T0327560

Project Finance For Construction & Infrastructure

Principles & case studies

Frederik Pretorius

Department of Real Estate and Construction,
The University of Hong Kong

Paul Lejot

Faculty of Law, The University of Hong Kong

Arthur McInnis

School of Law, City University of Hong Kong

Douglas Arner

Faculty of Law, The University of Hong Kong

Berry Fong-Chung Hsu

Department of Real Estate and Construction,
The University of Hong Kong

*All associated with the Asian Institute of International Financial Law,
The University of Hong Kong*

Blackwell
Publishing

Blackwell Publishing editorial offices:
Blackwell Publishing Ltd, 9600 Garsington Road, Oxford OX4 2DQ, UK
Tel: +44 (0)1865 776868
Blackwell Publishing Inc., 350 Main Street, Malden, MA 02148-5020, USA
Tel: +1 781 388 8250
Blackwell Publishing Asia Pty Ltd, 550 Swanston Street, Carlton, Victoria 3053, Australia
Tel: +61 (0)3 8359 1011

First published 2008 by Blackwell Publishing Ltd

ISBN: 978-1-4051-5127-6

Library of Congress Cataloging-in-Publication Data

Project Finance for Construction & Infrastructure: principles & case studies/
Frederik Pretorius...[et al.].

p. cm.
Includes bibliographical references and index.
ISBN-13: 978-1-4051-5127-6 (hardback : alk. paper)
ISBN-10: 1-4051-5127-7 (hardback : alk. paper)
1. Project management–Finance. 2. Infrastructure (Economics)–Finance. 3. Construction
projects–Finance. 4. Public-private sector cooperation. I. Pretorius, Frederik.

HD69.P75P724 2008
624.068′1–dc22
2007017022

A catalogue record for this title is available from the British Library

Set in 10/13 pt Trump Mediaeval
by SNP Best-set Typesetter Ltd., Hong Kong

FSC
Mixed Sources
Product group from well-managed
forests and other controlled sources
Cert no. SGS-COC-2953
www.fsc.org
© 1996 Forest Stewardship Council

For further information on Blackwell Publishing, visit our website:
www.blackwellpublishing.com

Contents

Preface

Project finance is not a recent phenomenon, nor has it offered any real or lasting controversy in academe's quest for meaning in finance. As a subject of study it is not particularly complex either – indeed, it is a practical subject, much of it simply outlining how finance is structured for particular ventures and why it is generally accomplished in a distinct way.

We believe this book will be a useful resource to academics who find themselves in particular situations which mirror our own, and may be explained as follows. A fundamental problem of many taught courses at undergraduate and graduate level in many disciplines is that students lack the luxury of studying applications in the field utilising a generic approach to the subject or its uses, as majors in that subject might do. There are good reasons for this, including modern trends to provide more and more interdisciplinary interpretations of subjects that were once considered entirely specialist in nature. Unfortunately, circumstances often require subjects to be taught by instructors who might not themselves have a similar interdisciplinary perspective. In the study of all aspects of projects in the built environment, for example, this is frequently the result of courses being taught by instructors with professional or technical backgrounds not necessarily associated with management or finance of large projects, whether in infrastructure development, power generation, mining, oil and gas, or some sector of real estate. Consequently such instructors may find themselves relatively unfamiliar with institutional environments that govern the economics of projects, and/or the design of business entities to facilitate the business objectives of project promoters/developers, and/or the manner in which national or international financial systems allocate finance to any such project. It may be argued that such courses may produce students entirely familiar with technical aspects of large projects, and some economic and financial considerations; but who remain largely unfamiliar with the commercial and institutional context within which such analyses may be conducted. In our opinion, this circumstance applies also to project finance.

In all, this situation is exacerbated by most generic finance textbooks, which are typically aimed at students of corporate finance, investment finance or financial economics. At best, they tend to confine project finance to a single chapter, or submerge the subject in descriptions of syndicated lending. Further, finance texts that are otherwise carefully crafted and pedagogically sound will often be country specific, usually to the United States, and from a corporate finance perspective lean heavily towards US corporate finance practice, corporate governance structures and problems, and the influence of particular economic institutions such as complex tax laws, such that underlying principles become skewed or obscured. This is also clearly reflected in a large proportion of corporate finance research. United Kingdom or Commonwealth derived applied finance books can be even more problematic, because so many are written by professionals associated with project delivery (whether architects, engineers, surveyors or lawyers) who may be far less familiar with commercial and finance concerns than specialist aspects of project delivery. The problem may be inevitable, given that so many international project finance techniques have roots in Anglo-American commercial practice, but this ought not to influence excessively a comprehensive treatment of the subject. This is well illustrated in the physical infrastructure sector by the treatment of the build-operate-transfer concept (BOT) as a project delivery mechanism or contractual arrangements governing physical construction, rather than primarily a model of funding for such projects based on principles of corporate or project finance. In this way finance and construction contracting arrangements become conflated or confused in the realisation of teaching programmes.

Given all such considerations, this book has a simple aim. It is written primarily to help non-finance students understand the economics and commercial aspects of project finance transactions. In essence we have set about this task by viewing project finance as having been deliberately mystified by academics, finance professionals and the specialist media so as to differentiate their views or products, each conjuring new jargon or acronyms for what are typically similar concepts or phenomena. Our response is essentially to make project finance transparent by presenting in accessible language the nature of such transactions and the context within which they occur: in each case the simpler the better. We present it as a facet of corporate finance, albeit in our view one illustrating in an interesting way where the focus of corporate finance research has been directed over the last three decades, a period that embraces a transformation in the multiplicity of financial sector activity. In keeping with the subject, our presentation is practical, and based upon first principles. As academics, we believe it is also important to present theoretical matters that are relevant to the topic, but our intention remains simply to demystify project finance. This means examining the context in which project finance has been viewed, whether as a way to overcome corporate or public sector balance sheet constraints, as a financing mechanism for infrastructure schemes, or as a special category of corporate finance. Project finance is all of these.

Of course, while project finance is based upon certain fundamentals, it continues to evolve in form and practice through new components or applications. For example, the widespread post-1980s drive to privatise government service delivery adds a further perspective to the broadening application of project finance principles. Many professionals from engineering, the built environment and law have become associated with the delivery of capital assets in private finance initiative (PFI) arrangements, public private partnerships (PPP) or similar such ventures. These initiatives have reflected problems typically addressed in the core project finance model, such as managing incentive conflicts and other risks. In our view, an important observation is that poorly conceived PFI or PPP arrangements may represent the flawed evolution of a generally sound model, taking place under political pressure for the public sector to devolve more and more risks to private sector agents. While aspects of PFI and PPP transactions are within the project finance domain, they may be conceived and constructed without due consideration of the high asset-specificity of project finance, which is crucial in its transactional integrity. For teaching purposes there is thus also significant importance to explaining typical PFIs and PPPs, making it clear that while privatisation initiatives may achieve objectives in line with the project finance model, they may often not meet the conditions required for successful project finance ventures.

The book is intended as a textbook and source for teachers and students, not a guide for researchers, although where appropriate it hopes to demonstrate clearly the practical relevance of academic research in the field. It is intended to be a post-entry level textbook for tertiary education, but one that assumes no specialist corporate finance or accounting knowledge. It seeks to introduce the vocabulary essential for an understanding of the relevance of these matters as the subject develops in the text. The book is targeted at filling an important void in finance teaching literature, namely a generic, non-national, non-locality specific project finance textbook for non-finance students. The main transferable skill to be imparted by this text is an analytical framework that can be applied to dissect and understand any project finance arrangement, and possibly the structure of most business ventures in general, using systemic and systematic analyses of those institutions and economics that surround these transactions.

A last note for academics contemplating using this book: the cases in Chapters 6–9 were prepared in general to support instruction using the case study method. Thus we have prepared teaching notes for the four cases covered in Chapters 6 to 9 – visit http://www.acrc.org.hk/ and follow the instructions.

Frederik Pretorius
Paul Lejot
Arthur McInnis
Douglas Arner
Berry F-C Hsu
Hong Kong, August 2007

Acknowledgements

We have so many people to thank for their generosity in helping us complete this project that it is impossible to do justice to all and inevitable that we would offend some by not mentioning them. So we have decided instead to identify those people and organisations without whom we could not even have started. First, we want to thank all members of staff at the Asia Case Research Centre (http://www.acrc.org.hk/) at the Business School of The University of Hong Kong for the fantastic intellectual support we enjoyed from them during the writing of this book. In particular Professor Ali Farhoomand, Director, and Ms Pauline Ng, Assistant Director, were highly motivational, and Mary Ho, Grace Loo and Alison Bate were inspirational under pressure. The Asian Institute of International Financial Law (http://www.aiifl.com) in the Faculty of Law at The University of Hong Kong provided a scholarly environment with distinguished visitors that helped discussion and kept us somewhat focused. We also wish to express our gratitude to the Hong Kong Research Grants Council which funded 'An empirical investigation into project finance (PF) arrangements in bank-dominated financial systems' (HKU 7127/03E), a project which led to many of the insights reflected in our book.

Our spouses and families absorbed an unfair share of the stress generated by this project, and deserve thanks. Very special thanks to Adrienne.

Frederik Pretorius
Paul Lejot
Arthur McInnis
Douglas Arner
Berry F-C Hsu
Hong Kong, August 2007

About the Authors

Frederik PRETORIUS is an Associate Professor in the Department of Real Estate and Construction at The University of Hong Kong and a Fellow with the Asian Institute of International Financial Law, University of Hong Kong. Dr Pretorius holds a PhD from the University of Hong Kong and an MBA and BSc(QS) from the University of the Witwatersrand in South Africa, and concentrates on project finance, real estate investment and finance, real options analysis and urban economics and development. He has experience in Hong Kong, Australia, New Zealand and South Africa as an academic and professional, in the corporate sector and as consultant for various activities and industries including regional economic development, real estate development, building, and mining and process engineering.

Paul LEJOT is a Visiting Fellow with the Asian Institute of International Financial Law, University of Hong Kong, and a Visiting Research Fellow at the ICMA Centre, University of Reading. Formerly an investment banker with a wide transactional command in structured debt, fixed income, regulatory capital, and financial restructuring, and with extensive experience in Europe and throughout East and South Asia, he is now engaged in research into financial market development and policy reform. His interests include legal and institutional aspects of financial market behaviour, regulation and development, transaction law and economics, and Asian regional financial policy. Since resuming an academic career in 2003, he has published widely on these topics and consulted with official organisations, in particular examining the legal and practical obstacles that constrain capital market development and effectiveness in Asia. His current work includes research and postgraduate course development in law and finance, notably legal influences on development, instruments, institutions and markets, as well as structured finance, regulatory arbitrage, and financial derivatives.

Arthur McINNIS is a Visiting Fellow with the Asian Institute of International Financial Law, University of Hong Kong, and Managing Director, International Law Institute (Hong Kong). A former practising lawyer in North America and Hong Kong, he currently teaches at The City University of Hong Kong, and has an international reputation which is tied to more than 60 publications in the construction, contracting and planning fields. Arthur is also the Honorary Legal Advisor to the Joint Contracts Committee, the body which has lately published the new Standard Form of Building Contract (Private) for Hong Kong. He is a founding member of the Centre for Infrastructure and Construction Industry Development (CICID) and a former co-director of the Asian Institute of International Financial Law (AIIFL) both at the University of Hong Kong. Arthur's expertise in the field of construction, projects and dispute resolution is diverse with long experience at both Baker, McKenzie and Clifford Chance. Arthur holds Diplomas in both Civil Law and Comparative Law from Sherbrooke and Dalhousie Universities respectively. His first degree in Economics and Political Science (Regina) was followed by an LLB degree (Sask). He also holds a Bachelor of Civil Law degree and an LLM from McGill University in Montreal and a PhD in law from Queen Mary and Westfield College at the University of London.

Douglas ARNER is Associate Professor and Director, Asian Institute of International Financial Law, Faculty of Law, University of Hong Kong. In addition, he is a member of the Board of Management of the East Asian Economic Law and Policy Programme and Co-Director of the Duke University-HKU Asia-America Institute in Transnational Law. Prior to his appointment at HKU, Douglas was the Sir John Lubbock Support Fund Fellow at the Centre for Commercial Law Studies at Queen Mary, University of London, a consultant with the European Bank for Reconstruction and Development, and Director of Research of the London Institute of International Banking, Finance and Development Law. He has consulted, lectured, co-organised conferences and seminars and been involved with financial sector reform projects in over 20 economies in Africa, Asia and Europe. Douglas specialises in economic and financial law, regulation and development. He is author, co-author or editor of eight books and more than 50 articles, chapters and reports on these subjects. He holds a BA in literature, economics and political science from Drury University, a JD (cum laude) from Southern Methodist University, an LLM (with distinction) in banking and finance law from the University of London (Queen Mary College), and a PhD from the University of London.

Berry F-C. HSU, BSc, LLM (Alberta), MA (Oregon), PhD (London), Barrister and Solicitor (Supreme Court of Victoria) is an Associate Professor of Law in the Department of Real Estate and Construction, and Deputy Director of the Asian Institute of International Financial Law, Faculty of Law, University of Hong Kong. He is the author of several books on the common law system, banking and finance, and taxation, including *Financial Markets in Hong Kong: Law and Practice* (Oxford University Press).

1

The Nature of Project Finance

Great projects fire the imagination. In conception and realisation, such ancient wonders as the Pyramids of Egypt, Aukor's temples, the stone city of Petra or Rome's revolutionary water courses have caused veneration and wonder in both the contemporary and modern viewer. They inspired emulation by successive kings and emperors, whether to honour a deity, subjugate a people or repel an invader, or to create a lasting economic infrastructure. More recent schemes such as the transoceanic Panama Canal or 19th and 20th century rail and water transport systems in Europe and North America have transformed the fortunes of national and global economies, and seemed to suggest unlimited scope for humans to transform the landscape they inhabit. Whether from commercial needs or the fiat decision of a ruler, it seems that great projects have been underway throughout recorded history, and a neglected curiosity that projects as diverse as the Stonehenge circle and the Inca road system were effectively subject to transaction costs and financing concerns.

What is most interesting is that the reasons for undertaking all such feats have changed little over the millennia. The typology of reasons that explain their construction remains relatively limited, although priorities have changed over time. Without putting too fine a point on it, these motives can be categorised as the demonstration of authority or reverence, the enhancement of security or the creation of new economic resources. They appear as devotional creations (temples, pyramids or cathedrals), schemes to manage the elements (great dams and irrigation schemes), economic structures (ports, transmission projects, canals, road or rail systems, mines, process engineering), and political (iconic public structures, developmental projects, the exploration of space, defence systems, and projects of prestige that demonstrate national power or emancipation). In an age of immense computational power the analysis of large-scale projects has become increasingly complex, and while our insight into the environmental, social and economic impact of all such schemes has developed immeasurably, it will remain inevitably incomplete. It is thus quite conceivable that future actions to address the negative impact induced by so many major projects might yet depend on the instigation of still more, even on a more demanding scale.

This book aims to provide a framework to comprehend large projects in the modern world, concentrating on the financing of projects. In considering project finance, we draw on several other disciplines. Of necessity we make use of an economics vocabulary, but we approach it with a somewhat unconventional view. We take traditional demand, supply and economics of industrial organisation concepts based on market equilibrium analysis as handed down from neoclassical thinking, but our view is that to understand the world of project finance requires a further vocabulary of market imperfections. Hence the economics we use largely reflects the world of friction in economic exchange, contracts, transaction costs, agency conflicts and economic institutions. Possibly the most conventional part of our approach is to corporate capital budgeting decisions – in the end projects are expected to be feasible in aspects that are important to their promoters and hosts. To our knowledge there is not yet a widely accepted methodology to assess project feasibility in both the public and private sectors other than extended discounted cash flow methodology and its derivatives, although we do not pretend that observed market prices are necessarily information-efficient. We also do not confuse desirability with feasibility – this is a matter of political, ideological or philosophical choice. We concentrate largely on private-sector aspects of projects, but it will become clear that it matters little if the principal shareholders in project companies are public or private. Irrespective of political ideology we consider that societies everywhere continue to value the economic application of their resources, so we broadly assume this convention has merit.

This introductory chapter has five sections. First, we present a capsule history of a famous project of the recent past in order to provide indications of the approach we shall follow to large projects and their financing, and perhaps to introduce the richness of the world of large projects. Thereafter we describe generally the matrix of arrangements within which modern projects are typically executed, concentrating on concepts around corporations and their structures and how this framework functions to govern projects, and specialised project-related companies. A reader with knowledge gained from previous study of economics (especially institutional economics and transaction cost economics), finance (particularly banking and corporate finance), law (particularly commercial law), or business and accounting will recognise many of the concepts we describe throughout the book, but our objective is to explain concepts sufficient for students in a range of non-cognate disciplines to become fully comfortable with the subject matter. We thus outline the corporate context within which modern projects are conceived and executed. Third, we introduce the nature of project companies, their typical business model and financial structure, and the economic nature of modern project company facilities (such as bridges, tunnels, power plants, transmission pipelines, and refineries). Fourth, we believe it will be challenging to make sense of as disparate a field as project finance, because it draws on such a wide range of disciplines. Thus we fall back on an old ally to help make sense of complexity, and identify and outline a number of systems theory concepts to guide our overall approach to complex matters as they arise in the rest of the book. Last, in Section 1.6 we present the plan for the rest of the book. In all, we wish to impart an approach to analysing the context of projects to identify where the likely economic, financial or political risks may be lurking, in order to consider how such risks are managed. No single concept here is radical but we prefer to think we are suggesting a novel way to approach the study of particular types of projects. In a practical sense, we consider this way of thinking to be appropriate for the analysis of any identifiable venture within a commercial context.

1.1   The world of projects today

Projects, projects everywhere. It would indeed be wonderful to write a book about the history, engineering, finance, and stories associated with mega-projects – each such story would be a gem: the Panama Canal, Suez Canal, Oresünd Link between Denmark and Sweden, the US interstate highway system, and many, many more. We simply do not have this luxury, so in order to spur early interest, we present a short narrative about an amazing project in terms of ambition, engineering ingenuity, and vision – the Channel Tunnel, developed in the 1980s and early 1990s linking France and the United Kingdom by a rail tunnel under the English Channel. It is ironic that this project has been hailed as a wonder of the modern world – and yet the finances of this magnificent project, in terms of the concepts covered in this book, remain an absolute mess, with several financial restructurings undertaken since it commenced operations in the early 1990s, with no real prospects of it being a successful infrastructure asset in economic terms for some time. An English Channel Tunnel has lain in the imagination of engineers from early Victorian times but in becoming a reality the project grew into a contractor's dream and an investor's nightmare. The project was conceived in the mid-1980s, in part as a high-profile cooperative venture between two long-standing national rivals, but has been viewed by successive British and French governments as a scheme that would not receive direct state capital investment or transactional support in either construction or operation. The selected design involved boring two rail tunnels and a third service tunnel between the southeast coast of England and a point in northern France, an undersea distance of around 34 km, and the creation of road and railway terminals and associated infrastructure (Figure 1.1). The plan was for the tunnel to be built, owned and used by commercial interests under a long-term franchise granted by the two governments.

Construction began in 1987 and the tunnel opened for commercial use in 1994. Work on development and construction was undertaken by TransManche Link, a company formed by a group of contractors and financing banks, and subsequently listed on the stock exchanges of London and Paris. TransManche Link's technical achievement is well-regarded, but its financial record was dire from the moment earth was first struck, and the company has been subject to repeated cash crises and financial restructurings. This can be ascribed to two main linked factors. First, the

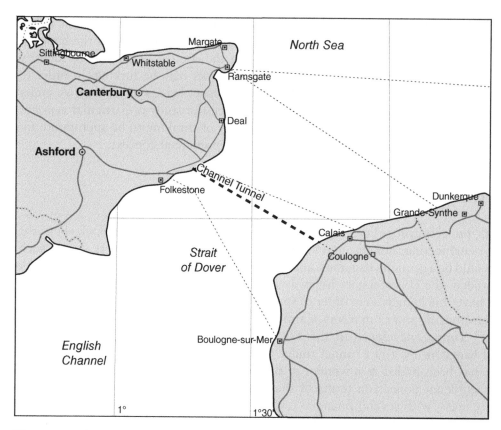

Figure 1.1   Geography of the Channel Tunnel.

attraction of being associated with a prestigious project blinded the sponsors and financiers to its true economics. TransManche Link was severely undercapitalised and poorly funded, and needed to be rescued by its creditors long before the tunnel opened to train traffic. Second, the sponsors were woefully optimistic in their revenue forecasting, and in particular neglected heightened competition from increasingly efficient ferry operators, which even today retain a significant portion of cross-channel traffic, and the more recent success of low-cost airlines.

In essence, the project's early financial modelling and strategy were wholly inadequate, and the scheme has never been financially stable. It should be noted that despite the financial ill-health of the tunnel builder/operator, both governments have expended resources in developing new infrastructure associated with the project, for example in city centre rail terminals, and together with its popularity among users this may ensure that the Eurotunnel project is 'too big to fail'. More technically, it may be significant that the project was developed prior to financial modelling becoming relatively sophisticated and integrated with the regulation of bank capital. One banker associated with the initial listing of shares has written about the casual approach taken to forecasting which were it to take place today, might be the subject of litigation given that financial representation and disclosure is regulated more closely than in the 1980s (Freud, 2006). Furthermore, it is notable that among the large syndicate of banks that provided loan financing at the project's inception, very few were US-domiciled and only one US bank took a prominent part in the transaction. Major US lenders were then among the most well-resourced in credit risk and project analysis, and almost all were formidably sceptical as to the project's viability. By contrast, European banks faced unsubtle political pressure to join the transaction.[1]

1.2 Corporations, finance and projects: important concepts

To be true to the objective of simplicity, we consider it appropriate to open this book by outlining the circumstances that underpin the concept later defined as project finance. In order to avoid inevitable confusion if this book is to be read in parallel with other traditional texts on corporate or project finance, or the international financial system, it is preferable that we make no assumptions whatever about the existing language of finance, and thus commence by placing this book in its proper context. We therefore request the reader to be patient as we consider several concepts that we deem too important to take for granted.

In essence this book is about the economics, and in particular the financing, of certain classes of projects. While the development and financing of historic projects will remain great narratives in social custom, economics and politics, not to speak of the stories of powerful and single-minded patrons, the focus of this book is thoroughly modern. We are quite selective in our choice and use of economic concepts, what is meant by finance[2] and how it is set in context, in order to make sure that the analyses in this book can be applied as clearly as possible to representative and real-world problems. Further, we are also particular about the meaning and use of *project*, because, despite the ubiquity of the concept, some projects are more prone to failure than others, as the politics that surround some ventures will show (think of the Channel Tunnel as but one example). In short, some projects are more equal than others (apologies to George Orwell).

In such cases as Eurotunnel we see that the venture required an enabling organisation, a form of business entity, to execute development and then operate the project. This indicates our point of departure: projects do not simply happen; they require the formation of such enabling organisations, which in turn require enabling institutions for the project entity to fulfil its objectives. To make quick sense of this notion, we present a fast tour of the corporate world and where projects fit in. We concentrate

---

[1] Author's recollection from direct experience.
[2] It has been said that there is no economics, there is only finance . . .

on projects that can be categorised as having commercial objectives, but in later sections will also focus attention on the institutions and politics of projects driven by issues of public policy.

1.2.1 Corporations, companies, and more: what is meant by project?

We commence with an attempt to circumscribe what is meant by project. It is correct to say that project finance refers to a particular family of financing mechanisms used to achieve objectives by very large, mostly multinational corporations, governments, banks, and developmental projects funded by regional multilateral organisations such as the World Bank or Asian Development Bank. This is not a mechanism that is in regular use by small and medium-sized enterprises, except in some cases in the specific industry sector of commercial property development. This fact requires that we approach project finance from the perspective of major, complex schemes and, to make the book useful, demands the introduction of many concepts and much vocabulary that is used commonly in such an environment.

In the first instance, our approach follows the meaning of project in business and corporate finance literature, where any and every decision to invest corporate funds in some business activity (i.e. the capital budgeting decisions of the company, or the corporation) may be viewed as a project, and is very often simply referred to as such. This view follows from well-established concepts in corporate planning, popularised by illustrious management personalities such as Chester Barnard, Peter Drucker, Russell Ackoff, Kenichi Ohmae, Michael Porter, Tom Peters, Peter Senge and other figures from the distant (and more recent) past. Any such narrative will tend to begin with *what* a company wishes to achieve, and *how* it may go about achieving it.

'What' is generally expressed as a corporate mission, a high-order, often abstract, statement of the purpose of the company. As many corporate annual reports will show, vision or mission statements are often artificial. Most true corporate purposes can be distilled to being the creation of wealth for shareholders, subject to the caveat that the means to do so is legal and considered morally acceptable. We understand stakeholder interests to be taken as part of legal and ethical considerations. Thus we see limits to being able to justify how any economy may function.

'How' the company sets about achieving its mission constitutes its strategy and, importantly, includes decisions about which products and services the company will create or procure, to whom it will market the products and at what price, all in line with the selected corporate strategy. Corporate, or business, strategy is a similar concept to business model, a term which we shall use in a somewhat revised context in following chapters.

More practically, however, how this mission further is seen to be achieved is through devising a corporate plan and setting and achieving of objectives, which are more usefully expressed in terms of the nature of business activities, or targets such as market share or product profitability. Once objectives and means (products, services) have been confirmed, a crucially important further variable in electing how to achieve those objectives lies in decisions about organisational structure, from which comes the famous organisational design adage, *'structure follows strategy'*. This is then a short description of what is in general known as the systems approach to planning and, more pertinently for our purposes, to corporate planning. The essential features of this framework are shown in Figure 1.2, where the overall plan is developed from left to right. Internal consistency can be assessed by reversing the direction of the process as if from right to left, and considering whether each stage in the process necessarily leads to its successor.

In working towards describing the context of project finance, it is important to understand that the way the adjective-noun 'project finance' is used in this book is as the product of one organisational design strategy that responds to a particular set of financial circumstances (this will be explained in later sections and chapters). There will also be more about the systems approach; in fact it is fair to say that many of the structures presented in later chapters draw on this elegant old theory.

Let us elaborate in more practical terms on how we use the term 'project' within a corporate strategy framework. Consider a hypothetical large corporation which owns or controls a number of subsidiary companies, and which may be organised into a number of divisions for managerial purposes. At the highest level there is an overall corporate mission or vision guiding divisional objectives. Ideally, each subsidiary contributes in some way to a divisional objective, and so on, following the notion of nested objectives in corporate planning. The corporation's structure (corporate level, divisions and subsidiaries, functional management or matrix organisations, and so forth) is also organised to achieve objectives (structure follows strategy).[3]

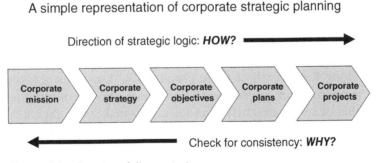

A simple representation of corporate strategic planning

Direction of strategic logic: **HOW?** ⟶

Corporate mission ⟩ Corporate strategy ⟩ Corporate objectives ⟩ Corporate plans ⟩ Corporate projects

⟵ Check for consistency: **WHY?**

Figure 1.2   Structure follows strategy.

---

[3] Organisational theory students will be aware of what a bewildering array of organisational structures there are – from bureaucratic to organic, and so on (Handy, 1999). What is important is that these forms do not just happen; for our purposes they are a conscious response to an organisation's strategic objectives. Of course, organisation structures often lose their effectiveness, contribute to organisational decline, and form the subject of painful restructuring.

Within this hypothetical structure, each subsidiary may be viewed as a project defined by what product or service it deals in. For example, suppose Global Consilium Corporation (GC) is a large multinational mineral and energy resources company that explores for and extracts mineral deposits and energy resources (coal, oil and natural gas), and refines and sells a range of metals and energy resources in the organised international commodities markets.[4] Assume further that GC is considering acquiring a company that has production and resource capacity in a metal such as platinum which it considers to be of strategic importance: let us call it Kalgoorlie Platinum and locate it in Western Australia. It could analyse this opportunity as an investment project and make a decision based on its capital budgeting rules, typically applying extended discounted cash flow analysis using a target weighted average cost of capital (WACC) (Chapter 3), conduct due diligence, bid for and acquire the company. If the acquisition is not successfully concluded, GC may of course instead consider starting a new venture to produce the service or good itself, but in either case the task is taken to be a project.[5]

In strategic terms, GC might decide itself to develop, rather than buy a similar asset but it nevertheless remains a corporate project. In similar fashion, a decision to invest in research and development of a particular technology or product is considered a project, a new model car is a project, and by similar reasoning virtually every corporate investment (and divestment) in effect becomes a project. So there are corporate subsidiaries that may be projects, and the subsidiaries themselves may be portfolios of smaller projects. Where a project is defined within a corporate entity depends almost entirely on what is appropriate for corporate strategy and structure, and what seems efficient for management purposes. But importantly, it also matters for financial considerations, as we shall see in following sections.

As an illustration, we may consider any well-known large multinational company. The scope and complexity of such companies' project portfolio are typically large – think of engineering companies like Fluor or Bechtel, mining giants such as BHP Billiton of Australia, Anglo American Corporation of South Africa, or Rio Tinto of the United Kingdom, or think of the many projects of telecommunications and port infrastructure conglomerates like Hutchison Whampoa and Cheung Kong Infrastructure of Hong Kong. Each of these corporations functions across the globe and has wide-ranging, complex project portfolios which may be fully understood by only a small number of senior executives and decision-makers. In fact, a commonly used idiom in modern corporate planning is that corporate strategy can be viewed as a portfolio of present projects and options on future projects (Copeland and Antikarov, 2001).

---

[4] The major soft commodity and metals markets are centred on Chicago or London, although oil and gas are traded substantially over-the-counter, or directly between buyers and sellers (Chapter 4).
[5] The symbiotic relationship between a powerful manufacturer and monopoly supplier and the strategic or contractual alternatives that then arise in production are typified by the 1930s case of General Motors Corporation and Fisher Body Inc. (Coase, 2000).

1.2.2 *Finance and corporate finance*

At this stage it seems appropriate to remind ourselves that this book is about finance. While we have been defining terms, we have also been purposely avoiding the term finance but can no longer do so.

In a narrow sense, this book represents a general introduction to one specialised aspect of corporate finance, known as project finance. This means corporate finance, as a concept that subsumes project finance, also has to be outlined briefly. Predictably, following Brealey *et al.* (2006), corporate finance concerns the financing of corporate activity. In the corporate strategy framework outlined above, this essentially means financing strategic investment decisions or if we revert to our project-centred framework, it may indicate the corporation's alternative choices in financing the acquisition or creation of a company/project it has identified as desirable. In general, it can be made this simple – given Global Consilium's decision to buy Kalgoorlie Platinum in Western Australia, *how does GC pay for it*?

At the most basic level it is this simple: GC pays with the proceeds of issuing equity or debt. Using a particle physics metaphor, there are only two fundamental sources of external finance from where all financial engineering originates, and these fundamental corporate finance instruments, or elemental financial claims, are *equity* and *debt*. Leases, in our view, are a specific form of debt where the transaction purports to achieve further objectives, which we will explain in Chapter 2. In essence all further transactional financial engineering derives from these fundamental contractual claims. Debt and equity are contributed to the company from external sources (they represent its liabilities), while internally generated funds (net retained cash flow over any accounting period) form a further source. Keep in mind that net internally generated funds technically belong to the company (and ultimately to its shareholders). However, this is not to say it is always easy to identify whether equity or debt is to be used, or was used in a particular application – a simple complication is illustrated by certain hybrid financial instruments such as convertible bonds, which may rank as equity or debt depending upon an external event, for example, the corporation's share price or lapse of time.

We do not intend to consider in this book more than the most fundamental corporate finance instruments, but it will become clear to students of financial engineering where this field of interest fits within corporate finance activity, and where further study of this interest will lead. A further observation about financial instruments: unlike investment finance textbooks, together with insurance, we consider a range of financial derivative instruments such as options, futures and forwards only in terms of their risk management uses. Although derivative instruments are extensively used in corporate finance and in project finance to hedge risks, we will suggest only their uses and not any technical features. Similarly, it will become clear to students who are interested in how corporate risks are hedged where this field of interest locates within corporate finance activities, and where to find inspiration for further study of this subject.

1.2.3 The project is a company

Given these points, we can turn to Figure 1.3, where we have illustrated Global Consilium's balance sheet as consisting of assets (all its subsidiary companies, or strategic projects, if you will), which are financed simply by corporate debt and equity – in aggregate the liabilities of the group. For the corporation we have arbitrarily chosen a capital structure of 69% debt and 31% equity (or a debt to equity ratio of 220%), which may be considered a fairly typical capital structure for a large diversified corporation, all things considered. Of course, recall that this is a consolidated balance sheet, so this final picture is the net accounting sum of each subsidiary company's assets and liabilities.[6]

The Global Consilium Corporation (founded 1929)						
Consolidated Balance Sheet at June 30, 2007[a]						
($ millions)						
LIABILITIES				**ASSETS**		
		$	%		$	%
Equity				**Fixed assets**		
				Corporate head office		
Issued ordinary shares (breakdown)	11,000			building	1,000	
8% Cumulative preference shares (non-voting)	5,000			Fixtures and vehicles, etc.	50	
Retained earnings (accumulated past profit/losses)	20,000			LESS:		
Minority interests	3,500			Accumulated depreciation	-150	
Shareholders' funds, ("net worth")		**39,500**	31	**Net fixed assets**	**900**	1
Debt				**Associated companies[b]**		
Long-term debt				Consilium oil	30,000	
Secured bank debt, due 06/2010	31,500			Consilium mining	30,000	
Outstanding bonds, maturity at 06/2015	48,000			Consilium jet engines	15,000	
				Consilium infrastructure and power	35,000	
Long-term office lease	400			Other consilium companies	11,000	
Net long-term debt		**79,900**	63	**Associated companies**	**121,000**	96
Current liabilities						
Commercial paper: 3 months	5,000			**Current assets**		
Short-term bank loan: 1 month	2,000			Incidentals, inventory	1,400	
Current liabilities		**7,000**	6	Other current assets	2,000	
				Cash and marketable securities	1,100	
Total debt		**86,900**	69	**Current assets**	**4,500**	4
Total:		**$126,400**	100	**Total:**	**$126,400**	100

Notes

[a] Please understand that we make no effort at all to reflect Generally Accepted Accounting Principles as practiced in ANY jurisdiction in our analyses. The detail reflected here is not typical of many consolidated balance sheets.

[b] In many (most) jurisdictions, associated companies will NOT be named at all, there will simply be a one-line item reading 'Associated Companies', with a total against it. We itemise associates to illustrate the underlying structure of corporations that have project company subsidiaries.

Figure 1.3 The corporate balance sheet as a portfolio of projects.

---

[6] As stated before, we would prefer to err on the side of caution with vocabulary. So, if we assume no accounting knowledge, we have to say – assets must equal liabilities, and the balance sheet has to balance.

It follows then that we view corporate activities as projects contributing to an overall strategy. However, for legal and regulatory purposes, and often in accounting, projects are typically incorporated within separate companies, or even adopt an alternative form of business. Nevertheless, this is unusual, for close corporations, partnerships or sole traders hardly ever feature at the level and scale considered here, although some atypical private companies may do so (think of the Olympia & York group, which was for a long time a private venture, and its 1990s Canary Wharf development in London's Docklands). For simplicity, we assume that every project in Global Consilium's portfolio is incorporated as a wholly-owned subsidiary; or it could be itself a listed company, majority-owned or controlled by Global Consilium. In corporate matters every project's legal structure (a company or other business entity) matters greatly for legal and regulatory purposes. We shall also see in later chapters that the technical details of incorporation, shareholding and control all influence the reasons for adopting project finance as a strategic choice to finance a venture. A further dimension to consider here is that corporate law and business regulation differ between jurisdictions, so an additional layer of complexity is introduced when international projects and their incorporation are being considered. Following this, we wish to emphasise that the legal vehicle, or structure chosen to execute a project, plays a central role in the concepts presented in this book.

In the stylised corporate strategy framework presented above, a project company may be viewed as a company created or acquired to execute some strategic objective. Assume, for example, that GC has decided that platinum will be critical to the process of developing better catalytic converters so as to reduce pollution from fossil-based energy sources. Suppose, for illustration, that this objective is translated into the practical corporate objective *'to increase Global Consilium Corporation's share of world platinum sales to 30% by 2015'*. Assume further that Global Consilium Corporation creates Global Consilium Platinum Ltd (GCP), a wholly owned subsidiary, for the sole purpose of achieving this objective. Of course, there are many ways in which Global Consilium Platinum can go about achieving its objective. One way, for example, is to acquire controlling shareholdings in a sufficient number of existing platinum mining companies all over the world until it reaches the target market share. Another option may be to locate a large platinum deposit and develop its own platinum mine. Global Consilium Platinum then becomes a company with one asset only, a platinum mine, with one narrowly defined commercial activity, to operate the mine and extract and sell platinum. In the vocabulary of this book, Global Consilium Platinum is a company with a single asset, the mine, possibly with a sole shareholder in GC – this context explains the term project company as the term is used in the field of project finance. A critical further observation is necessary – Global Consilium Platinum may have been created as a company, but the mine still has to be developed, which carries much further significance, as we see in later chapters. This essentially means that GC, the single shareholder in Global Consilium Platinum Ltd, is also the mine's developer – in project finance terms also known as the sponsor or promoter.

The essence of what has broadly become known as project finance are the financial, regulatory and legal mechanisms that have developed around such single asset project companies and their financing. In practice, Global Consilium Platinum may have other shareholders (sponsors or promoters), and may even become a public company, but for present purposes it is a single asset project company with one shareholder-sponsor in the parent Global Consilium Corporation. Most project finance ventures since the early 1970s have been industrial or resource development projects such as oilfields, transmission pipelines, extractive mines, process engineering, hotel and resort development, but project finance has also been used extensively in infrastructural schemes, often in collaborative arrangements between governments and the private sector (public private partnerships (PPPs), private finance initiatives (PFIs), build-operate-transfer (BOT) schemes in sectors such as electricity generation, telecommunications, transportation infrastructure (roads, tunnels, bridges, railroads, water treatment facilities), hydroelectric projects, port facilities and container terminals, latterly also air-cargo terminals, and more. It has been applied widely in both developed and developing countries, including Asia Pacific, Africa, the Americas, and Europe – it is certainly a technique that is proven. But with this familiarity have also come problems, mostly from attempts to relax the fundamental project finance model in some applications such as PFI and PPP schemes – we will also draw attention to some such developments.

1.3   The project company business model

We have now introduced enough terms and presented sufficient context in order to move towards defining what is meant by a project company business model within the broader field of corporate finance. We described above what could be characterised as a typical project company; that is, a company that owns and operates a single economic asset such as a mine, toll road or oil refinery. In economics terminology these are described as highly specific assets, generally taken as capital assets that have only one economic function (or at most a small range of functions). For example, economically an oil pipeline can practically not be used for much besides a pipeline, unless significant additional investment is made in order to change its function. Typically mines, oil refineries, energy transmission pipelines, chemical process engineering plants, power plants, certain marine vessels (purpose-built liquefied gas carriers, for example), roads, bridges, tunnels, certain seaport and airport cargo handling facilities all exhibit high asset specificity – these assets certainly cannot easily be redeployed to some other function, or have the flexibility, say, of a personal computer in an office environment, or a delivery vehicle in an urban setting (more about asset specificity in Chapter 2).

Application of highly specific assets in a commercial setting raises further considerations about the risks surrounding their use. Not only are they typically capital and scale-intensive, they also tend to utilise established and low-risk technologies within mature and well-understood industries, where risks are clear and well understood, and with well-delineated demand and supply chains. It may be argued that deploying assets such as these might not be risky, but that asset specificity itself is the risk and is at the heart of the problem of irreversibility (sunk capital) briefly considered in Chapter 3.

A useful further observation may be introduced to categorise typical project company assets, based on the way in which they generate earnings for their owners. We can think of two classes of assets, namely process-based assets and stock-flow assets. Process-based assets use raw material or basic inputs, which they convert through an industrial or chemical process, and sell the output – similar to the functioning of many industrial companies. In a project company context this may be illustrated by an oil refinery or electricity generating facility, where raw materials form a significant share of the facility's continuing cost function. Stock-flow assets are more complex. A pipeline scheme requires in its completed and operational state only limited further inputs (facility management, systems input) to generate revenue from the service it sells (product volume transmitted through the pipeline). Ongoing inputs are comparatively less important as an element of the system's cost function compared to most process-based facilities. The pipeline is said to represent a stock-flow asset: its value as a capital asset is represented by the total stock of services that can flow during its life or until it is unable to generate a service for which there exists demand (it may be economically depreciated, or there may simply be a substitute for the service it offers). Water treatment facilities, bridges, roads, tunnels, mines and oilfields may be conceived of similarly.

Project finance, therefore can be seen as a financing mechanism, developed around business ventures with assets that exhibit high-asset specific characteristics, often seen most clearly in relatively predictable revenue streams, within well-understood industries and often involving few technological uncertainties, as identified above. Project companies may often be financed by a parent group – there is no reason why not – in which case we may consider their financing to be simply consolidated into the corporate balance sheet under normal corporate finance conventions. But such project companies are often not financed with corporate finance, but instead are financed using what have become known as project finance principles. So what then constitutes project finance? Here, we take as a definition the narrow meaning accepted among all financiers, and which is constituent with the institutional approach that underlies this book.

Project finance thus represents a form of non-recourse external debt funding of an identified scheme, carrying defined claims to its revenues, assets or contractual rights (such as purchase contracts or third party insurance provided other than by the project sponsors), and without contractual rights or non-statutory claims in relation to the debt against the project sponsors or shareholders. The funding will almost always be provided to a company established solely for the purpose of owning the project, and

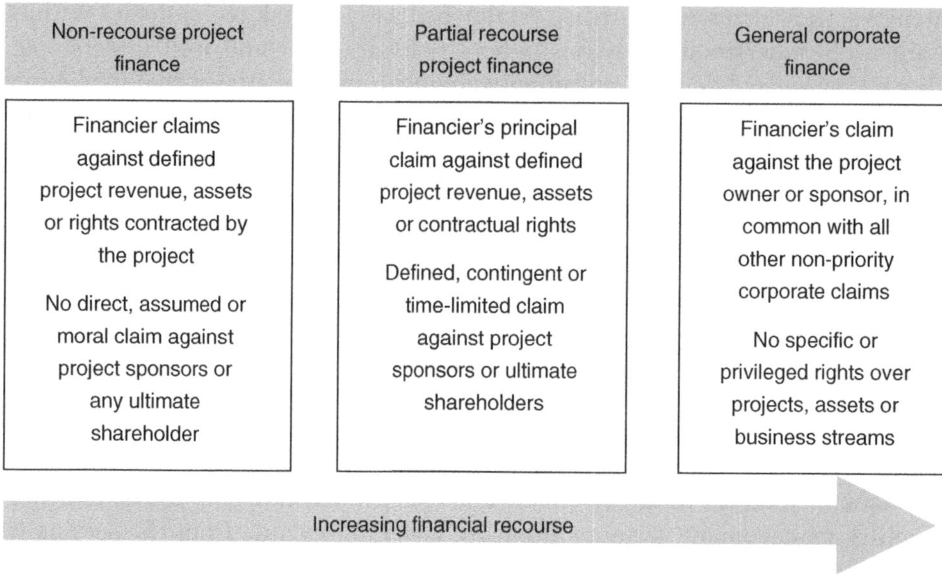

Non-recourse project finance	Partial recourse project finance	General corporate finance
Financier claims against defined project revenue, assets or rights contracted by the project No direct, assumed or moral claim against project sponsors or any ultimate shareholder	Financier's principal claim against defined project revenue, assets or contractual rights Defined, contingent or time-limited claim against project sponsors or ultimate shareholders	Financier's claim against the project owner or sponsor, in common with all other non-priority corporate claims No specific or privileged rights over projects, assets or business streams

Increasing financial recourse →

Figure 1.4   Degrees of financial recourse.

entering and servicing its funding liabilities. That this is distinct from other common forms of financing can be seen in Figure 1.4, which characterises the three elemental types of finance which interest us here.

Non-recourse financiers of projects, such as bank lenders or investors in project bonds, obtain claims only against the net assets of the project, usually including any rights that the project vehicle acquires from third parties. If the project falls into disuse or becomes a commercial failure and unable to service its debts, then the financier cannot look for recourse to the project's sponsors or shareholders, even if it is with them alone that negotiations then begin to find a solution to the project's problems. At the other extreme, general corporate finance will see funding being provided to a company or group – such as the sponsor of a project – without specific contractual rights over any specific part of its activities or assets. This is by far the most common form of external funding.

Between these extremes lies a form of project finance in which a degree of recourse is granted to the financier, perhaps for a limited period, or on certain quite specific terms, and is widely known as partial-recourse finance. Projects that take time to construct or mature may use each of these financing variants at different times, so that at inception when the project assets amount to no more than a pile of engineering blueprints, it may be cost-effective to induce lenders to provide short-term land or construction finance to the project sponsor. When the project is properly formed, it can be funded with partial-recourse finance, and when all is completed and revenue has begun to grow then non-recourse finance may be attractive to both lenders and sponsors. We discuss in Chapter 2 the sequence of financing and phasing of transactions common to complex projects, but the basic elements are always simple, and intended to isolate and define the costs and risks of funding the project for its sponsors and financing investors.

Thus there are distinct features that make project companies somewhat different from a group that is corporate financed. These involve separate incorporation, capital structure, and management discretion over company assets (there is only one asset). Project companies are typically legally separate (standalone) capital-intensive single asset companies, often classed as a special purpose vehicle (SPV) or special purpose entity due to their lacking any activities not associated with the project. The project company or SPV is the entity through which are channelled all the project's contractual matters – typically also the SPV is the project company itself. They also have common capital structures, in particular concentrated equity ownership (no more than 5–10 sponsors usually own all the equity in a project), together with higher leverage in their capital structure than with frequently observed established companies.[7] This typical capital structure with relatively high levels of debt in relation to shareholders capital serves several purposes, each forming the subject of later sections in the book. (See Figure 1.5 for what a single asset project company balance sheet might look like.)

| **The Baguio Power Company (incorporated in the Republic of the Philippines 1999)** | | | | | |
| Balance Sheet at June 30, 2007 | | | | | |
Liabilities		($ thousands)	**Assets**		($ thousands)
Equity			**Fixed assets**		
Issued ordinary shares:					
Global Consilium Corporation	175,000 35.0%		Power generating facility at Subic Bay (at cost)[1]	2,600,000	
Bagatelle Projects D&C Inc	175,000 35.0%		Furniture, fittings, equipment, vehicles, etc.	5,000	
Hydrex Turbines Inc	150,000 30.0%		LESS:		
Shareholders' equity		500,000	Accumulated depreciation	(1,000)	
			Fixed assets		2,604,000
Debt					
Long-term debt			**Current assets**		
Secured bank debt, due 06/2022	1,400,000		Incidentals, maintenance inventory	5,000	
Capital leases: generating equipment	600,000		Other current assets	310,000	
Long-term office lease	10,000		Cash and marketable securities	5,000	
Long-term debt		2,010,000	**Current assets**		320,000
Current liabilities					
Commercial paper: 3 months	114,000				
Short-term bank loan: 1 month	300,000				
Current liabilities		414,000			
Total debt		2,424,000			
Total		2,924,000	**Total**		2,924,000

1. The land at Subic Bay is leased from The Philippines Government at a neglible annual rent of US$1 per annum.

Figure 1.5   Typical project company balance sheet.

---

[7] However, relatively high leverage can be common in particular industry sectors, for example, in trading-related businesses or utilities, and may change with long-term interest rate cycles.

We need to make further observations about project company capital structure, however. An ideal project finance arrangement would also aim to achieve a further characteristic with respect to its debt – it would aim to structure as non-recourse debt. This means that the providers of debt finance to the project company have no recourse to the project sponsor. In the event that the project company defaults on its debt the providers have recourse only to the project company and the project company's assets. Lenders have no recourse to shareholders or sponsors in the event that the project company defaults – hence non-recourse. This means that banks in practice are exposed to a different form of credit risk with project companies because the project company and its assets are alone devoted to the servicing and repayment of the debt. This is distinctly different from project companies financed with corporate finance, where banks may share recourse to all corporate assets, depending on the seniority of their loans. It would be incorrect to characterise the risks associated with lending to a project company as necessarily greater than that of a broader corporate claim, but they are distinct, and will tend to alter as the project matures. We refer to non-recourse debt as the ideal debt arrangement in project finance, in that project sponsors from the outset attempt to negotiate it as such. But non-recourse debt is seldom fully achieved in practice; it is most often limited recourse debt, which allows banks some recourse to sponsors or third party guarantors in the event of default or project failure, and which allows a degree of influence for the financiers in any subsequent negotiations, in effect as a control device to draw the sponsors to the negotiating table.

A further characteristic of typical project companies is the extent to which their activities are regulated by contracts entered into by the SPV. Much managerial discretion is curtailed through contracts – meaning that the company by contract is managed as a one-asset company only, with limited if any executive discretion over surplus revenue that may be generated during operation. Often this is brought about through stipulations in the debt contract, which may prescribe the way that the project company is to distribute its free cash flow (Chapter 3) – typically free cash flow is first applied to servicing and repayment of debt before any dividends may be declared. Another important reason for intensive contracting is to manage risks associated with the company's ability to service and repay its debt – recall that typically these companies are highly indebted. These contracts typically cover parties in the whole supply chain through to the purchasers of the project company's output – so we have long-term supply contracts covering raw materials and other inputs used in processing, such as energy, and also long-term contracts to purchase the output of the project – such as the platinum that will be produced by Global Consilium Platinum Ltd. The extensive use of supply and demand contracts functions to stabilise project company net earnings, or cash flow, in order to assess what its safe debt capacity is (safe debt capacity is a relative term, of course; financial risk can never be eliminated entirely). We consider below further how to conceive of project companies using systems concepts.

1.3.1   Why project finance?

There are matters to settle before we continue further. A fair question to ask is why use project finance as one special form of corporate finance rather than using corporate finance for all corporate projects? After all, companies may enter into what are often termed project funding loans, which tends to indicate the purpose for which a business or a group obtains a loan – in other words it is not a general business loan in which specific terms (loan covenants) restrict the use of proceeds. It may be obtained specifically for the purchase and development of real estate (as in a development and term loan), for the expansion of an industrial plant's capacity, or to finance a strategic acquisition. Such loans may also be large, and may also have to be provided by a syndicate or group of lenders acting in concert. The defining difference between these and project finance loans – regardless of their purpose – is the nature of the claim given to the lender – corporate term loans provide recourse to all group assets, and without specific recourse to particular project company assets as outlined above as we are describing business activities at the major corporate level, why should such enterprises consider a special kind of arrangement such as a project financed standalone project company? The multinational corporation (MNC) has the widest possible array of financing options available to finance any feasible chosen project, so what makes it consider using project finance? In principle we can isolate at least three reasons: two arise from financial imperatives, while the third relates to corporate governance. These are corporate consolidated balance sheet constraints, public sector financing constraints, and the contractual discipline and focus that project companies impose upon their managers as a result of funding with project company debt.

1.3.2   Balance sheet constraints

Balance sheet constraints are often cited as a primary reason for using project finance to finance the procurement of large capital assets, and while opinions differ, it mostly remains at least a valid consideration. First, remember that although there are important developments in financial accounting and international accounting rules that may be converging, there is still no single internationally accepted general accounting practice. The financing decision may be affected by accounting rules as formulated in certain jurisdictions, and the consequences of capital investment decisions for a commercial group's balance sheet. Mostly the argument centres around the costs and benefits of achieving off-balance sheet treatment for external financing – the ability to raise finance while not having to state fully such liabilities on the corporate balance sheet. It would then appear to readers of financial statements as if a corporation was less indebted than the true figure. To illustrate: suppose corporate accounting rules prescribe that if a corporation owns less than 40% (say, for example) of another company, and it is not contractually or formally liable for that company's liabilities (say, as a guarantor), then it may not be required to consolidate the associate's debt into its corporate balance sheet – the full amount of the associated company's debt is thus 'off-balance sheet'. This example represents an economically justifiable reason for not consolidating the associated company's debt into the corporate balance sheet as, after all, it is not formally liable for any of the liabilities.[8]

Suppose further that GC is faced with many strategic investment projects, and that it simply cannot afford to allocate a large amount of capital to any one investment project. Recall the case of Global Consilium Platinum, and now remember that GC cannot finance this new development from group financial resources, because Global Consilium Platinum is only one of its many projects. However, by the treatment explained above, if it owns less than 40% of GCP equity, and it is not liable for any of GCP's liabilities, it need not consolidate GCP debt into the group balance sheet. This means if GC begins a joint venture with others to develop the mine, as long as it observes the shareholding and recourse rules, then a joint venture is feasible. Now assume two other companies are potential partners, Rambo Civil Engineering and WA Railways, which agree to construct civil and mechanical engineering works and a railway line to a nearby port. Each company is subject to identical accounting rules, and all face financial constraints as a consequence of limited corporate resources compared to plentiful investment opportunities. If these companies together form Global Consilium Platinum, each owning less than 40% of the shares, minimise the absolute amounts they invest (equity), and maximise GCP's debt, then they control GCP and none of them will need to report GCP's debt on their consolidated balance sheets – if the debt claim lacks recourse to any of them. The way to achieve this is through making the project company alone liable for its debt. This is the stylised private sector project finance company model, and readers will observe these characteristics in the hypothetical project company presented in Figure 1.5. Lenders agree to such arrangements under comprehensive conditions, including removing virtually all operating discretion from the project company, as explained in Chapter 4.

1.3.3   Public sector budget constraints

As companies in the private sector are balance sheet constrained, so the public sector is often also budget constrained. In a case analogous to the commercial example explained above, national, regional or local governments face many spending demands with limited resources, including the necessity to invest in costly infrastructure, whether roads, rail facilities or public buildings. Whatever our ideological views of its desirability, the world seems for the moment to favour a public finance model where the provision of public infrastructure is no longer solely a public sector financed activity; governments everywhere are adopting various forms of private sector partici-pation in the provision of public infrastructure. Most examples of such cases (and

---

[8] This example describes the nature of one particular accounting rule. We generally caution readers against the quest for off-balance sheet finance as a good reason for financing decisions.

possibly the most desirable from many viewpoints) are modelled on the characteristics of typical project companies as described: concentrated equity investment, high non-recourse debt, and a large single-purpose asset. Any equity partners may be subject to a 40% equity accounting rule, but different rules may apply given public sector involvement while any project debt is non-recourse – so neither the public sector nor the private sector is liable. Important here is to recognise that while public and private sector accounting may differ in objectives and applications, the public sector may have just as great an interest as commercial entities in appearing to be underleveraged (implying a lower budget deficit) or efficient (signifying revenue maximisation).

The essential feature of public private sector ventures is that while capital investment in the asset is made solely by a private sector project company, the asset is ultimately intended for public use, so the project company will not own it forever. Depending on the stage of the project life cycle, private sector participation in financing public sector infrastructure assets is usually facilitated by transferring rights to a private sector project company to build the asset or operate the asset for an agreed term, and then return custody to the public sector. This is the well-known build-operate-transfer (BOT) model (and its numerous parallels), used widely in the development of roads, tunnels, power stations, rail projects, pipelines and other landmark projects all over the world. In the final analysis, the public sector obtains its infrastructure without drawing directly on limited public resources, and the private sector has temporary control for profit of a single-asset project company.

The central agreement between the private sector and the public sector that facilitates this form of collaboration is the contract which gives the private sector the right to build and/or operate the asset as a project company, including financing it. Such contracts are commonly referred to as concession agreements which set out details of the agreement between the parties, including the term of the agreement, regulations governing charges that the project company might levy on users, how changes in charges might be regulated, the government's undertakings with respect to restrictions on developing similar facilities, and the project company's responsibilities for managing the facility and handing it over in good condition. We give some insight into the nature of concession agreements in Box 1.1.

The potential to draw private sector finance into projects in emerging economies using the principles explained above has not gone unnoticed with developing states and multilateral organisations such as the World Bank, Asian Development Bank and Inter-American Development Bank. In the last two decades infrastructure projects of all kinds have been completed in Asia, Latin America and Africa with private sector participation in public infrastructural development. In many cases national or regional governments have participated as equity contributors together with international corporations such as Bechtel, Fluor and Enron (*that* Enron, yes), while debt has been provided by promoters, governments, development agencies, and very often by

Box 1.1   Concession agreements.

At the centre of all infrastructure projects with a government or government regulatory body as principal, there is typically a formal concession agreement. Normally, this is a right granted by a public authority to a particular person or corporation stipulating the rules under which they may be allowed to build and operate a public facility such as a power station, bridge or tunnel. In order to encourage private financing in building infrastructure and to minimise the public burden, a concession agreement may be granted to a private sector agent who then assumes the borrowing risk to provide the financial resources to develop a particular infrastructure facility. Concession agreements typically differ with different infrastructure assets, because the nature of risks will differ – for example, while a power purchasing agreement may be at the centre of the feasibility of developing a power plant in that it may stipulate volume and prices, it is unlikely that such a demand guarantee would be found in the agreements to develop a toll road. Concession agreements are an important mechanism to facilitate project finance, because they typically outline several regulatory and operating rules, often including prices for services generated by a facility, price adjustments over time, the term over which such rules are in operation, and possibly may further include taxation incentives as well as other privileges. A concession agreement normally stipulates where the private sector receives a concession from the state, for example a franchise to operate a tunnel for ten years, in return for its building and financing the particular facility. The acronym for this type of arrangement is BOT, which stands for build, operate, and transfer. The content of a BOT agreement may incorporate terms granting the promoter the privilege to design, finance, and construct the project, providing the promoter with ownership and operation rights over the infrastructure, specifying the party responsible for the operation of the infrastructure, and the ownership of the infrastructure reverting to the grantor after the expiry of the concession period.

international banks. In many cases the World Bank and other multilateral agencies have learnt in parallel through co-financing, similar to that of a part-guarantor of third party contractual performance, in order to encourage the participation of international lenders and other investors in developing country projects. It has been argued that often such projects test institutional development in host locations – we return to aspects of such arguments in Chapter 4.

1.3.4   Corporate governance

Corporate governance may be the least prominent of the three reasons that support the use of project finance principles to fund assets but has few opponents, particularly amongst bankers and corporate investors. It has been recognised for some time that conflicts of interest arise between the claims of shareholders and professional managers of companies, more so with quoted companies with widely dispersed shareholders. A principal conflict arises over the use of free cash flow, considered to be the cash flow generated by the corporation in excess of that which it can profitably invest (in attractive projects). Under these circumstances, it has been shown that managers with no or little interest in the company may devote free cash flow to non-essential activities that tend to benefit themselves instead of the company – think of executive jets or luxury cars. One mechanism to prevent such behaviour is a corollary of external debt, which induces managers to focus on generating sufficient cash flow to service and repay debt, and allows little discretion over free cash flow. In many cases, higher leverage would also signify more complex contractual requirements in the terms of the debt, including restrictive covenants (Chapter 4) that provided a direct means to constrain managerial action. Moreover, the best candidates for high debt are companies that can be given a narrow business focus so as to not distract from company objectives, as with leveraged acquisitions of conglomerates in the 1980s, or today's infrastructural targets of private equity funds. The best candidates for high leverage are arguably single-asset companies – project companies – with specific resources in well-known industries, supporting reliable revenue generation and using proven technology. We return to these matters in Chapter 2.

1.4 The project cycle

We explained that project finance is characterised by the generally specific nature of project assets. We also explained that corporations may choose to purchase an existing project company, or to employ a new project company to develop a project asset, all dependent on the economics of the choice and strategic objectives. A familiar concept in project finance literature is the project cycle, a mechanism that we use extensively in Chapter 4 to identify risks that may arise in the project finance model that might require risk management action. We introduce the project cycle concept here to illustrate the fundamental differences in risks associated with different times in the project's life and to hint at how such risks may be managed.

For now, let us return to Global Consilium Corporation and its subsidiary, Global Consilium Platinum. Suppose GCP is considering the development of a new platinum mine, following a breakdown in negotiations to purchase an existing operational mine. Under these circumstances GCP has to go through a typical five-phase life-cycle of a project facility before generating revenue from the sale of platinum. Of course, it is also possible for Global Consilium Platinum to negotiate entry at a different phase in the life of a project under separate development by other sponsors. In any event, the five-stage life-cycle model of project companies is presented in Figure 1.6.

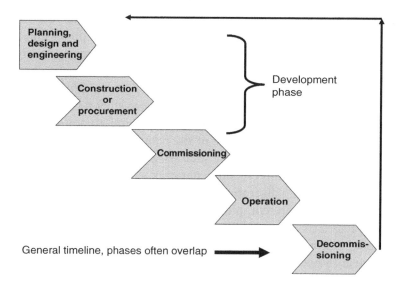

Figure 1.6   The project cycle.

The stages in the project cycle are summarised as follows:

1. Planning, design and engineering ⎫
2. Construction (or procurement)   ⎬ Development phase
3. Commissioning   ⎪
4. Operation   ⎭
5. Decommissioning

The first three can be grouped loosely as the development phase, which for present purposes we may treat as one phase. Each phase carries particular risks, which may or may not be quantifiable, and which all have implications for the project finance model. A critically important fact is that no operating project company exists before the development phase is successfully complete. This means that if development is unsuccessful, no revenue generation is possible, which is hardly attractive. It requires no leap of imagination to realise that even if conceived properly, the risks to the success of a project company are substantial in the development phase (even if manageable), because it is here where the project company either procures a working asset or not. We mentioned that the project finance model evolved around single-purpose assets, with well-developed technologies and in well-understood industry sectors. Transactional familiarity and standardisation in revenue generation also facilitated the development of risk management in the context of many typical projects, which include varying risks associated with the project's maturity. In the rest of this section we outline the context within which the project company functions in each phase of its life so as to appreciate how risks may vary between different phases. As in later sections and chapters, we change the sequence of phases in order to emphasise the importance of the development phase. We therefore discuss the nature of the operational phase first, followed by the development phase, and then make some concluding comments about decommissioning.

1.4.1 Operational phase

Concern with the operational phase in a project life cycle allows us first to conceptualise the project vehicle as a going concern, to use an old accounting term. Recall from the last section that the operational project company will typically have a high leverage ratio, as illustrated by the fictional Baguio Power Company's (BPC) balance sheet in Figure 1.5. BPC is of course another Global Consilium brainchild, and for explanatory purposes, let us suppose that BPC owns a US$2.5 billion, 2000 MW coal-fired electricity generating plant built on a 50 Ha site next to a deep sea port near Manila, the capital of the Philippines. A dedicated coal terminal was developed as part of the power plant's supply chain arrangements, complete with stockpiling, coal handling and road infrastructure to service the plant and link it to Metropolitan Manila's road network.

BPC is a continuous process system, turning coal into electricity on a large scale and at low unit cost for sale to a diversified regional customer base. For practical purposes BPC's electricity generating process may be viewed as fully automated, consisting of four large combined cycle turbine-generators with all necessary technical control equipment, driven by steam produced by a continuous coal combustion process. At BPC's transmission exchange facility, its electricity output is metered formally and fed into the receiving distribution grid. To manage, operate and maintain this process facility BPC has in place a proven team of operations managers and highly competent technical and maintenance professionals from an international pool of project operating teams. BPC's operation can be presented as centred around a continuous electricity generating process (Figure 1.7).

Baguio Power's operations illustrate what we categorise as a process-based single-asset project company. Suppose also that BPC has a power purchasing agreement (PPA) with the Manila Electricity Distribution Board, owned by the Metropolitan Manila authorities, which buys some 80% of its outputs for distribution to its customers. BPC has high leverage and we assume that a major proportion of its cash flow is committed to debt service and repayment. Generally, operational risk is more broadly defined as any and all risks that may cause the project company financial distress during operation. A high proportion of debt in the project company's capital structure requires some observation about financial instruments, and the terms of debt contracts in particular. While long-term supply and purchasing agreements may seek to address certain market risks, many debt contracts will introduce additional operational risks in the form of interest rate volatility, and international debt and input supply and off-take contracts may involve exchange rate risks. We return to these in Chapter 4, but for now suffice it to say that the critical operational objective

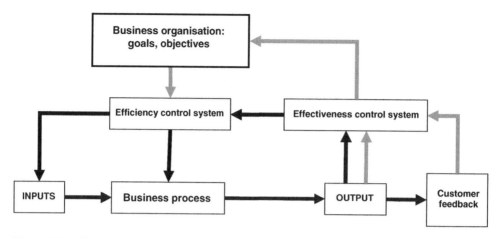

Figure 1.7   The Baguio Power Company – a process-based project asset.

of, and risk to, the project company is inadequate or unstable cash flow and consequent financial risks.

We wish further to highlight three categories of risks that emerge from considering BPC in the operational phase, namely input risks, process risks and demand and distribution (customer) risks. However, we remain mindful of the project's lenders' position first, and so emphasise those risks which threaten the project company's ability to generate sufficient cash flow to service and repay debt. If a large proportion of cash flow is committed to debt service and repayment, the project company would prefer cash flow to be stabilised as far as is possible in order to manage this commitment. We may disregard for now process risks as BPC deploys proven technology; thus its operational risks are concentrated in input supply and supply chain risks (coal), and demand for and distribution of electricity it produces. Both demand and supply are subject to market risks – taken to be volume availability and price volatility.

The essence of BPC's operational risk management is aimed at stabilising supply, demand and price risks by entering into long-term input supply contracts, matched on the demand side by long-term off-take agreements, with both agreements allowing for volume and price adjustments. Prospective financiers may often require matched agreements to be agreed in principle and in place before contemplating lending to a project company, and established formally prior to disbursing funds. With stock-flow type project companies, on the other hand, it may be argued that most risks are concentrated in demand for the service generated by the project asset, say a toll tunnel, but off-take agreements may be unfeasible with such projects but project financers nevertheless commit to lend, usually based upon certainty of traffic forecasts. Two such cases are dealt with in Chapters 6 and 7.

Generally, the project company attempts to minimise its net exposure to operational risks by shifting them to suppliers and customers, and so make cash flow predictable throughout operations, regardless of the behaviour of these market-determined cost and output variables. This allows revenue to be dedicated first to debt service and repayment, and thus gross cash flow becomes the principal determinant of the project company's debt capacity. Think of it as follows: if there were no debt, much of this risk would disappear at a stroke – high debt imposes a range of disciplines on managing the project company in its operational phase. And also, in the world of finance every contract has a probability of not being honoured, including long-term supply and off-take agreements made with reputable suppliers and governments. Not all risks can be managed in a world where humans are fallible or cannot devise perfect contracts, and where the unpredictable often happens. Thus we also see that the world of project companies is further characterised by a spectrum of third party guarantees to compensate for the cost of contractual failure – these are frequently required as preconditions to lending to the project company.

1.4.2 Development phase

While the basic project finance model and the operational phase in a project company's life cycle may be read together as characteristic of typical project companies as going concerns, the project development phase undoubtedly presents the single biggest obstacle to the creation of a successful project, if only for its complexity. Furthermore, it invariably takes a long time to conclude; for example, it may take 3–5 years to develop an electricity generating facility using fossil fuel, and developing a large-scale deep-shaft mine can easily take ten years. If we consider the scope for costly mistakes and unplanned events during project planning, design, engineering and construction, it is to the credit of those responsible that many more financially disastrous projects are not recorded. The complexity of these transactions is legendary, not least because of the scale and capital needs of many projects (think of the complexity of building the Panama Canal). Economically, there is no viable project until the asset produces according to plan in terms of capacity, volume and quality. Before it produces according to plan, there is no cash flow to service debt, or pay dividends to shareholders – there is nothing. This is a project promoter and banker's nightmare, and this stark fact has thus also brought about a fundamental strategic response by project company lenders. We draw your attention to this strategy shortly, but first we embellish the circumstances that surround the development phase.

Of course, before considering the development phase and its pitfalls, one has to assume that the fundamental economics of the project are sound – there is a demand for the service or output generated by the project asset, and it can be satisfied profitably given the associated supply chain and off-take risks. In all, this remains a corporate capital budgeting decision, and whether it represents a strategic investment decision (such as a growth option in real options terminology), or an investment opportunity that satisfies all current corporate investment criteria is not considered here; we assume that the corporate decision to develop the project is justified economically and financially (we present insight into these decisions in Chapter 3). Thus development risks surround decisions about planning, design, and engineering the project, and the directly associated risk that the asset will cost more than it is worth. This explains amply the critical importance of controlling development and construction costs throughout the development phase – particularly if kept in mind the multi-billion dollar scale of the investments. With these assumptions, we briefly consider risks in planning, design and engineering project facilities, thereafter we briefly consider construction risks, and then briefly present accepted mechanisms to manage both these risks.

The fundamental approach to risk in the project development phase is avoidance, whenever possible. Because most of the technologies employed in projects such as electricity generating, chemical process engineering, road and rail development, or tunnelling are well understood, designing or choosing appropriate technologies and engineering solutions for facilities to perform to target outputs and desired quality is commonplace among experienced professionals. There are, however, always informational problems surrounding broad planning of the project facility itself that could lead to serious errors in assumptions about geography, geology, infrastructure, supply chains, distribution chains and many further physical project facility requirements. Dedicated roads, railways and other physical infrastructure may be essential for project success, just as dedicated rail and port development may be required for Global Consilium Platinum or port facilities for Baguio Power. Despite involving experienced professionals and adopting best practices, careful planning and attention to detail, each project is unique, has a different location with its own geographical and institutional circumstances – and thus carries non-replicable risks. It is a challenge to plan new projects without making errors, simply because of the scope and complexity and far-reaching impact that any scheme will have on its physical and economic environment. Possibly somewhat more serious, there may be assumptions about institutional arrangements that are simply mistaken, often in elementary factors such as operating licences and regulatory arrangements.

Project companies have at their disposal a range of options to address planning, design and engineering risks. It is important to understand that such companies as Royal Dutch Shell and British Petroleum have resources commensurate to plan most facilities that they may consider developing. So they conduct much of their planning, if not all, in-house. Other project companies may require planning to be outsourced to a separate professional firm, as their project-based activities may be insufficient to justify economical employment of project planning teams. There may also be the case where the corporation and its project company may be faced with a project that they simply have no previous experience with – it will be an extremely irresponsible project company that does not acquire all possible planning resources. Thus in the world of corporations and project companies there is a range of planning activities that may be outsourced based on the companies involved, their projects, previous experience and budgets.

While there are relatively fewer risks in facility design and engineering than in overall project planning, the construction phase of any project generates further complexities and problems that no planning can fully anticipate given bounded rationality

– the inability to foresee perfectly and plan for any and all future circumstances and contingencies that may arise during the execution of large, complex, capital intensive construction projects, possibly in remote locations with poor infrastructure and problematic logistical facilities (as may be the case with Global Consilium Platinum in remote Western Australia, more than a thousand kilometres from the closest port and with little usable transport infrastructure). Imagine supervising the construction and assembly of scores of thousands of components, and the logistical challenges associated with such endeavours under extreme geographical circumstances. And from a project customer's view, there loom the customary construction industry project risks of cost overruns, missed delivery dates and quality risks, all with potentially serious effects on project feasibility. Despite the fact that cost overruns, missed delivery dates and quality risks are well understood professionally and intellectually, they are viewed in the construction industry with a combination of depressed inevitability and hope that the current project will be an elusive exception. Economically, the worst case for the project company is to have sunk budgeted funds into a fixed and irreversible capital investment, but find the project is incomplete and has output quality problems, for once construction has commenced, under most circumstances withdrawal from the project is rarely a feasible option. It is thus unsurprising that project companies take cost, delivery and quality targets seriously, if not by their own imperative then as a consequence of external professional advice, and frequently also by industry analysts and providers of construction finance.

Fortunately project companies are not passive recipients of construction industry characteristics. They act strategically to manage construction risks, mostly through procurement systems (or contract strategies, as these are also referred to) and financial instruments such as performance guarantees. Risks of cost overruns can be managed through contractual means such as guarantees of maximum costs, or incentives to save costs are achievable with target cost contracts with sharing of savings and overruns, depending on where responsibility is located. Further, in all construction projects there are disputes about delays, quality and costs, so that construction contract dispute resolution is now highly sophisticated. In the world of large projects, however, design and construct contracts are commonly encountered as a mechanism to conclude transactions for large project facilities. Under such arrangements project companies typically invite bids from pre-qualified reputable and experienced design and construct companies, and then engage the successful firm to plan, design and engineer the project facility, in addition to managing its construction. The intention is simply that the design contractor hands over a completed facility with all ancillary works on a particular date, and to an agreed budget at a required level of quality. Such arrangements may include target costs, guaranteed maximum costs or gain-pain sharing agreements, but in essence the principle is simple: the project company employs a design and construct company to deliver a completed and fully commissioned project facility, and only this party is responsible for those actions under the core contract. If this sounds too good to be true, it often is, but it does allow clearly the identification of the contractual performance requirements of all parties, and in certain industries (energy, chemical process engineering) such design and construct companies have grown into giant and respected corporations, and include companies like Bechtel, Fluor and ABB Asea Brown Boveri. Plus, such arrangements typically include third party performance guarantees, and post-contract maintenance arrangements to manage quality. Construction contracting is notoriously complex, but is populated by astute advisors acting for both counterparties.

A last and important point has to be made about the different project phases and differences in risks in the various phases. Recall that the operational project company is often highly leveraged, but also note that the project lenders technically invest long term in the project company once it is commissioned and operational – that is, once the development phase, including commissioning, has been concluded successfully. This is somewhat of a half-truth, though, because it is normal for lenders to provide commitments that they will fund the operating project company on the condition that asset is successfully constructed and commissioned. In this way lenders manage development phase risks by technically not being exposed to these risks at all.[9] But this is further accompanied by appreciating that there is also a profitable lending opportunity in providing finance to the project company to develop the facility. Lenders do take both opportunities by legally separating the two phases, and providing two sequential loans. First, a short-term construction loan is agreed between the lender and the project company (sometimes the design and construct company), to be repaid by the expected completion and commissioning of the project facility. Further, the project company's parent will often guarantee repayment of the loan (or the design and construct company if it is the recipient of a construction loan). The successor loan, providing long-term project company debt, is conditional upon successful construction and commissioning. In this manner, if all goes well, a lender manages to book two large loan transactions in sequence, often with the same counterparties, both with attractive transaction fees earned from arranging the facilities. Further, as construction loans are inherently more risky than loans to a revenue-generating project company, interest margins earned on construction loans will also be higher, relative to the duration of the two transactions. There is more to this, however, and we return to it in Chapter 4.

1.4.3   Decommissioning

Over the last two decades decommissioning of facilities has become a matter of substantial concern, even more so as the reality of the planet's environmental state has moved from society's fringes to mainstream social, political and economic agendas throughout the world. Societies are now entering a new era where accountability for pollution and contamination is demanded and will become increasingly difficult to escape. For example, in some important test cases banks have been held jointly liable

---

[9] Be sure that there are also many cases where banks are less risk averse than this statement implies. Also, banks also lend for project formation without firm take-out commitment.

with corporate borrowers in damages for land contamination. We argue that decommissioning risks are most productively considered in planning and design, by choosing technologies and processes that prevent physical contamination and pollution that will be costly to remedy (given imperfect information about environmental impact now or in the future). For example, while it may have been the best environmental option for Shell to dispose of its disused Brent Spar production platform in 1998 by sinking it in the North Sea, public outrage at that option indicates what may become customary with all environmentally sensitive corporate decisions. We cannot imagine the circumstances that will surround the decommissioning of industrial facilities even one decade from writing. As information becomes more generally available, our understanding of the effects of pollution and contamination grows as never before, and problems identified that were once never considered, as with the use of asbestos insulation until the latter part of the last century. In all, our point of departure is that decommissioning critically represents a matter of corporate values, which relates to the attitude with which decisions are made in the present. As the saying goes, '*. . . as the circle of knowledge expands, its border with ignorance increases . . .*'. Because these are complex issues, we view pollution and decommissioning more philosophically as part of our discussion of institutions in Chapter 4. It should be noted that increasing attention to ethical investment in developed markets may induce more responsible corporate behaviour in these respects for reasons that are entirely driven by share price concerns.

1.5 Systems concepts and the project company

We alluded above to the fact that we will borrow selected systems theory concepts to form a framework to consider risk and project companies, and we now turn in this direction. As with other theoretical concepts in this book, we will concentrate on those systems ideas that are immediately useful, and will thus sacrifice many of the intellectually appealing aspects of systems theory. However, be sure that there is nothing as practical as a good theory, and so it is with systems.[10] For our purposes, think of systems theory merely as an analytical tool – a way of ordering complex environments for further inquiry, be it our ecological, socio-economic, technological, commercial, political, institutional or social environment. One of the great applications of systems thinking is its ability to facilitate comprehension of complexity in our environment. Since the late 1980s there has been something of a renaissance in enquiry into complex systems; think of research at the Santa Fe Institute in New Mexico into the new science of complex systems, much of it inspired by concerns for

---

[10] Systems theory is likely to enthuse some readers who venture to read more about it – there is an exciting journey ahead for those who do read more (Senge, 1990; Checkland, 1999; Blockley and Godfrey, 2000).

the environment, climate change and complex social systems.[11] For our purposes, however, the system we will be dissecting is the project company as a business venture, also because many of the risks it faces are determined by the company in its technological, geographical, socio-economic and institutional environments.

An early note about complexity will be useful. What do we mean by complexity? There is no single definition or description that does justice to the width and depth of this concept. We may have technologically complex machines, built by industrial processes that display relatively little organisational complexity (for example, a car assembly plant). The companies that build the vehicles, however, may exhibit higher complexity, but of a different kind – social and organisational complexity. Alternatively we may have a technologically relatively low complexity project, say a toll road, needing to pass scrutiny by a range of stakeholders in a politically and socially complex process. Immediately we see here more than one side of complexity: technological complexity, situational complexity, process complexity, or institutional complexity. In Chapter 2 we return to transactional complexity as one particular concern with this term.

First, however, we need to introduce selected systems theory concepts as useful vocabulary to make sense of organisations, products, processes, and more. We introduce three concepts which we consider important to facilitate initial comprehension of complex business and project arrangements. We introduce the fundamental notion of emergent properties and systems functions/objectives, and then discuss the important concept of systems boundaries – critical to understanding risks in project company environments.

1.5.1   The first principle of systems: emergence

We have to start with a somewhat technical definition of a system. For our purposes a system is seen to have at least three properties.[12] The first is that all systems have an emergent property. Kramer and DeSmit (1977) define a system as:

'. . . a set of interrelated entities (parts), of which no subset is not related to any other subset. This means that a system as a whole displays properties which none of its parts or subsystems has.'

As an example, a bicycle is a set of parts but when assembled, a transportation system emerges as the assembly's higher order function. Analysing each individual

---

[11] We highly recommend the Santa Fe Institute's work as an important entry point to understand the problems of complexity as it most certainly applies to social, ecological and open systems generally. A very accessible entry is found in Waldorp (1992). This is a wonderful counterpoint to the frustrations that many experience with the logic of the scientific method when complexity is evident.

[12] Many systems theorists expand the set of essential requirements of a system. We are not going to bore the reader with the many differing views; we believe that we need only convey the essential principles to use systems concepts fruitfully for present purposes.

part (entity) of the bicycle in isolation cannot explain the emergent property of the whole – it is necessary also to consider interrelationships between parts. It follows that parts of the system may also be systems; this is the origin of the term subsystem. So an entity, or a part, can be a subsystem. The bicycle's pedals, cogs, gears and chain collectively form the locomotion subsystem (requiring human pedal power, though); the brakes form a subsystem, and in this case it may also be viewed as a feedback and control system, which allows an operator to adjust the system's behaviour according to what may be desirable, or required by an external prescribed norm.[13] It may not be socially acceptable or legal to travel at 160 km/h in a city on a bicycle.

A durable idea that has entered and left fashion with social scientists a number of times is that all forms of human organisations can be viewed as systems, including business organisations (of which project companies are a subset). We represent a typical organisation as a system in Figure 1.8. This could be any human organisation, such as a school, a statutory body such as a city's road traffic authority, an industrial company, a state or provincial government, a regulatory institution such as a central bank, or a stall operator in a street market anywhere in the world.

This representation views a business organisation as a system which transforms inputs (raw materials, energy, labour) through some business process (possibly utilising capital equipment), and produces an output – in the case of a commercial enterprise this will be a product or service to be sold to customers, such as electricity from the hypothetical Baguio Power Company, or road access from a toll tunnel. However, the business organisation cannot simply produce products or services in some abstract way: it is part of the economy of some society, and will be influenced by how it fits into a wider economic system. In essence, any single business output is itself part of

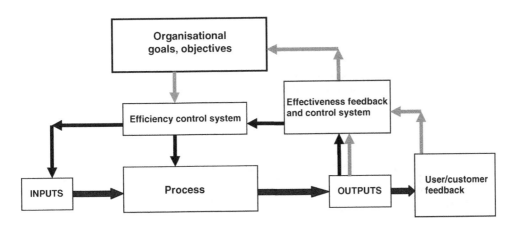

Figure 1.8 A simple business system.

---

[13] The study of control/feedback systems is termed cybernetics – The 'Terminator' is a cyborg, short for cybernetic organism.

another organisation's inputs; similarly the business may use another organisation's outputs. In this way a national and the international economic system are integrated in a complex matrix of input-output relationships: natural gas is piped to a power plant, used to produce electricity, used in manufacturing of vehicles in Germany, which requires transportation infrastructure, to transport food to markets, and to transport maintenance personnel and equipment to manage the power plant and pipeline, and so on. In ecological systems we have the most amazing and complex nested input-output phenomena – think of the richness of life along a river system as an example. Often the word holon is also used in this context, broadly indicating connectivity of parts within subsystems within systems, from which some property emerges.

In the wider economic system, any business organisation in a well-functioning market economy will be subject to forces and constraints which will affect its functioning. From an economic perspective there will exist economising constraints that may result from various sources: costs of inputs, price and volume of outputs, process costs, or users' satisfaction with the service. In typical business systems there are two feedback and control mechanisms, simply termed efficiency and effectiveness control systems. In Figure 1.9 we present these as separated subsystems, since not all their feedback signals are necessarily common. We may simplify the efficiency control system by presenting it as primarily being concerned with economising – appropriate use of resources, waste elimination, and so on. The effectiveness control system is concerned with the business entity in an organic and dynamic economic environment, where business organisations respond to their environments in order not to be rendered superfluous. In order to do so, the business system requires further basic

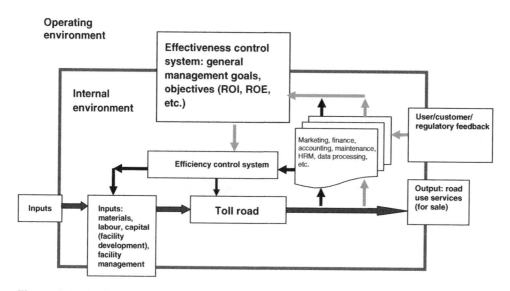

Figure 1.9 A slightly modified business system.

inputs – goals and objectives, and feedback mechanisms to control the system according to the goals and objectives. This is the effectiveness control mechanism, but we show that an important input into the effectiveness control mechanism is efficiency – in most circumstances efficient use of resources remains a critical concern.

1.5.2 *The second principle of systems: function, purpose or goal*

Before considering systems objectives or functions critically, we need to take a step back to consider emergence again. In nature, we may say we see that a system's function emerges – it has come about. One emergent result of forested hills is soil preservation and ultimately flood control, but humankind may have come to realise a little late just how important this function is in the complex input-output system of a river basin. There is a difference between emergent functions in a natural system and conscious human decisions to design and engineer systems to have some function we desire – in systems design terminology, the objective of the system being developed (more correctly the intended objective) is the emergent property we desire. Thus, the second fundamental principle follows from the above definition, and that is that a system (and therefore a subsystem) has a function, or purpose or goal. In natural systems it is from the interaction between parts and subsystems that the function of the system (the whole) emerges; for example, an ecological system sustains life, amongst other things. With design of human organisations and physical machine systems such as a bicycle, the process is mostly reversed in that the emergent property is defined first (what is the purpose of this organisation?) and then it is designed, developed or engineered to achieve this function or objective.[14] Subsystem interactions and the system's emergent property are often not directly observable, for example in human organisations or biochemistry, but particularly in physical systems such as machines they are typically clearly observable, as with our bicycle above. We may term a bicycle a hard system, because its function is clear, the interaction between parts and subsystems is clear and predictable and can be carefully engineered for efficiency; its behaviour will likely be as intended, and so on. We may also say this is a purposeful system. Some hard systems can be engineered to an exquisite degree, as the manufacturers of the world have shown with cars, consumer electronics, computer hardware, and so on, but it seems that we have only started learning with more complex hard systems (like applications software).

For analytical purposes, most business literature considers systems theory to be an appropriate model to analyse businesses as human organisations, and we therefore also consider it appropriate to apply it to project companies. When viewed as a system, a business enterprise would have some intended function or purpose reflecting the intention of its creators (as difficult as this may sometimes be to imagine). Recall our

---

[14] We must also immediately say that many human/social organisations evolved historically into their current forms and were not consciously designed (some political systems, for example).

rendition of Global Consilium Corporation's corporate strategy/planning machinations in Section 1.2, and that this process relied explicitly on a statement of corporate objectives – for our purposes, these become the system's objectives. Cynics might say that any business has only one objective, and that is to make money, but this loses sight of the fact that all business organisations ultimately have some function in highly complex societies (remember, by definition a society's culture includes its economy). Until a better metric of effectiveness is devised, money unfortunately continues to be the principal measure of a business' success in fulfilling its objectives.[15] Businesses are social systems created by humans to pursue some objective in the economy, but because humans are purposive and human motives and incentives often cannot be directly observed, it means interactions between a business's functional parts and subsystems (marketing, finance, accounting and administration, human resources, production, general management) are not always predictable, and to a significant extent how the system itself ultimately may behave is not predictable. Human systems are often also described as soft systems, which cannot be finely engineered to achieve well-defined purposes with the expectation of much success.

The concepts of a system's function or purpose, and that of control systems are closely related. Because unpredictable disturbances may enter a system which may cause its actual performance to deviate from its intended performance, feedback about system performance and adjustment mechanisms are necessary. Feedback/control systems must generate signals to allow the system to adjust or compensate for disturbances, and it therefore follows that subsystems, depending on their complexity, may in turn have further control systems. In systems development, the integration of subsystems may in fact be a matter of integrating a number of nested control systems, as may be the case with a chemical process engineering plant. Figure 1.9 may well represent for our purposes a toll road leading from one city to another, built and operated by a private project company under a BOT arrangement with an appropriate transport authority, with the expressed hard objective of providing users of road transportation vehicles access to high quality, safe and convenient road transportation at an affordable cost. For the project company, one control metric may require that this be provided in a profitable manner to shareholders, while another control metric may be to beat low accident targets to comply with its operating concession. The system (the toll road) is designed, engineered and built to achieve the planned objectives, or functions.

1.5.3   *The third principle of systems: systems boundaries and systems thinking*

An important principle surrounding systems planning, design, engineering and operation of a business system such as a project company concerns subsystem interaction

---

[15] Perhaps accounting systems may still develop to broaden this metric, but we are still grappling for proper multi-purpose metrics that can reflect accurately sustainability.

and systemic behaviour. It is generally accepted that when designing (or developing) a system to achieve a particular function or purpose, and each subsystem is designed to be as efficient as it can be or each part of the system is optimised, it is likely that the system itself will under-perform – it may not be as effective as it could be. The very best braking system for a bicycle may make it too heavy; the very best mechanical systems in a building may undermine feasibility; the very best technology in a manufacturing facility may add no value to customers while undermining unit cost competitiveness; and the very best individual players may not combine into the best football team. This is the difference between efficiency and effectiveness in systems thinking: efficient parts may not integrate into an effective whole.

In system design and development optimising individual parts is known as reductionism. This follows from the first principle, the fact that in systems the sum of the functioning parts is less than the functioning of the whole – often also described as synergy. For example, note in Figure 1.9 that we identify two control systems: an efficiency feedback, and an effectiveness feedback system. For our purposes, we note that a business system consists of hard and soft subsystems. Efficiency could be viewed as being concerned with how well the system is performing its hard functions – how energy efficient is it, how low are unit costs, is the quality as planned? Effectiveness may be viewed as incorporating its soft, human systems and the interface with the hard systems – how effective is the project overall in achieving the project company's objectives? It may be efficient in achieving its output, cost and quality targets, in a hard sense; but overall it may not as profitable as had been planned for. Thus it may be viewed as not as effective in achieving its goals as hoped for. Thus some adjustment may be required – change in inputs, throughput, quality, pricing, marketing. In all, we may consider that effectiveness is concerned with overall performance, and efficiency with the specific performance of subsystems. Of course, we have to alert readers again to the fact that there are no rules about when efficiency and when effectiveness becomes the measure – it will depend on a system in its environment where such boundaries are defined. We start dealing with these concepts in the following subsection.

Reductionism in systems design and development can undermine a system's intended function, and systems thinking is fundamentally non-reductionist. Systems thinking fundamentally requires that we firstly regard reality in terms of the effectiveness of the whole and not optimality or efficiency of subsystems or parts. From these first principles almost all of systems theory is developed, and for our purposes we will isolate a further important concept that is relevant for analysing risks and the project finance company. We need to introduce the third systems concept, namely the system's boundary, which is intended to define where the system under analysis ends and where the system's environment begins. While our last thoughts on reductionism urged us not to be reductionist, view the world systemically and avoid optimising subsystems performance, we also know that in practice, to use Checkland's words (1999:60):

'... *Cursory inspection of the world suggests it is a giant complex with dense con-nections between its parts. We cannot cope with it in that form and are forced to reduce it to some separate areas which we can examine separately'.*

Unfortunately we have to be reductionist on occasion to achieve practical outcomes, but this is forgivable only if we think systemically first. This is the context within which we attempt to understand and use the concept of a system's environment and its boundary when analysing risks in project companies as systems.

1.5.4 System boundaries and the system's environment, open and closed systems

In general, a systemic view considers the economy, ecology, and human institutions as a set of interlinked systems with increasing complexity. In order to make sense of a system's behaviour and how feedback mechanisms may have developed (as in eco-logical and biological systems), or which feedback mechanisms may be appropriate when developing business systems or other human organisations, it is necessary to define as clearly as is possible the system's boundary, which effectively separates the system from its environment. As we extend our view outwards from a given system we consider those things that influence the behaviour of the system, termed the sys-tem's environment.[16] A rich systems perspective will include more of these concerns in an analysis of any system within its wider systems environment and at various levels of aggregation while a narrow(er) systems perspective attempts to isolate systems from the richer systems environment possibly for closer analysis, problem identification and possibly intervention (as may be the case with project companies or other business organisations). In essence we wish to concentrate more on the logic of a business system's boundary and environment here, because insights into what can be achieved with risk management depend in great measure on identification of the boundary of the venture under analysis. This is a most critical concept in under-standing the limits to what risk management and risk mitigation can practically be expected to achieve.

In essence, a system's boundary often depends on what the particular purpose of a system's analysis may be. Of course, if a system has a boundary, it implies the bound-ary is between it and its environment. First then, we narrowly envisage the system boundary as between the system and its environment. A *closed system* is a system which does not interact with its environment – one could say the environment does not exist in the case of a closed system. One could envisage a solid-state electronic component such as a semi-conductor as a closed subsystem – its output signal depends

---

[16] Paradoxically the levels of complexity may also increase as the level of aggregation decreases, as we may observe with particle physics and quantum theory, and complex biochemical processes such as the functioning of the human brain.

entirely and directly on one input signal only, and its control mechanism (if any) is independent of the environment. In contrast, we speak of an *open system* when a system does interact with its environment. Defining an open system's boundary can often be a confounding problem, and what's more, it happens that boundaries may be defined differently for different purposes or, more frustrating, boundaries may not be stable. We may explain an open system as one where there is a set of entities outside the system, which does not belong to the system, but could influence the system or could be influenced by the system. With human organisations it is to be expected that system and subsystem boundaries may not be stable, that entities outside the system/subsystem may influence its behaviour, and this is in fact one of the risks to manage or mitigate in project finance arrangements, particularly with international ventures. We also have to deal with the concept of open and closed systems when a system's boundary is under consideration.

In order to illustrate the nature of open and (somewhat) closed systems and their boundaries in project companies, consider again the Baguio Power Company. The Baguio Power Project is a 2000 MW coal-fired electricity generating plant located next to a deep sea port close to the city of Manila in the Philippines. BPC is a continuous processing system and produces electricity at large scale and low unit cost. As part of this massively capital intensive facility we have numerous process control systems that control the rate of burning, steam distribution, emissions, and so on (the hard systems), including control mechanisms that manage technical goals for the project (overall efficiency of factor inputs). Together these describe the business system's internal environment. In a narrow sense we can represent the internal environment and its boundary as illustrated in Figure 1.10, itself an expanded version of Figure 1.9.

For practical purposes BPC's electricity generating process may be viewed as fully automated. At BPC's transmission exchange facility, the electricity it produces is metered formally and fed into the receiving electricity distribution grid. In Figure 1.10 we see that the plant operations management team is supported by the normal business functions that provide efficiency and effectiveness information for control purposes, such as accounting and administration, facility management, data processing, and so on. This information function effectively informs both the effectiveness and the process control systems. Together with the electricity generating system itself, this is presented as the system's internal environment.[17] The effectiveness control system is part of BPC's internal and operating environments, and it reacts to BPC's relations with GC's corporate objectives for BPC, such as return on invested capital (RIC), return on equity invested (ROE), and other corporate objectives. From here it controls the internal environment through the process control system. This is to illustrate that BPC's management obtains signals from the operating environment

---

[17] For the sake of economy, this curtailed description neglects altogether the detail of electricity generating plants' operations.

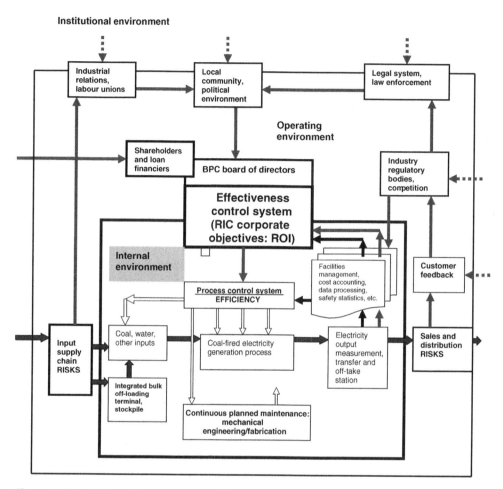

Figure 1.10 BPC coal-fired power project: internal and operating environment.

and adjusts objectives if required, which would likely require effectiveness and efficiency objectives to be altered accordingly. In business systems, we thus differentiate the system's relevant environment into what we described as the internal environment, possibly containing the production process itself and the normal harder functions such as accounting and administration; the operating environment, concerned with the business in its competitive environment; and the institutional environment, the regulatory bodies, laws, norms, customs, etc. in the society in which the business is located.

This characterisation of a business's environment into internal, operating and institutional environments is useful because each of these indicates an increase in the level of complexity in the environment, as well as a decreasing ability to exert influence over the environment. Each level is also associated with a higher noise to signal ratio – less certainty about information. For example, typically a well-managed business would have good information about its internal environment, and possibly less specific information about competitors and other factors in the operating environment. The institutional environment requires special attention, because there can be excellent information with well-developed institutions, or very poor information with weak institutional environments. Of necessity there are some overlaps in this categorisation, but the framework explains how one could conceive of a project's environment and where the boundaries between the different levels of environment may be. At some stage in analysing a system, it becomes clear which factors may affect the project's operating environment and lead to financial risks, and it also could become clear which risks can be managed or mitigated and which cannot. It is clear that finding the boundary(ies) between the project company and its various categories of environment is important to understanding where risks originate and how they may be approached from a managerial perspective. In Chapter 4 we return to the project company's environment as an analytical mechanism to help identify risks in the project company's various levels of environment, in order to conceptualise risk management activities. We also devote part of Chapter 4 to the institutional environment, where the project company is least able to influence its environment, and may be most vulnerable to being influenced.

1.6 Plan of the book

Recall that the definition of project finance introduced in Section 1.3 implies that at least three interdependent elements should be included in the analysis of project financed ventures. These are:

- the economic/business unit under study (the project company including its business model and environment, earnings/cash flow, and risks)
- the nature of the assets deployed by the business unit (highly specific assets)
- the business unit's financial structure (with particular attention to the business unit's debt, and its servicing and repayment), including agency conflicts in finance and management of the project.

In order to consider the context within which any project company is created and operates, an unstated fourth element also matters. In our view a useful analytical framework should facilitate an integrated analysis of any project finance venture, and facilitate identification and explanation of decisions surrounding the first three elements when consideration is given to the institutional environment of any project finance transaction. Integrated analysis is also intended to facilitate forming of ideas about planning, design, engineering, construction and financing project financed ventures. All this implies that a particular viewpoint for such a framework is required, and we apply systems theory and institutional economics concepts as general paradigms to consider the economic and financial nature and structure of project companies, and systems theory concepts in general to maintain functional logic for the presentation where required. All of what follows in this book is either an elaboration of or application of concepts introduced in Sections 1.2 to 1.5. The overall aim of our book is thus to produce a text that details an analytical framework drawing on applied institutional economics, which facilitates analysis of the logic which underpins the structure of generic project financed arrangements. Our approach is based on an analysis of the institutions that govern project financed transactions, the economics and agency risks of costly contracts (transactions), and risk management concerns derived from these analyses.

The framework outlined above provides the logic for the book's structure. Chapter 2 introduces the world of complex transactions. We use basic institutional economics concepts to build an image of complexity in single, large, unusual economic transactions, as is the case with large single-asset project companies. In the process we introduce the fields of transaction costs, contracts and concepts of agency theory and incentive alignment as part of the costs of transacting. We point out where the project finance model with high asset specificity and typically high indebtedness locates within the corporate finance environment. Many concepts identified in Chapter 2 require a further institutional economics context, which is made more explicit in Chapter 4 where we consider the importance of institutions in concluding complex transactions successfully.

Chapter 3 represents an entirely conventional approach to the three most important financial decisions in typical project financed companies. We firstly consider the fundamental capital budgeting decision faced by any corporation that is considering investing in a facility such as a road tunnel or other typical project. Thereafter we consider the financing decision associated with such a decision, and we also consider the financial circumstances that surround the construction phase of such a venture.

Chapter 4 considers broadly how financial risks originate from the project finance business model. We outline generally approaches to risk management, and present introductory explanations to a range of risk management mechanisms including a number of important financial instruments such as options, futures and swaps. The emphasis of this chapter is, however, on explaining an approach to risk identification based on systems concepts, and we do not attempt to provide the detailed machinations of any particular risk management approaches. In Chapter 4 we also present a compact introduction to the institutional context within which project companies function, and indeed within which all commercial activity is conducted. We identify important institutions in (normally) well-functioning economies, and point out their relevance to identification of risks that face project companies, as well as their relevance to the success of project ventures, particularly in an international context. We note the importance of government, legal, corporate, political and regulatory institutions that influence project ventures, and also consider the risks generated by unstable or poorly developed institutions.

Chapter 5 considers a further category of complex transaction, namely PFIs and PPPs, and a range of transactions with catchy acronyms. We present these as well-intentioned variants on the idealised project finance arrangement motivated by public sector budgetary constraints, but with important differences. While the typical project finance transaction is based on a narrowly defined business objective, a single highly specific capital asset and relatively clear risks, the PFI and PPP structures seem to test increasingly the logic of the project finance business model by structuring transactions around less well-defined assets with more complex income generating characteristics than typical project ventures (schools, hospitals).

Chapters 6 to 9 present case studies, with both general and specific purposes. Application of the frameworks introduced in Chapters 1 to 5 in the text will be illustrated through a number of project cases from a number of world regions at different stages of political, legal, regulatory and financial sector development. The intention is that while each project case will have a general integrative theme, it will also be used to illustrate one or more specific project finance principles highlighted in the text.

Chapter 6 outlines the circumstance surrounding the recent development of the Sydney CrossCity Tunnel, a toll-road tunnel that provides an alternative route to surface road users in the busy and traffic congested Sydney central business district. This case introduces decisions concerning project special purpose vehicle structuring considered in Chapter 3, and risk identification concepts in Chapter 4. It further hints at concerns with PFI and PPP concepts.

Chapter 7 also considers a toll-road tunnel, this time under Hong Kong Harbour. We present the Western Harbour Crossing, a project that forms part of the infrastructure created around Hong Kong's international airport. This case is concerned with two particular financial decisions raised in Chapter 3, namely the capital budgeting decision and the financing decision. In particular, it is concerned with the estimation of the scale of a loan decision pending a hypothetical decision to purchase the West Harbour Crossing SPV by investors.

Chapter 8 explains the institutional problems and conflicts surrounding the infamous Dabhol Power Plant close to Mumbai in India. Focusing on concepts outlined in Chapters 2 and 4, this case relates a range of problems from project company risks associated with political opportunism to issues with enforceability of third-party guarantees in power purchase agreements, to problems that may be described as a lack of credible corporate commitment by project promoters.

Chapter 9 returns to PFI and PPP arrangements outlined in Chapter 5. We present the problematical attempts to raise capital to modernise the London Underground system through private sector participation. The case explains problems associated with the nature and state of the system's assets, which draws on asset-specificity concepts introduced in Chapter 2.

Key concepts

The following concepts are considered sufficiently important to memorise as key vocabulary for use in subsequent chapters.

Corporate finance
Emergent properties
Financial structure
Internal environment, operating environment, institutional environment
Non-recourse, limited recourse finance
Project companies
Project finance
Systems model, closed system, open system

2

Complex Transactions

In Chapter 2 we step back from the broad concepts of project finance and the project company business model. Here we return to basics, and introduce ways of thinking that allow the elemental analysis of commercial projects. These centre on the idea of complex transactions, and in particular how they are viewed in economic terms. Much contemporary project finance literature accepts many such concepts without question, as if controversy took no part in commercial analysis or judgement, but familiarity is by no means a cloak that we wear with comfort. This chapter will thus put into context the essential vocabulary used throughout the rest of the book, but it is important to remember that one purpose here is to simplify terminology and not join controversies that lie behind some of these matters.

We draw on insights from an eclectic array of analysis including the economics of industry, institutional economics, transaction cost economics, law and economics, information science, aspects of behavioural theory of the firm, agency issues in finance theory, and property rights considerations. Some ideas are drawn from contemporary game theory and evolutionary economics. We use new institutional economics primarily to explain that organisations exist for reasons of transaction costs, taking the long-established view that any commercial organisation represents a vertically integrated alternative to what otherwise might be a collection of separately defined activities. But these notions are also important to maintain clarity in categorising ideas that straddle the analysis of both transaction costs and institutions. This chapter thus presents a review of the concepts required to expand understanding of the vocabulary used in Chapter 1 and elsewhere in the book, and helps in particular to explain where to locate the project finance model within a corporate finance framework, and why. More will be said about institutions in Chapter 4, especially the practical concerns of promoters of project finance in international markets.

The chapter is structured as follows. First we introduce what is meant by transactions and how we use the terms transaction cost economics and the expense of transactions. Our point of departure places transactions centrally in a simplified approach to the economic analysis of project finance. Rather than base our assumptions on perfect information, optimal contracts, profit maximisation or a rationalist view of economic welfare, we take a less abstract and expansive view of human economic behaviour, including its fallibility and a continuing desire to increase economic efficiency or reduce waste – in this case with transactions. In Section 2.1 we explain the fundamental ideas of transaction costs as used in the rest of the book, and introduce the concept of transaction governance. Section 2.2 presents a more formal model of elements of transactions, drawing mostly on concepts from financial contracts. This model forms the basis for analysing how cost is incurred in transacting. Section 2.3 then considers the influence of delegation to agents of economic functions and activities, and the impact that it may have on the cost of transacting through the necessity to control and monitor those agents. In Section 2.4 we (re)state fundamental principles of incorporation and corporate finance, to develop context for Sections 2.5 and 2.6, where we relate the concepts to project companies and project finance, specifically using transaction costs and agency theoretic logic to explain the project finance model.

2.1   Transactions, the cost of transactions, transaction cost economics, and projects

It is a neglected notion that while profit and revenue maximisation are held to be established human traits, so also is a tendency that is a corollary in terms of behaviour. This is the wish to economise in all kinds of endeavours, commercial or otherwise, or wherever possible to reduce loss or wastage. Economic incentives appear to be at work equally at both ends of this welfare spectrum; increase gain, certainly, but try to reduce loss as a fundamental part of the same equation. In a sense this explains why we choose transactions as the anchor for much of the analysis presented in this book, because market economies are concerned with the exchange of goods and services, and the efficiency with which such exchanges take place.

We thus open discussion of transactions from the premise that economising in costs of any kind is as powerful a motive as profit seeking. In our framework we use exchange and transaction as synonymous, but we prefer transaction as being more useful in the way it captures the completeness of both the process of bargaining and the resulting bargain itself. Of course, transactions may be based on simple goods and services or the most intricate projects. A simple transaction may involve the purchase of an amount of fuel oil; a complex transaction could be the acquisition of an oil refinery by Global Consilium Corporation. If Global Consilium decides to construct a new refinery instead of buying an existing facility it essentially acquires a virtually identical asset but employs a principal contractor (as its agent) for future delivery, rather than engage with a seller company, making the bargain somewhat more complex. To risk labouring the point, transactions for the purchase of services can also be simple, say buying the right to drive through Global Consilium's toll tunnel, or may be a transaction for a far more complex service, such as appointing a bank to advise on corporate financing strategy and arrange a large loan facility. There are countless ways in which goods and services are categorised to create a typology of transaction types (and sophisticated typologies such as the international Standard Industrial Classification (SIC) are widely used) but we imagine instead a simple typology as presented in Box 2.1.

This typology is self-serving, and allows us to identify two important points. First, some transactions (e.g. for industrial commodities or household foodstuffs) occur frequently, will be standardised and usually take place in markets in large volume. These goods are largely undifferentiated and relatively homogeneous. Others, such as transactions for consumer durables, occur less frequently but in sufficient number for the goods involved similarly to be largely identical, although they may be technology intensive, as with digital cameras or personal computers. For practical purposes, these goods may be considered largely homogeneous, with only minor functional differences. Other transactions will occur infrequently and are likely to be highly capital consuming so as to lead to scale economies; an example would be the transfer of industrial process plants. These bargains will typically involve non-standardised or heterogeneous products, although standardisation in subsystems (e.g. in power turbines) and processes (such as oil refining) will produce scale economies in the design, engineering and delivery, and is always an objective. In addition to being inherently complex systems composed of several subsystems, such facilities are usually also situationally complex.

Box 2.1   Basic transaction categories.

Consumer and household goods and services, such as food, transport, etc. (some essential, some not).

Consumer and household durables, such as household appliances.

Industrial production facilities, such as manufacturing facilities, electrical power plants, oil refineries, mines, roads (capital goods).

Industrial commodities, such as electricity, iron ore, platinum, copper, standard electrical components, microprocessors.

Business and personal services, such as banking and financial services, medical, consulting services including marketing, design, information systems, professional services such as accounting.

and so on.

A further point we observe from Box 2.1 is that a transaction typology of this nature may be constructed based on the information set that is part of a buyer's expectations, and a seller's representation of the good or service at the time that the bargain is made. Thus some categories of goods are transacted where the product or service is *information-simple*, as with a quantity of crude oil purchased on the spot (cash) market. In this case, the transaction would also be expected to be simple because the information set needing to be matched by the transacting parties is simple, has become standardised over time and is subject to high dealing volumes. Other transactions may be described as *information-complex* or *information-intensive*, such as the consensual acquisition of one company by another – the expectations of the buyer or the representations of the seller can be defined with far less certainty than the sale of a simple commodity, nor can it be said clearly what may be an efficient exchange of transaction information.

Two concerns emerge from considering the context within which transactions take place and the information complexity of the good or service being exchanged. First, with information-complex transactions such as the acquisition of a company, the actual case is also expected to be complex: it is certain to involve extensive due diligence and other legal and official requirements. Second, information may be skewed among the transacting parties – one begins the bargain knowing far more about what is being exchanged than the other. In these circumstances information is said to be asymmetrically distributed, and we may also expect the transaction process to be complex because so much information must be uncovered and representations about it tested or verified. If nothing else, it may take Global Consilium Corporation quite some time to conduct due diligence and assess the company it hopes to acquire, and problems of all kinds may be revealed by the process before the transaction is formally concluded, not least since they will affect the final terms of the contract of sale, and the means by which the purchase price is financed and paid.[1]

No leap of imagination is needed to conclude that the more complex the information set to be put before contracting parties, or the more unusual or heterogeneous the subject product or service, so the more demanding it is to arrange and complete a transaction. We term this the process of transacting. The more complex the process of transacting, the more costly it is likely to be to execute – regardless of the basic value of the good or service under exchange. The least expensive and most efficient process of transacting discrete homogeneous goods occurs in well-functioning markets (markets need not be perfect for this to occur). But complex, one-off project facilities are not purchased in open markets with many buyers and sellers. The transaction process involved in acquiring an oil refinery or bringing on-stream a deep underground

---

[1] In market practice, sensitive acquisitions may involve a forensic audit but it is far more usual for the sake of despatch for a buyer to conduct more modest due diligence over a limited period, highlighting particular issues, and in executing the deal to rely upon conditional contractual warranties and representations made by or on behalf of the seller.

mine will be both intricate and expensive. It is fair to say that while there is no continuing market for the purchase of new oil refineries per se, markets do exist in a more generic form in elements including subsystems, and design, engineering and management skills that form part of any new refinery.

The organisation and management of transactions is a focus of the new institutional economics. Note that in this discussion, there is a conceptual distinction between out-of-pocket expenses associated with completing a project or transferring its ownership between buyer and seller, and the bundled transaction costs inherent in any single project, from the simple to the most technologically demanding. The two ideas are related, but the first – the expense of concluding a bargain – is relatively trivial, even if large in relation to the purchase price, whereas transaction costs are both important in their consequences and fundamental to the way in which a project comes to fruition. Remember also that the most complex industrial project can involve modest contractual expenses in its formation or sale, but will tend to be associated with high transaction costs. This can most easily be seen by considering the case of the sale of a complex and capital intensive industrial facility – think of nuclear power plants or a pulp and paper mill, for example. Everything about the facility will be formidably complicated in its engineering and construction; all aspects of operations are multilayered and protracted to bring about. No-one could contemplate replicating the working of the facility with individual contracts that each defined a single indivisible task in the process because the result would be unwieldy beyond all measure. In certain industry sectors a semblance of such laborious work is undertaken; for example, the manuals for large commercial ships are usually a multiple volume explanation of every function on board the vessel.

If we can describe a single project as standing in place of an infinite number of contracts, then we would characterise that project as being associated with high transaction costs. Relatively closed markets for projects are likely to have comparatively low transaction expenses in relation to their financial value. In each case, the out-of-pocket expenses of locating a suitable project (search costs), information gathering and due diligence are comparatively low because buyers and sellers are mutually known and highly informed. But we would expect the transaction costs associated with any modern industrial process to be substantial, simply because the project represents an agglomeration of many thousand discrete functions that could each in theory be undertaken by bargaining in an open market.

Information thus plays various roles in all transactions. It is an inherent characteristic of the good or service being bought or sold that an information set (from the simple to the very complex) relating to the good's functional characteristics must be known for the sale to be possible. Second, the information set associated with the subject process and its complexity in the execution of a sale must also be exchanged. Third, with more complex or heterogeneous transactions, information about the functional properties of the subject good and its creation is often asymmetrically distributed. In devising transaction governance mechanisms (rules to guide parties in executing a transaction) all these three matters are crucial and will be important to each contracting party.[2] Simple, standardised goods and services are easily transacted in volume in markets (in this sense approaching what is meant by markets in a theoretical sense). Information intensive goods and services occur less frequently, require more developed processes, including legal engineering and other intense professional input, and may be concluded only over an extended period using highly structured contractual forms, perhaps with dedicated organisational forms such as special purpose vehicles (SPVs). No standardised goods have developed, or standardised procedures evolved, to facilitate these types of transactions. In a different context, Rescher (1998) provides a useful way of thinking about how complexity may affect the costs of transactions – '*complexity is the inverse of simplicity. The latter is a matter of economy, the former a matter of profusion*'. Complexity thus increases the cost of a transaction – 'costly contracting' is a term encountered in law and economics whenever process complexity is involved.

2.1.1 Transactions, contracts and information

This approach allows us to broaden our view of transactions, and introduce further useful vocabulary. It has become customary in economics and finance literature to use contract terminology to appraise and summarise the processes that form the conduct of transactions. There are good reasons to revert to the language of contracts, because this is exactly what a transaction represents – an economic exchange among two or more separate parties. In its simple form one party delivers a good or provides a service to another at an agreed price, where the buyer requires and the seller represents that a good has certain characteristics or a service contains certain functional elements.[3] Using the information set metaphor, then in order to contract successfully and fairly the parties must agree on the information set embedded in the good or service, and the exchange can then occur. Verbal contracts may suffice for simple goods and services, and the overwhelming number of transactions in everyday economic activity are verbal and take place instantaneously for most practical purposes. However, as goods and services become more functionally complex, more information-intensive or the transaction process becomes intricate and time-consuming, the transacting parties will adopt written contracts to manage that complexity. In essence we can contend that sound contract, assuming it is written,

---

[2] Transaction process has been variously referred to as transaction methodology, procurement methodology, procurement systems, and more, depending on which industry the analysis is conducted in.

[3] It is wise to emphasise that use of the language of contracts in economics does not imply a legal interpretation of transactions – it remains metaphorical, although the distinctions have become abstract in the writings of some academics. An example is Richard Posner's (1992) proposition that the common law seeks to achieve economic efficiency. Posner means judge-made law and assumes behaviour that may not be observed in the real world, for example that individuals seek to maximise their personal satisfaction and are aware of what that implies. In Posner's context efficiency refers to legal solutions that are optimal among the parties to a dispute rather than for the world at large.

is physical evidence of the information set that is exchanged between transacting parties. But contract terminology allows a finer analytical scalpel to be applied to transactions because formal written communication of buyer expectations, seller representations, and the expectation of enforcement of the agreement all become subsumed in the analysis of transaction processes.

The wider identification of how and where costs are incurred in complex processes is facilitated by the principles and vocabulary of contracts as phenomena. So a simple contract will generally be implicitly inexpensive to complete (purchasing a quantity of fuel oil), while a complex contract may be costly and take time to conclude (developing a new refinery). A further point becomes relevant with complex transactions: exchange is not always instantaneous and complete – it may entail continuing commitments by a series of parties, or complex performance assessed over extended periods. In these cases in addition to defining the subject of the exchange, contracts written to facilitate a complex transaction will contain procedures to manage its realisation, including the transaction subject, method of execution, and what the contracting parties will do in the event of changes in circumstances compared to those imagined at the time of contracting. These are often referred to in the finance literature as other states of nature. Complex transactions require rules to guide the actions of parties in unforeseen circumstances or in contingent conditions, as will contracts that govern performance over many months or years, such as long-term supply or service contracts.

Such complex contracts may be characterised as private systems of law or rules specifically to govern the behaviour of parties to specific transactions.[4] We refer later to transaction governance to describe the actions required of parties to a transaction, how those actions are monitored during the term of the contract, and the rules for contingencies that may require specific actions by any party. We also point out how this differs from corporate governance, a distinction which is useful when hierarchies of contracts are considered. These circumstances become more meaningful when we describe a firm as a nexus of contracts (Jensen and Meckling, 1976), or more expansively as a nexus of treaties (Williamson in Aoki *et al.*, 1990) because much of what happens within a firm is about transactions, or elements of transactions, their governance, and the conduct of the entire business entity (corporate governance) within wider legal and social institutions.

It remains for us to place transaction cost economics in perspective as a field of theoretical pursuit, because it considers the concepts we have discussed so far in a somewhat more abstract way and at a higher level of aggregation in the economy. Williamson (Aoki *et al.*, 1990) describes transaction cost economics as relating to assessing the efficacy of alternative ways of conducting similar transactions,

---

[4] Here, we use private law not in any formal legal sense. Private international law deals with the way that jurisdictions govern the enforcement of contracts across borders, and is always a concern for professional advisors to major international projects.

explaining why transactions are located, for example in markets or in organisations, or how they occur (possibly with various different organisational forms and with elaborate written contracts). In all, it is proposed that the way economies, sectors of economies, and even individual transactions are organised may be viewed as attempts to minimise transaction costs; for example, to reduce the aggregate costs of transacting in a sector, or the cost of a particularly complex transaction. At the level of sector or the aggregate economy, this is a useful approach to considering the role of economic institutions. Efficient institutions reduce the costs of all the millions of transactions in a particular industry, and the many billions of transactions that make up an economy, and are thus of immense political importance in any society. Efficient institutions are extremely important to the cost of conducting transactions internationally, but what is included in a state's institutions differs from country to country, even city to city, and in each case both formal and informal institutions need to be appraised. And institutions in some jurisdictions are more equal than others, as you may have guessed. We will return to institutions in Chapter 4.

Let us return briefly to Chapter 1, where the nature of project finance is discussed. The analysis of transactions presented above allows us to consider how we may categorise a project finance transaction. First, of course, acquiring a large-scale industrial facility such as an oil refinery is a capital intensive transaction, and the facility itself is expected to be somewhat information intensive because it involves the aggregation of thousands of bought-in and purpose-manufactured subsystems – but as these things go, oil refineries have been around for a long time, and are unlikely to comprise wholly new technology although some of the subsystems may be novel. We may therefore agree that while an oil refinery (or power station, toll road, or container terminal) is moderately information intensive, in that it is possible to define and specify its functional requirements as facilities with some accuracy (the goods or services to be delivered), it is unlikely that there will exist significant asymmetries in the information held by those involved in the refinery's development about their respective functional duties or requirements. It is nevertheless also unlikely that such facilities are homogeneous, because there will be critical differences in the location, capacity, and other design requirements, which may introduce important informational problems. However, because of the scale, capital intensity and management resources required to develop such facilities, the transaction process to acquire a new refinery is expected to be particularly complex. Thus, although the facility is of moderate inherent complexity, the transaction process required to bring it to realisation is expected to be complex and management intensive, with a large number of contracting and subcontracting parties engaged in interdependent actions in the delivery process. At every level, considerable potential exists for poor exchange of information sets. In institutional terminology, writing a contract for this transaction (the facility and its delivery), and agreeing rules to govern performance, and manage the actions of those involved following contingent events or unforeseen contingencies during execution is expected to be difficult – and expensive.

2.2 A more formal approach to disaggregation of transactions

While we considered to some extent the complex nature of a capital intensive industrial facility and its procurement process in Section 2.1, we have yet to consider the fact that the purchasing party has to finance the transaction. We alluded to the fact that debt forms the largest source of finance in the capital structure of typical project finance ventures. Debt, of course, is created by a transaction whereby one party borrows from another – for a period and under conditions agreed at the outset, including interest costs. This steers our discussion towards financial instruments or financial contracts. We plan to use vocabulary developed around financial transactions as a framework to amplify the concepts of Section 2.1 because it will allow the introduction of agency theory easily and further highlight informational problems in a highly relevant context. In all respects financial transactions are exactly as conceived in Section 2.1. Fortunately this is a field that attracts much research, leaving us with a significant body of insight to draw upon to consider further the complex sphere of contracting that characterises all aspects of project finance.

2.2.1 Elements of transactions

Before we introduce agency theory to the analysis of transactions, we first outline two concepts to help order our thinking. The first is describing transactions formally, using transaction cost economics vocabulary, and the second is presenting a practical approach to disaggregating transactions into stages, based on Greenbaum and Thakor's (1995) approach to financial transactions. Transactions may be viewed as comprising the following cost elements, with each present to a greater or lesser degree in any transaction, depending on its nature and complexity. There is a sequence suggested in the way these elements are presented in Table 2.1, which is generally the case in most transactions when objectives are clear and all goes according to plan. In more complex transactions there may be extensive overlapping and iterations between elements or stages in the process.

These stages are an effective way of ordering thinking about transacting. The first stage, the *search for counterparty*, refers to the search one party has to undertake to find a second to transact with – the party that will provide the good, service or facility that is to be acquired. The search may be simple, as with a commodity or other well defined good, or it may be complex in the case of a more specialised good, for example a complex component in a numeric controlled machine. Auctions or closed-form tendering are mechanisms to search for counterparties. *Screen counterparty*, the second stage, requires each party to assess the other's capability of completing the transaction satisfactorily – the due diligence requirement that parties undertake when transacting, sometimes required by regulation or corporate governance in complex transactions such as corporate acquisitions. Again, the screening process may be simple in the case of a well-known counterparty that has transacted often and with good outcomes, or it may be complex in the case of a new or unknown

Table 2.1 Anatomy of a typical transaction.

Phase	Nature of Activity	Transaction Cost	Agency Cost
1	Search for counterparty.	Search for potential counterparties	
2	Screen counterparty	Cost of information about and confirmation of standing of counterparty	
3	Negotiate, write contract (can be verbal, or implicit)	Bargaining, bonding arrangements and their costs	Conflicting incentives, bonding costs
4	Execute and govern contract (bond, monitor, enforce performance, mechanisms to facilitate negotiation of changes/ disagreements/disputes and deal with unforeseen circumstances and contingent events)	Cost of administration, costs of renegotiation	Monitoring costs

counterparty, or in demanding prequalification exercises that form part of tendering for complex construction projects.[5] In this case it often involves costly efforts to communicate both expectations and representations. As with searching for a counterparty, it is clear that information plays a role in the simplicity or complexity of these stages.

The third stage, *contracting*, refers to the process of agreeing what is being transacted and how it is to be delivered (if the good or service is not in existence or delivery involves performance over time, how this is to be monitored, reported, remedied, actions and decisions prompted by contingent events, and all other variables). Preparing written contracts to govern complex transactions is specialised and time-consuming, as anybody who has experienced well-drafted contracts for complex transactions will testify. This process can be extremely detailed and complex, as may be the case with many project finance transactions. The fourth stage, *execute and govern*, has a more literal meaning when we consider contracts with detailed rules about performance. In such cases the concept of *transaction governance* becomes clear, because it is concerned with managing ongoing aspects of the transaction, under the guide of rules that were agreed in the preceding stage.

This laboured presentation of the search, screen, contract, execute framework highlights a simple reality – every stage of every transaction is expected to locate some-

---

[5] It is common for screening risks to be mitigated with third party contractual commitments such as performance guarantees or bid bonds.

where on a continuum of complexity in each of the transaction elements – from simple to complex. North (1990) suggests that *'the costliness of information is the key to the costs of transacting'*, and so it is with each stage in the process. The associated costs of conducting every transaction element and the aggregate cost of the transaction are thus expected to be a direct function of its complexity, its informational nature and the distribution of the information between the transacting parties. But each element is also subject to the effects of economising actions: repeated transactions facilitate learning and function to eliminate previous errors, and so common (even standardised) procedures evolve that reflect efficient solutions to transaction problems. Thus the overall costs of a transaction comprise the costs of transaction elements, plus the cost of the institutions in the economic system that supports enforcement of the transaction, the shared cost of maintaining and developing further the economy's transaction infrastructure (institutions), including courts and procedures for enforcing agreements, and creating and regulating standards and professions. More about these concepts appears in Chapter 4.

2.2.2 Dimensions of transactions

The elements introduced so far in this chapter are seen to be influenced by the nature of transactions, which we discussed in Section 2.1. While we concentrated there on general characteristics, here we wish to introduce further concepts which influence the costs of transacting. These concepts come into play when we consider agency costs and corporate governance matters. Essentially following Williamson's (1985) conceptualisation in *Transaction Cost Economics*, the nature of transactions is described by two principal phenomena, namely the critical dimensions and the behavioural assumptions of transactions. We describe the nature of each concept in more detail, and the role that information plays in each case soon becomes clear.

2.2.3 Behavioural assumptions in transactions

In transaction cost and institutional views of economic activity, the central assumptions in transacting are that information matters and humans are fallible. These two factors are summarised in the concepts of bounded rationality, information incompleteness and opportunism. These are phenomena we encounter frequently in activities beyond economic exchanges, certainly in interpersonal exchanges at many levels, but a particular vocabulary has developed to describe these phenomena and their influence on transactions and their complexity. The three concepts are interdependent, but we can make sense of them separately to indicate their underlying nature.

First, *bounded rationality* proposes that there is a limit to the human capability to formulate and solve complex problems, as originally articulated by Herbert Simon in 1947. Bounded rationality does not suggest that humans do not pursue rational behaviour, but that *'human behaviour is intentedly rational, but only limitedly so'*

(Simon, quoted in Aoki *et al.*, 1990, p11). Bounded rationality occurs in any transaction, but is expected to increase with its complexity and thus may be best demonstrated by a large and complex transaction, for example the development of a deep water natural gas field and associated recovery terminal. When required to create governance mechanisms for this complex transaction, and take account of a range of associated contingent conditions, physical and economic risks, and states of nature (each with a spread of consequences), it is clear that the task is inherently too extensive to conceive comprehensively when the contracting parties enter the transaction. Further, even if the parties knew for certain all future states of nature and contingencies, and could fashion such a transaction, information may be distributed asymmetrically, or may not exist at all – we are said to have information incompleteness or imperfection. Under these circumstances the contract to evidence this transaction and its governance will be incomplete; there will almost certainly be circumstances that arise during its life that were unanticipated at execution, and then may involve additional costs in negotiation and contracting. The phrases bounded rationality and incomplete contracts are frequently used to describe this situation.

Second, to humanists, opportunism is a far more unattractive behavioural characteristic than bounded rationality, but the first fact to recognise is that it is also intimately related to information about a transaction, the transaction parties, and the distribution of this information. Williamson (1985) describes opportunism as '*self-interest seeking with guile, and as making self-disbelieved statements*' – both these phrases are euphemisms for misrepresentation of fact. Misrepresentation may be not disclosing all facts pertaining to the state of an asset, telling half-truths, or simply lying. The word guile is extremely unflattering – according to Webster's Dictionary it means cunning. More about consequences of opportunism when agency theoretic concepts are introduced. However, it also has to be recognised that opportunism may not be *ex ante* (before the fact) intended, but may be lucrative *ex post* (after the fact), for example if an unanticipated circumstance arises which favours one party hugely if it defaults opportunistically (against the spirit of the original transaction). It may thus be difficult or impossible to distinguish *ex ante* honest people from the dishonest, but a related problem is that opportunistic behaviour can occur both *ex ante* and *ex post* where performance over a period is required – and this circumstance is aggravated further when a condition of bilateral monopoly is present (described in Section 2.3).

2.2.4 Critical dimension of transactions

We combine the behavioural assumptions considered above with the nature of assets, goods and services being transacted, referred to as the critical dimensions of transactions, which yields a rich environment within which we may analyse all transactions. Let us briefly deal with the critical dimensions of asset specificity, uncertainty and frequency, and with what is termed bilateral monopoly. Asset specificity is the most important of the three and receives more attention.

Asset specificity is a simple yet extremely powerful concept and of central importance to any discussion about project companies and project finance. It can be explained as the degree to which an asset can be redeployed in different applications without sacrifice in productive value (Williamson in Aoki *et al.*, 1990). The more specialised an asset, the harder it is to redeploy and the higher is its asset specificity. It is clear that some assets, such as desktop computers, have wide application potential in all kinds of economic activity, as do trucks and certain industrial machines. A desktop computer can be used without modification by a retailer, in a bank, or by an accounting office. Such assets may be labelled homogeneous: they approach the characteristics of commodities and may be thought of as interchangeable in a productive process without causing disruption. These assets have low asset specificity. Other assets have value in specific uses and not in others. Small electric motors and petrol engines have a huge range of applications from lawnmowers to chainsaws, are produced in large volumes at relatively low cost, and can be described as having low asset specificity. However, a power generation turbine can only produce electricity in one application, while a diesel engine might be used in a diesel electric rail locomotive or in a marine application to propel a ship. In each case the applications are fewer and less diverse than those of a desktop computer. The turbine has higher asset specificity than a small electric motor, and the fact that it can be expected to be far more costly is not incidental, given that fewer are demanded, and the applications in which they are used also are more restricted and specialised, or more specific.

If we return our focus to project finance transactions, we noted in Chapter 1 that project finance transactions, such as for infrastructure or transmission pipeline development, were often highly asset specific. Returning to Baguio Power Company (BPC) introduced in Chapter 1, BPC owns a coal-fired electricity generating facility, which sells all its output to the hypothetical Manila's electricity authority for distribution to final consumers. This plant can produce only electricity (this is specifically what it was designed for), only using coal and only at that particular location (producing electricity efficiently requires large-scale plants at fixed locations, preferably with port access), and sells a very large proportion of its entire output to one client, probably under a contractual arrangement. The principal asset, the power plant, exhibits extremely high asset specificity. While it could be moved to another location or converted to consume oil or gas, the cost of modifying the plant in theory would be high and likely to make such options uneconomic.

An important concern arises with asset specificity in more complex transactions, for example when two or more parties enter into a transaction jointly to pursue a project venture which may extend over a significant period of time – known broadly as the field of relational contracting. The degree to which either party may be concerned about its vulnerability in such a relationship depends on how specific the assets are that form part of the transaction, and also that their specificity may be increased by the joint venture. Williamson (1985) outlined four ways in which asset specificity can be identified with reference to relationship-specific investments. The first is *site specificity*, where a buyer and seller co-locate with each other, usually attempting to minimise input costs. A power plant may be co-located with a marine oil terminal, for example. As with BPC, once sited the assets in place are effectively immobile. The second is *physical asset specificity*, for example when one or both parties to the transaction invests in capital equipment and machinery that involves design characteristics specific to the transaction and which have a lower value in any alternative use. This may be the case where an aero-engine manufacturer invests in equipment to make a limited number of special engines with characteristics specific to an application, say a small fleet of powerful short-take-off and landing cargo aircraft. Such engines may have alternative uses but none will be as valuable, because their specificity means they are not close substitutes for other aircraft engines. This concept also reflects partnering agreements as initially conceived in scale-intensive industries such as car assembly, whereby original equipment manufacturers (OEM) and their suppliers could jointly benefit from dedicated subassembly design and manufacturing that is integrated with OEM requirements.

In an industrial supply context, a capacity expansion made by a supplier or subcontractor especially based on the prospect of selling a significant volume of products to a particular customer explains what is considered to be *dedicated asset specificity*. If an investment was made in additional capacity under an agreed contract, but the contract was terminated prematurely, the supplier would be left with significant excess capacity and higher fixed costs. A last category of asset specificity recognises that human capital can also be specific, thus *human specificity* refers to investments in human resource skills that may be an essential component of any specialised form of joint venture activity where learning-by-doing processes are involved. For example, our aircraft engine could require specialist maintenance skills, which may be specific to that engine design and may be only a partially transferable skill, to use modern parlance. The technician undertaking aircraft engine maintenance, and the manufacturer may both be exposed to this form of human specificity.

Frequency and complexity, as two further transaction dimensions, were alluded to in Section 2.1 and in the discussion of behavioural assumptions. Both are related to asset specificity and particularly relevant in the context of project finance (Figure 2.1). Complexity requires little further elaboration, as it is intimately related to bounded rationality and the informational requirements of transactions, but frequency requires confirmation of earlier observations. As explained, when products are well known and standardised and transaction volumes are high, it is likely that such transactions will occur in markets – the appropriate institution for high volume trading of commoditised goods and services. However, with infrequent complex transactions of high asset specificity, specialised structures may be required to govern the transaction, and will most certainly be needed in the development of Global Consilium's offshore gas field project. It follows that the cost to create a governance mechanism to manage such complex single transactions is

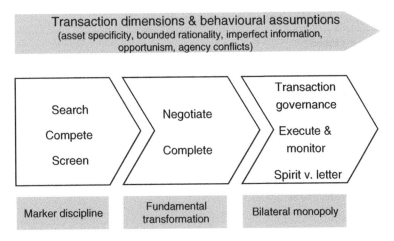

Figure 2.1 The process of transacting.

expected to be high and as with any industrial activity, such high initial costs may be reduced (written down or amortised) with repeated use of a basic governance structure. If we consider the next phase of Global Consilium's gas project, bringing the lifted gas to wholesale distribution, it may be expected that this phase will require a different and cheaper governance structure, because gas sales occur frequently and in volume, most probably in organised international commodities markets.

A last phenomenon we have to outline for present purposes is a critical concept known as the bilateral dependency condition (or the bilateral monopoly condition). This state is possible with any transaction where ongoing obligations are agreed – for example, with a long-term maintenance contract that Global Consilium Jet Engines might enter for the specialist jet engines it sold to Acme Air Cargo. Crucially for our analysis, Global Consilium may have competed with other engine builders to supply this plant, and the final sale price may be highly competitive. This will not necessarily mean that the benefits of competition or favourable pricing exist *ex post* contract and that market power is neutralised. Competitive conditions in the supply of the asset undergoes what Williamson (1985) describes as a fundamental transformation into an *ex post* contract state characterised as a bilateral monopoly (bilateral trading) arising from the monopoly supplier-monopsony buyer arrangement created by the contract's fundamental transformation. It reflects a closed, exclusive *ex post* contract condition between two parties where ongoing obligations exist – the fact that Global Consilium supplied the engines under competitive conditions is one thing, but will it honour the commitments to maintain the engines according to the rules that will govern the ongoing phase of the transaction? Amongst other things, Acme's expectations about Global Consilium's future behaviour depends critically on the information it holds about Global Consilium – its previous experience with other suppliers, Global Consilium's reputation, and so on – all providing scope for

Table 2.2 Assumptions underlying transaction cost economics.

Critical dimension of transactions
Asset specificity
Uncertainty/complexity
Transaction frequency and the fundamental transformation

Behavioural assumption of transactions
Bounded rationality
Imperfect information and incomplete information
Opportunism

Source: following Douma and Schreuder (2002).

hidden information, misrepresentation and opportunism.[6] In all, asset specificity and informational concerns are at the centre of a large proportion of activities that are subject to managerial decisions in everyday business, and have a most critical influence on strategic decisions. Asset specificity also has particular implications for property rights considerations under bilateral dependency, to which we shall return. We summarise behavioural assumptions and critical dimensions of transactions in Table 2.2.

2.3 The influence of agency on transaction costs – agency costs

Our interest in agency is subsumed into the wider scope of the economic analysis of contracts. We are therefore interested in the economic analysis of agency problems, not agency as a phenomenon in certain forms of contract law. In modern business few transactions are conducted solely between principals – without a doubt the vast majority of transactions in developed economies have some kind of agency characteristics. Managers are the agents of shareholders in corporations that are not owner-managed; employees are agents of their employers; borrowers are agents of debtors; tenants are agents of landlords; lessees are agents of lessors; fund managers are agents of investors; and so on *ad absurdum*. Agency analysis quickly becomes complex in institutional relationships that involve many parallel contracts and/or multilateral contracts, and it is therefore no surprise that there is a considerable literature on agency theory, or principal-agent theory (Tirole, 2006; Laffont and Martimort, 2002). Our foray into agency theory is modest but serves to elaborate the influence of agency on transaction costs and the structuring of transaction governance mechanisms in project companies. Because agency is typically subsumed into the wider realm of

---

[6] It gets more complicated than this – with transactions that require both parties to perform ongoing obligations, such as with operating leases, there exists symmetrically the risk that either counter-party could exploit the bilateral monopoly condition.

contracts, we thus also are able to draw on much of the vocabulary already introduced in the following discussions.

In its simplest form, agency theory is concerned with the relationship between two parties, whereby one party (the agent) contracts with another (the principal) to act on its behalf in defined circumstances. While analysis of bilateral agency problems is in itself exciting, we propose that for the most part the fundamental nature of agency theory based solely on bilateral relationships can be extended into quite complex multiple agency situations, without losing sight of more complex transaction aspects such as corporate governance and project finance. To avoid confusion, for present purposes we assume agents and principals are legal persons (governments, companies, or other forms of enterprise) and their representatives, while we consider defined circumstances to mean given transactions (explicit and implicit) within a commercial context. In reality there are many grey areas, which explain why agency theory is so fascinating.

The first concern in developing a vocabulary to analyse agency problems in project finance (and other transactions) has to do with behavioural assumptions, and in this respect we have to repeat those associated with transaction cost economics, namely bounded rationality, incomplete contracts, and opportunism, all as outlined above. But there is an important additional emphasis, and this concerns self-interested behaviour of parties to a contract. We assume a party to a contract acts in its own best interest when concluding the transaction, and thus that tensions that emerge in transaction governance occur only between the incentives and motivations of principals. When one party acts as agent for another, and if both parties seek to maximise utility, it is commonly accepted that the agent may not always act in the best interests of the principal, but instead favour its own interests – which fund manager is not interested in performance bonuses, possibly more than in maximising the return to their investors? This introduces a potential conflict of interest, or incentive conflict, which may result in an agent benefiting at the principal's expense by acting opportunistically upon the conflict. Every principal-agent contract aims to align and manage such conflicting incentives. Crudely, and somewhat depressingly for idealists, unless incentives between agent and principal are perfectly aligned, there is no such thing as a perfect agent.

Thus we expect conflict between the incentives of the principal and agent, with opportunism a further distinct possibility, in addition to human conditions such as incompetence, negligence, or a regrettably short attention span. Even excluding bounded rationality, opportunism and human fallibility, a true alignment of interests between agent and principal can be achieved only with perfect information, complete contracts and perfect markets – but we have already concluded that these conditions are found only in fiction. Agency relationships extend an additional layer of considerable complexity to an analysis of original transactions between principals that may be complex to begin with.

The presence of agency conflicts has important implications for transaction costs and the form chosen for transaction governance mechanisms. Since it is generally impossible for the principal or agent to ensure that its counterparty will make decisions that are optimal with respect to its own interests, in agency relationships the principal will incur costs to manage agent behaviour as an integral part of transaction governance. These are generally referred to as bonding costs (typically incurred by the principal) and monitoring costs (typically incurred by the principal). In some contracts both parties will incur monitoring and bonding costs. In operating leases the principal may will maintain an asset, while the agent must take care not to abuse the asset. Both parties' actions have to be bonded and monitored (Pretorius *et al.*, 2003).

There is an important additional information cost in agency relationships, which concerns the costs of uncovering the nature of incentive conflicts inherent in the type of transaction that the agent has been appointed to conduct. Following convention, we thus assume the costs of agency relationships as having three elements: monitoring expenses by the principal (or the agent), bonding costs typically incurred by the agent (but in some contracts involving both parties), and the additional costs of uncovering incentive conflicts inherent in the type of transaction. However, there is also a residual loss – the general welfare loss to the economy from the cost of all transactions (Jensen and Meckling, 1976). Minimising this overall loss is a public policy issue that involves institutions concerned with regulating the economy. These issues are touched on in Chapter 4.

An obvious additional point should be made, which is that agency costs may be incurred through the primary contract as well as a related agency contract, so it is often useful to identify those incentive conflicts that require bonding or monitoring that are generated from the nature of transactions, and consider who then experiences the incentive conflict.

So how may we explain easily what may be categorised as bonding costs, monitoring costs, and costs of uncovering the nature of incentive conflicts that arise in agency arrangements? Research into agency costs in typical financial contracts such as equity, debt, or leases has gone some way towards establishing a typology of principles towards this purpose, but of course every type of transaction involving an agent will have different conflicts. Let us consider the meaning of bonding and monitoring costs in the context of leasing the electricity generating turbines in a power station, a common financing mechanism in power projects.

First we need to uncover the nature of incentive conflicts that may arise generally from lease transactions (which also explains why this is in principle an agency transaction). Following Pretorius *et al.* (2003), leases are defined as financial contracts whereby an equipment owner (the turbines) grants an exclusive right of use to another party (the project company) for a defined period in return for a defined stream of payments whilst retaining ownership of the equipment during the term of the contract. Note that this is essentially a purpose-specific credit contract: instead of lending money to a corporate borrower for general purposes, the provider of capital provides it in form of an asset but it is nevertheless a credit transaction. The separation of rights that characterises the transaction and the conflicts that they generate, in addition to the agreed schedule of payments, form the source of information to be processed by the lease parties prior to executing the transaction. Information about the nature of the leased asset informs the desirability of the transaction, while information about counterparties informs expectations about their respective behaviour during the term of the lease including potential default and other risks. The form of the transaction (the lease contract) including its payment schedule are together the outcome of the information exchange between the project company and the lessor-manufacturer (the turbine supplier).

For present purposes, we concentrate on two incentive conflicts in lease transactions: asset-specific bilateral monopoly exploitation, and residual value expropriation. In order to understand the nature of the residual value expropriation incentive, we restate elements of the transaction. The project company takes possession of and installs the turbines, and is committed to a schedule of payments. In exchange it obtains the rights to any surplus income (over the lease payments) generated from using the equipment over the term (known as residual returns). The schedule of payments compensates the turbine manufacturer-lessor by generating a return on its investment in exchange for temporarily assigning the rights to its residual income, while retaining rights of ownership and consequently the equipment's residual value upon maturity of the lease term. The party that decides how the equipment is used during the term (the project company) has control over the asset for the lease period, but is not the residual value claimant and is directly responsible for its actions during its period of control only to the extent that the effect of its decisions on changes in the equipment's residual value can be detected. For example, if a lessee contains the cost of using the equipment over the lease term, it directly increases the residual income over which it has full rights for the term of the lease. The project company therefore has an incentive to economise on maintaining the turbines during the term of the lease, ideally to where the residual income is maximised over the full period of the term given that the quality of services generated by the asset is likely to diminish with poor maintenance. Project company decisions that adversely affect the residual value of the asset but go undetected, such as under-maintenance, are characterised as actions that directly expropriate residual value from the lessor.

Although this scenario conveniently does not consider various different forms of leases (Box 2.2), it is nevertheless useful for the purpose of explaining the nature of bonding and monitoring. For now, let us shift attention to the lease of a different asset by Baguio Power Company, say office space in a Manila central business district building. We may describe the nature of bonding in agency transactions from answering the question: how can a landlord make sure that a prospective tenant will protect and act in his best interests, namely by not expropriating the property's residual value, which requires proper care of the property and sound maintenance during the term of the tenancy agreement? Needless to say, a tenant may damage the property during

Box 2.2 A primer on leases: general characteristics (Pretorius *et al.*, 2003).

The function of leases as financial contracts is to facilitate separation of ownership of an asset from its possession and use in return for an agreed stream of payments. Whereas this characterisation purports to project a clear image of the transaction, in practice leases tend to be highly regulated in most jurisdictions, particularly for financial reporting and taxation purposes. Regulatory constraints generally limit to a significant degree the flexibility that the general definition of leases provided above suggests. Complications introduced by regulation, accounting treatment or taxation will be avoided in this brief categorisation.

Drawing on typical equipment leases for present purposes, two particular characteristics that form part of the definition of leases are employed to identify typical lease categories. First, joint consideration of the term of a lease and the economic life of the asset leads to the categorisation of leases as operating leases or finance leases (also often referred to as financial or capital leases). This categorisation arises as a consequence of considering four variables: the term of the contract, the economic life of the asset, the present value of lease payments over the term and the fate of the asset upon maturity of the contract. Briefly, the lease is generally categorised as a financial lease if its term approximates to the economic lifespan of the asset; the present value of lease payments approximates to the purchase price of the asset; and ownership of the asset is (or may be) transferred to the lessee at a favourable consideration upon termination. The resemblance of financial leases to secured debt contracts is widely recognised, and most tax regulators treat financial leases as liabilities for purposes of financial reporting, with a corresponding asset declared on the lessee's balance sheet. The conditions for financial leases to be treated as off-balance sheet financing have become onerous, but operating leases are generally still treated off-balance sheet under certain circumstances. Alternatively, a lease is normally categorised as an operating lease when its term is shorter than the asset's economic life or the present value of lease payments is less than the purchase price of the asset; that is, the lessor does not recover its investment during the first lease period.

Although lease transactions typically stipulate the term as a condition of the contract, and the lessor's return on investment may be closely associated with term to maturity, lease terms may also be contingent upon the exercise of options contained within the lease. Leases may contain cancellation options or may be non-cancellable and may involve or dispense with penalties for exercising any such option, as with prepayment penalties in small-scale loan contracts. Most financial leases are non-cancellable, while options to cancel are common in operating leases except for certain types of assets, notably commercial real estate (Senn, 1990). If the economic life of the asset exceeds the term as with operating leases, the asset's future residual value on maturity of the term is critical in the assessment of the contract's attraction to its parties. Where a positive residual is expected at the end of a financial lease, the right to the residual is frequently decided with the lessee granted an option to purchase the asset at an agreed price, while operating leases often specify a conditional option for a further term also in favour of the lessee (Brealey *et al.*, 2006). The agreed price generally takes the form of a method to determine the price at termination, but is often at a favourable consideration – for example, a bargain purchase option. Options to purchase an asset at maturity are less common in operating leases; similarly, options to extend financial leases are uncommon.

The second set of variables, the payment of rent and other cash flow considerations, is concerned with the financial purpose of the lease to lessee and lessor. For practical purposes, leases provide enterprises with the opportunity to finance their capital requirements by obtaining external financing for defined assets from specialist intermediaries. These entities often invest in an asset in order to conclude a lease transaction with a particular counterparty, as is common practice with equipment lease finance intermediaries, or in specific classes of asset, as with capital asset lessors. A consideration implicit in this categorisation of financial characteristics of lease contracts, however, is that a prospective lessee may choose to finance the asset with an alternative source of finance, commonly by borrowing or issuing shares to purchase the asset. Typically, the corporate decision associated with operating leases is between leasing and buying as alternatives to fund capital assets; while its decision in relation to financial leases is between leasing and borrowing as alternative contracts to finance capital assets.

the term of the tenancy or abscond without payment. The typical answer in a tenancy agreement is that the tenant is required to signal a commitment to the transaction, usually in the form of a sum held in trust – a deposit or a bond – which is at risk in the event of tenant default or opportunism.

The nature of bonding is thus that the agent risks a financial loss in the event of opportunism or some other occurrence that results in a loss for the landlord. Bonding agents to their undertaking can take many forms, but is always intended to crystallise a financial loss to the agent in the event of non-performance or opportunism. An important form of bonding is through risking reputation (my word is my bond), but it is also no mystery that loss of reputation can be extremely costly in the (mis)conduct of many modern commercial activities. Remember also that there may be regulatory penalties for abusing agency relationships – from being barred from practising a profession all the way to a prison sentence. In commercial leasing, acting opportunistically will likely cause the surrender of a deposit and a loss of reputation, which all else being equal may translate into a higher price or larger deposit for future leasing.

Word of default gets around, and is part of the information set available in a well-functioning market.

The loss of residual value as discussed earlier provides the point of departure for a useful explanation for monitoring (both as an agency cost and more widely). Recall that we stated that those decisions by the lessee that adversely affect the residual value of the leased asset and go undetected, such as under-maintenance or asset abuse, are characterised as actions that directly expropriate residual value from the lessor. Monitoring aims to prevent such losses by detecting them sufficiently early to institute action under the transaction governance mechanism (the lease contract). Actions may require remedies or more, but another important matter relates to the nature and complexity of the leased asset. Returning to the commercial property example we may thus ask a symmetrical question to that posed with bonding: how does a landlord ensure detection of asset abuse early enough to minimise residual loss? In turn, the typical solution in a tenancy agreement is for the landlord to inspect the property regularly for evidence of abuse or under-maintenance. Although evidence of abuse in commercial properties may be relatively easy to detect by inspection, this is not the case for complex machines. Equipment such as aircraft engines or power generation turbines may require complicated equipment and measurement to detect abuse early in the asset's economic life. Monitoring is a collective term for all activity aimed at ensuring compliance during the term of a contract, including measuring the quality of a product, the state of a machine, compliance with restrictive covenants in debt contracts, and also measuring complex human performance. In essence, as the requirement to bond performance is not without cost, so also is performance monitoring.

2.4   Corporate finance context

We have now introduced all the concepts needed for a transaction cost interpretation of the project finance model. To commence, in Chapter 1 we briefly identified how corporate finance and project finance are related, and will revisit that analysis in more detail to isolate the particular nature of project finance. For those readers who are unfamiliar with or need a reminder of essential concepts of the nature of corporations and their regulation (as practised in most jurisdictions with market economies), we state for quick reference the principles of incorporation and corporate finance, to develop context for Section 2.5.[7] We make no assumptions about handed down knowledge, because these are fundamental concepts that have to be clearly understood, especially to appreciate agency conflicts in corporate finance and how project finance arrangements help to manage such conflicts.

---

[7] Goode (2004) provides a good introduction to modern limited liability incorporation under English law.

2.4.1 *Corporate finance instruments*

The main innovation represented by the modern company when it became generally available in nineteenth century England and the United States[8] was to allow the simple creation of a corporate body as an ongoing legal person in order to conduct business in its own right, separate from its shareholders (its owners). Typically a company is created in law as a legal entity (a legal person), with a device known in many common law jurisdictions as a memorandum and articles of association consisting of a statement of the nature of the business it will conduct (its purpose), and how the company will be governed (we may say these jointly form the company's constitution). The articles set out details of the company's authorised share capital (the number of shares that may be issued). The authorised share capital may be issued (sold) in several offerings over time. It may also be increased periodically subject to existing shareholder approval. Authorised and issued shares are referred to as outstanding shares and in the case of public companies these may be widely held and/or traded on a securities exchange. In certain jurisdictions, outstanding shares may be bought back and cancelled, or held by the company as authorised but not outstanding, sometimes known as treasury shares.

The articles of association set out the rights and duties of shareholders, who may be those who agreed to contribute capital by subscribing for shares in the company, or subsequently purchased outstanding shares privately or through an exchange. Technically the shareholders control the company by having the right to appoint directors, who together oversee management and vote on decisions about company affairs. In some cases, the constitution will provide for several classes of shares with differential voting rights but while this remains common there is a modern tendency to undervalue publicly quoted companies with such shareholding structures.

Shareholder rights typically include sharing in profits if a dividend is declared. It may also include restrictions of shareholder property rights, for example to prevent the sale of shares by a shareholder-manager for a minimum period (lock-up periods), ensuring that an important individual shareholder-manager who may have been instrumental in the company's formation remains committed to its future despite selling shares. With respect to corporate governance, the articles stipulate the duties and liabilities of the executives and office bearers of the company (by regulation there are typically also prescribed officers, such as a company secretary).

The aim of incorporation has always been to afford the company the right legally to conduct commercial activities in its own name, for example by owning property, entering into contracts or litigation, and to limit the ultimate liability of shareholders for the company's debts only to the nominal value of their shares at issue. It also sought to provide continuity through the creation of a distinct legal entity, so that unlike a partnership, the company could survive the demise of its shareholders or

---

[8] Earlier models of specific incorporation date back much further (Arner, 2007).

allow each owner to withdraw from the company with a sale of shares. Thus a company may borrow money in its own name, without shareholders generally being liable to repay such loans in the event of later corporate financial distress. This last notion is important for the study of corporate finance and project finance but is a condition that applies mainly to large public companies. For the most part, shareholders and managers of small and medium scale enterprises (SMEs) are often the same and lenders will require owners to guarantee repayment of loans to their company, or provide collateral in the form of their own assets as security for loans to the company. This does not violate the principle of limited liability but confirms that lenders, when uncertain as to corporate credit risks or the costs of their appraisal, often look for collateral that is simpler to value when lending to a small company. It also finds comfort in drawing a material commitment from a person interested in the success of the company, and by bonding an owner-manager, it is able to overcome its information and opportunism concerns with lending to a small enterprise with little reputation.

Given these points, we illustrate again Global Consilium Corporation's consolidated balance sheet in Figure 2.2. In brief, corporate finance instruments are simply

The Global Consilium Corporation (founded 1929)
Consolidated Balance Sheet at June 30, 2007 [a]
($ millions)

LIABILITIES	$	%	ASSETS	$	%
Equity			**Fixed assets**		
Issued ordinary shares (breakdown)	11,000		Corporate head office building	1,000	
8% Cumulative preference shares (non-voting)	5,000		Fixtures and vehicles, etc.	50	
Retained earnings (accumulated past profit/losses)	20,000		LESS:		
Minority interests	3,500		Accumulated depreciation	-150	
Shareholders' funds, ("net worth")	**39,500**	31	**Net fixed assets**	**900**	1
Debt			**Associated companies** [b]		
Long-term debt			Consilium oil	30,000	
Secured bank debt, due 06/2010	31,500		Consilium mining	30,000	
Outstanding bonds, maturity at 06/2015	48,000		Consilium jet engines	15,000	
			Consilium infrastructure and power	35,000	
Long-term office lease	400		Other consilium companies	11,000	
Net long-term debt	**79,900**	63	**Associated companies**	**121,000**	96
Current liabilities			**Current assets**		
Commercial paper: 3 months	5,000		Incidentals, inventory	1,400	
Short-term bank loan: 1 month	2,000		Other current assets	2,000	
Current liabilities	**7,000**	6	Cash and marketable securities	1,100	
Total debt	**86,900**	69	**Current assets**	**4,500**	4
Total:	**$126,400**	100	**Total:**	**$126,400**	100

Notes

[a] Please understand that we make no effort at all to reflect Generally Accepted Accounting Principles as practiced in ANY jurisdiction in our analyses. The detail reflected here is not typical of many consolidated balance sheets.

[b] In many (most) jurisdictions, associated companies will NOT be named at all, there will simply be a one-line item reading 'Associated Companies', with a total against it. We itemise associates to illustrate the underlying structure of corporations that have project company subsidiaries.

Figure 2.2   The Global Consilium consolidated balance sheet.

the various transferable and other contracts used by companies to raise finance, often forming the greater part of their liabilities. Recall we have chosen a capital structure typical of a diversified industrial group of 69% debt and 31% equity (a debt–equity ratio of 220%). Assume Global Consilium's shares and bonds (publicly traded debt) are available to investors in all the major financial centres, including New York, London, Frankfurt, Hong Kong and Tokyo. Recall also that this is a consolidated balance sheet, so this final picture reflects the net accounting sum of each subsidiary company's assets and liabilities, plus assets technically owned by the corporation (such as its principal office building). For some elaboration, we also added a few other instruments, such as preferred stock or cumulative preference shares. Legally, preferred stock is currently classified as part of shareholders' capital and reserves rather than a long-term liability, and while less frequently used than ordinary shares (equity) it may have special uses in certain situations such as financing of acquisitions.

2.4.2 *Property rights and corporate finance instruments*

We begin our transaction cost explanation of corporate finance by reconsidering the fundamental instruments (or claims on a company), equity and debt, and the property rights that are vested in these claims. The company is contractually obliged to return the funding contributed to it as debt and equity to those who made these contributions, or their successor claim-holders. Thus, investors in equity and debt have financial claims on the issuing company that have to be settled by it at some future time, or upon a winding up. These are thus company liabilities. Should the company become insolvent, the order of priority of repayment (seniority) of claims is prescribed variously by company law and other statutes in the place of incorporation, and by valid prior agreements between groups of claimants and the company. Shareholders technically own the company, often referred to as equity interests, and as a contract equity ownership aims to bestow at least two clear rights. First, as owners, shareholders have the right to oversee decisions in principle about the company's business activities and have the right to decide how the company is governed, insofar as the law allows some flexibility, in each case by exercising the right to vote generally attached to shares.

Thus shareholders exercise the right to oversee the company through voting to appoint board members, who may in turn appoint professional managers who may not be shareholders, or may choose to sit on the board in their own capacity as shareholder-owners, or choose even to manage the company themselves (Jensen, 2000). For our purposes, this right to make decisions affecting the corporation's business or governance, usually restricted to ordinary shareholders, is taken to mean those shareholders with voting rights.

Second, in terms of property rights shareholders have a claim to the net balance (the residual) in case of liquidation (forced or voluntary) after all other claims have been settled (determined by statute or contractual seniority). Shareholders are thus often referred to also as residual claimants – they purchase an equity interest in the company at a quoted price, which allows them the right to become involved in decisions about the company's affairs, and to share in dividends (if an interim residual is declared in this form), otherwise they own a claim to the overall residual upon liquidation. They risk capital in exchange for the right to oversee decisions and participate in a residual claim.

These descriptions follow the general pattern of modern common law jurisdictions, and we choose this model largely because of its popularity in project finance cases, even in those involving sponsor shareholders from other jurisdictions. It is common in advanced civil law jurisdictions for large companies to be managed by two separate boards of directors usually known as a managing board and a supervisory board, and the latter explicitly represents the interests of a wider range of stakeholders than is the case in the common law corporate model, including those of current and former employees. Directors of Dutch or German companies, for example, have prescribed responsibilities to broader interests than those of their British or US counterparts as described in this chapter, who are (for the present) answerable primarily to shareholders.

Debtholders, on the other hand, have a claim senior to that of shareholders in the event of liquidation. Also, among the holders of debt claims as a whole, debt contracts such as bonds are usually unsecured claims, while other claims that enjoy security rights (by a charge over fixed assets, say) may rank ahead of all others within the general class of liability holders. There is often confusion about the meaning of *securities*, which are tradable debt or equity claims, and *security*, in the sense of collateral used to make secure a claim. In many jurisdictions, certain claims enjoy statutory priority. This often includes unpaid salaries and taxes. The modern view of debt is that the lender effectively acquires the company, while the equity holders have an option to buy the company at the real price of the outstanding debt.

Global Consilium's balance sheet illustrates a shareholder class called non-voting preferred cumulative shares, which is a claim carrying certain attributes of both debt and equity claims. As an example, these may be contractually entitled to a fixed annual coupon or interest payment based on their nominal value, and may or may not be traded on an exchange. They are equity-like in the sense that they may bestow rights to the residual upon liquidation, although contractually they may not carry rights to participate in decisions by voting at general meetings of the company. Although preferred shareholders earn a dividend that is predetermined in the same way as with debt, the directors can choose to withhold its payment. However, if these shares are cumulative then all dividends owed and outstanding to preferred shareholders will rank ahead of the rights of ordinary shareholders.

As a financial contract, debt can be fashioned into any number of forms, limited only by the tolerance of borrowers (issuers) and their shareholders, and the ingenuity and selling capability of financial intermediaries acting on their behalf. Later chapters will describe particular instruments, but we mention conventional variables that distinguish most debt contracts, which include a periodic interest rate and its method of calculation and payment, the final maturity, drawdown and repayment provisions, seniority and any provisions for collateral. Long-term debt instruments are usually treated as one of two categories, term loans and bonds, where the major difference has historically been that bonds are usually tradable in a public market. We present, in Appendix 2.1, a discussion of various long-term debt instruments which includes main elements of modern long-term loans and bonds.

2.5 Incentive conflicts in corporate finance

The classic theory of corporate financial structure argues that with perfect capital markets, the way that a company is financed is immaterial to its value – this is Modigliani and Miller's (1958; 1963) famous irrelevancy proposition. What matters instead are the company's investment decisions. Financial instruments represent only claims on uncertain net cash flows generated by the company, which is unaffected by financial structure. This view has been extensively modified in three respects (Milgrom and Roberts, 1992). First, the incentives and behaviour of various parties in corporate finance and governance influence net cash flows and the risk of financial distress. Second, professional managers are assumed to know more about the state of the firm and its opportunities than shareholders or creditors such as bankers (since there is information asymmetry), and with no equity interests in the company the managers have incentives to misrepresent these prospects, particularly for personal gain. Third, financial claims confer rights in decision making and control, and these rights are exercised by their respective holders to control the conflicts inherent in the incentives of managers. These rights have become the subject of substantial research using the concepts outlined in Sections 2.1, 2.2, 2.3 and 2.4. It quickly becomes clear that these conflicts are generic problems that arise from incentive conflicts in agency relationships, and consequently investors are seen to monitor constantly the financial decisions made by managers for signals about the company's prospects that they would not otherwise know directly.[9] In SMEs managed by their owner, conflicts caused by a variety of claims often remain at shareholder level, but these are no different to or less severe than the more public conflicts in the governance of large corporations.

We will discuss the principal agency conflicts and show how costly these can be to companies, using agency theory and the transaction cost concepts introduced earlier. We complete the section by presenting a parallel explanation of how the project finance model manages each main agency conflict in corporate finance. To keep the rest of Section 2.5 manageable, we concentrate on three critical incentive conflicts arising from financial claims in corporate governance, namely those between shareholders and professional managers, those among shareholder-managers and debtholders, and those within shareholder and debtholder groups.

---

[9] There is a large body of research into the classic theory's inadequacies, confirming these incentive conflicts and their consequences as agency costs of finance (Tirole, 2006).

2.5.1   *Incentive conflicts: professional managers and shareholders*

With some highly visible exceptions (Microsoft, News Corporation, Virgin Group), large corporations are managed by professionals who rarely own controlling shareholdings – indeed, in less aware companies, managers may have no equity stake at all. Professional management's objective is often ideally reduced to a single aim: to maximise over time shareholders' wealth, taken to mean company value.[10] The main incentive conflict in companies where shareholders do not own significant stakes in the company is seen to come from separation of ownership of the company from control over the company.[11] Concerns include control conflicts over free cash flow generated by the company, and about the nature of the company's strategic decisions.

Free cash flow is generally defined as cash flow in excess of the amount that can be reinvested to sustain a business in a way that is profitable compared to other uses. While this is not intended to suggest all managers behave in the way outlined here, it is nevertheless a risk that when companies controlled by managers generate large free cash flows, then management will have an incentive to pursue activities that misallocate that resource for their own benefit, for example by making economic investments in uneconomic assets such as corporate jets and luxury cars, or by promoting unwarranted schemes for performance-linked compensation. Technically free cash flow belongs to shareholders, and should ideally be returned to shareholders if management cannot profitably invest the marginal amount in the core businesses of the company, for example by increasing dividends or (if the law allows) by buying back outstanding shares. The conflict in such circumstances is thus between controlling costs and maximising shareholder value on the one hand, and on the other enjoying rents in the form of the benefits of untrammelled managerial control rights over free cash flow.

Conventional corporate governance assumes that such uses of free cash flow are monitored by boards of directors as the agents of shareholders. The directors have supposedly ensured that the interests of shareholders, and perhaps of other stakeholders, are represented in the firm's major decisions. However, the creation of a board of directors may drive the issue one level deeper: what disciplines the behaviour of the members of the board, and who monitors the monitors? Milgrom and Roberts (1992) point out that only ownership interests will ensure that managers do not expropriate shareholder investments, but also that those with relatively small shareholdings have little incentive to bear the costs of monitoring management or the board's diligence. Nevertheless, while directors monitor management's performance, agree

---

[10] Some jurisdictions require managers of large companies to acknowledge in their objectives the interests of other stakeholders including employees and consumers, or environmental concerns. This may become a more widely accepted aspect of company law compared to the traditional Anglo-American model described here.
[11] Credit for first formal analysis of the problems of separating ownership from control goes to Berle and Means (1932) (Berle and Means, 1991).

its remuneration, discipline and replace management in the event of poor performance, they may often inadvertently condone the actions of management. The owners of ordinary shares can only realise the current value of their claims on the company by selling their shares (through a securities exchange, or privately in some cases) which underlies a market for corporate control. This abstract market can provide a further monitoring function on corporate management, on the basis that the ultimate sanction against underperforming companies is their acquisition and subsequent restructure. In such cases, the extent of agency conflicts between shareholders and their representatives (directors and management) takes on depressing proportions.

In attempts to prevent takeovers, directors have sanctioned the most ingenious defences to reduce the attractiveness of their company to a predator. Following Milgrom and Roberts (1992), these have included the infamous poison pill which functions to increase the cost of acquisition by increasing present shareholders' claims triggered if a single entity acquires a large shareholding. Other devices have included scorched earth policies which lessen a company's value to a bidder, even if it reduces shareholder value simultaneously; staggered boards, where only a fraction of directors must seek re-election each year; and super-majority rules, which require as much as 90% of votes to effect a control change. A major point of this discussion is that there is not only an agency conflict between shareholders and managers, making directors effective agents of the shareholders is necessary to combat this. The fact that directors may have the power to waive such defences gives credit to corporate governance activists' views – such defences should not exist. Milgrom and Roberts (1992) point out that shareholders with large amounts at risk play an important part in disciplining management, because such shareholders have little interest in uneconomic investment that may function merely to reward managers at their expense; while, on the other hand, such takeover bids introduce the potential for substantial gains to large shareholders. It may also be feasible for shareholders to litigate against managers and/or boards to protect their interests; or if represented on the board, they may be able to dismiss or replace poorly performing managers. There is little doubt that the threat of replacement functions as an effective management performance incentive.

In cases where it is difficult to organise a sufficiently large block of shareholdings to distribute the costs of monitoring efficiently, a substitute monitoring mechanism is to replace equity with sufficient debt to achieve efficient monitoring by the lenders, thus effectively delegating monitoring to the lenders. Banks, in particular, have been characterised as specialist delegated monitors often able to obtain contractual undertakings from a debtor company over aspects of general performance (Diamond, 1984).[12] The extent of such control may be the result of competitive market forces and in particular the relative negotiation strength of debtor and creditor, with, other factors

---

[12] Other writers suggest more narrowly that only the granting of collateral by a debtor fulfils this function.

being equal, a more leveraged company needing to concede terms to its bank lenders that offer a greater degree of control of information and activity. A number of models attempt to explain this phenomenon, primarily based on the fact that management and investors are asymmetrically informed. Managers will usually be better informed about the firm's prospects than passive investors – known as signalling models. For example, higher debt levels are seen as signals to the market of better earnings prospects, because higher expected revenues suggest higher ability to service debt, and the market responds by assigning a higher value to the firm. Managers of firms with low expected revenues avoid taking on higher debt, however, because it is more likely to lead them into bankruptcy (Milgrom and Roberts, 1992).

2.5.2 Incentive conflicts: shareholders and debtholders

We have gone to some length to outline how serious can be the conflicts of interest between shareholders and board members and professional managers, that is, their agents responsible for managing the company. Conflicts also bedevil the relationship between providers of debt finance and shareholder-managers (the company). It is best to start the analysis by confirming the agency theoretic nature of the relationship between the company shareholder-management and its lenders (for convenience, providers of debt are simply called lenders) (Jensen and Meckling, 1976). Lenders expect the company to use debt finance diligently and in accordance with the transaction's commercial terms, which may specify how the loan proceeds are to be applied, and repaid at the loan's maturity. These expectations are of course tempered by bounded rationality, incomplete information, and expectations of opportunism. The company will be expected to commit itself to the requirements of a loan agreement, and subject itself to monitoring of a lesser or greater degree, depending on its credit standing and default risk, application of the funds, and capital structure. Regardless of the diligence of lenders in drafting the agreement or devising monitoring arrangements, management will usually be far better informed about the firm's prospects than lenders; again, managers of the company and lenders, in this case, are asymmetrically informed. Under these circumstances capital structure can affect company value by influencing incentive conflicts between the company and lenders, thus influencing the probability of default and bankruptcy with their associated costs. For illustrative purposes, it may be proposed that lenders consider agency risks separately from default and bankruptcy risk, but that all these contingencies are fully reflected in loan pricing. The primary agency concerns of lenders in assessing the incentive conflicts with the company revolve around the choice of investment projects. Two important concerns arise with management's control over financial resources, in particular its debts, which are management's incentive to take excessive risks (also known as the asset substitution problem), and the problem of underinvestment. Note that the company directors have no general duty to act in accordance with the interests of its lenders, unlike the obligation they have to shareholders, so the rights of debtholders will need to be expressed contractually.

2.5.3 Excessive risk-taking

An important agency conflict that occurs between shareholders and creditors is known as 'excessive risk taking' (a.k.a. 'asset substitution'). All following Milgrom and Roberts (1992), when a firm is heavily indebted and approaches financial distress, owners (and their manager agents) may have incentives to make risky investments because owners receive all the residual income if successful, while lenders bear the principal part of losses if returns turn out to be low or negative. The parties' interests are in direct conflict: an increase in risk to creditors is matched by a symmetrical increase in expected return to shareholders, while in extreme cases risky investments may even resemble a gamble where the costs have already been incurred. However, Milgrom and Roberts (1992) further point out that lenders expect the firm's owners to be tempted by excessive risk-taking under financially distressed circumstances, and consequently will lend less and/or increase the cost of borrowing to the firm. Thus the shareholders will bear the cost of this expectation, and in turn will motivate shareholders to commit themselves not to take excessive risk, for example by agreeing to restrict shareholder/manager discretion through loan covenants (see Milgrom and Roberts, 1992, p. 495).

2.5.4 The under-investment problem

The underinvestment problem is somewhat more subtle, and also arises when a company is deeply indebted, but in this case it may have profitable investment opportunities that are entirely within its normal risk profile. At first glance this does not appear to be a real problem, indeed fortuitous. Milgrom and Roberts (1992) explain how this circumstance reveals an important dilemma: on the one hand, servicing outstanding debt is the firm's immediate contractual obligation, but this may deplete available cashflow to the extent that new investment opportunities have to be foregone. Typically these circumstances would signal that a company requires additional external finance, preferably equity, so that it is able to take its opportunities. On the other hand, Milgrom and Roberts (1992) point out, while additional equity may be difficult to raise, additional borrowing also poses a problem, because the first lender is unlikely to share the benefit of its claim's seniority given the company's existing leverage and the higher credit risk that more borrowing might entail. They point out further that it may also not be rational for a new lender to extend debt to the company, because any new claim would rank junior to the existing debt, and the additional value created by new lending would first benefit the senior lender, possibly by an improved credit rating for a bond issue. As a result, additional finance may not be forthcoming, and the attractive opportunities may be foregone. The fact that value is effectively lost due to an 'overhang of debt' results in an outcome which is in neither the company nor its creditors' best interests. Commercial transactions have evolved limited ways to overcome such problems, usually involving agreements among lenders to share collateral, or purchase money mortgages, where the new lender obtains a prior claim over a capital asset acquired with its loan. These solutions are not necessarily easy to operationalise, however.

On first glance, being able to renegotiate debt contracts when a company experiences financial distress appears to be advantageous to all. Therefore it might seem that all the parties concerned would want to ensure that such negotiations are practicable, for example, by concentrating debt among relatively few lenders, as pointed out by Milgrom and Roberts (1992); but, they further point out, the error in this logic is that suggestions that renegotiation is possible act to shield shareholders (and managers) against the consequences of bad commercial outcomes, may reduce their incentive to act diligently on the company's behalf, and may reinforce their incentive to engage in risky activities. With numerous and diffuse debtholders, the threat of debt overhang without the possibility of renegotiation may encourage the company to restrict its borrowing in the first instance; and conversely, when debt claims are concentrated, making renegotiation more practicable, incentives may be created for managerial misbehaviour (see Milgrom and Roberts, 1992, p.496). One common commercial result is that highly leveraged companies will be subjected to onerous contractual requirements as part of loan agreements, in such a way as to minimise the scope for managers to act recklessly.

2.5.5   *Conflicts within shareholder/debtholder groups*

It would be amiss not to identify conflicts that occur within shareholder and lender groups. To start with, these problems are not clearly categorised as agency problems, but result from group dynamics where members pursue their own interests, but also have common interests that require monitoring to ensure best outcomes. Consider first the shareholders as a group. The group's interest is that management and company performance be monitored diligently. However, conflicts arise when one individual shareholder bears the full cost of any monitoring efforts, while the ensuing benefit is shared by the remaining members – all of whom thus have an incentive to become free riders. With highly dispersed shareholdings in large companies, individual shareholders are in no sense highly motivated to incur the full costs of diligent monitoring because the results are so widely dispersed. Unsurprisingly, there is empirical evidence suggesting that concentrated share ownership leads to more effective monitoring of company management and performance than dispersed shareholdings (Milgrom and Roberts, 1992).

The free rider problem is less relevant where a group of banks is lending to a company within a known hierarchy of claims, because it is likely that the scale of each claim will provide every lender with sufficient incentive to be responsible for monitoring. One advantage of the syndicated loan structure is that it creates scale economies for the company in reporting to lenders, since one bank acts as agent for the others and is a conduit for the company to provide day-to-day information to the bank group as a whole. However, with bond transactions the concern is often similar to that with widely dispersed shareholders, because bondholders have historically been relatively inactive investors compared to bank creditors, and unwilling to act as specialist monitors of the debtor. With bonds issued by highly leveraged companies, monitoring is often enforced by restrictions on managerial discretion, for example through covenants that restrict the right to issue new debt, require the company to meet financial performance objectives or even restrict the company from investing in new business activities. More generally, conflicts have arisen between banks and bondholders in cases of corporate distress, particularly when there is disagreement about the severity of a company's financial plight. Lenders with seniority in claims may prefer liquidation and a curtailing of losses while bondholders may prefer to be indulgent and hope that the company will trade out of its distress. An alternative explanation is that, except in the US domestic markets, bondholders traditionally favour anonymity and thus prefer not to engage actively in debt renegotiations. However, this may now be a fading motive.

These circumstances may require intervention by the courts, or the use of insolvency legislation giving companies temporary relief from their creditors (such as bankruptcy workout procedures (Chapter 11 provisions) in the United States). In certain cases since the 1990s, financial regulators have acted informally to promote negotiations among large creditor groups.

2.6 Transaction costs and agency – theoretic logic of the project finance model

In order to conclude this wide view of transaction costs, incentive conflicts and agency problems and their manifestations in the two fundamental financial contracts, debt and equity, we abstract several principles from earlier analyses in this section to highlight the logic of the project finance model in corporate finance. First, recall from Chapter 1 that project finance companies exhibit features that distinguish them from conventionally financed companies. This would include adopting a legally independent project company, supported by concentrated shareholdings (between one and three sponsors customarily own all project equity) and ideally acquiring non-recourse debt, owning and operating a single purpose industrial asset, and governed by extensive contracting with many prescribed rules over managerial discretion. Each of these project company features, informed by the logic developed in previous sections, will be elaborated upon. However, what is very important is to understand that these four features function together in the project company approach to addressing conflicts in corporate finance. In several circumstances more than one feature will seek to deal with the same conflict and similarly one feature may address several conflicts. The integration and discipline imposed by the project finance model on corporate governance are inescapable, and we are almost ready to suggest that the model has an air of elegance for its internal consistency.

2.6.1 High ownership concentration

We begin by showing how concentrated ownership helps address management–shareholder conflicts. In terms of corporate governance, concentrated ownership facilitates efficient monitoring of both directors and managers. It has been pointed out that shareholders with large investments can be critical in disciplining management, and with project companies there is little scope for professional management conflicts that arise from their having control of free cash flow. Project companies have no such free cash flow as the rules of governance will define how cash flow generated is to be applied. Further, project companies usually assign narrow objectives to operational managers, which are facilitated by the asset-specificity of the project assets. This allows the delineation of clear performance indicators, and clear mechanisms for their measurement, so that early indications of derelict performance are likely to be evident in operational or financial performance. The more specific management's targets, the more their scope for opportunistic behaviour is correspondingly smaller.

Incentives between project managers and shareholders are further aligned because the principal shareholders are likely to be the ultimate employers of the operational managers of the project facility, and only seconded for particular periods. In addition, there are further constraints on managerial behaviour – a high proportion of debt in the project company's capital structure will involve restrictive covenants controlling and prescribing how the facility is to be managed, including important forms of bonding such as lock-up periods for equity participants and rules for cash dispersal. From our earlier analyses, we know that banks are often characterised as specialist delegated monitors and the propensity for this to be the case will tend to reflect the specificity of the project and its financial structure.

Several additional project company features function to address conflicts that may arise between shareholders, particularly those with interests in early phases of a project company's lifecycle, such as contractors. By insisting on equity participation by promoters or developers as a form of bond, their commitment to the project is achieved, while some form of continuing equity participation may be a prerequisite to the granting of facility maintenance contracts.

2.6.2   High asset-specificity

We continue with the second feature, namely specificity of the project's assets, and point out how this feature complements capital structure decisions in managing the incentive conflicts associated with conventional corporate finance. Some of the lessons drawn from the increasing unpopularity of many major western conglomerates in the 1980s and in their subsequent break-up now resonate in project companies. At an earlier time, diversification had been thought to have risk management attractions, but this was a phase in corporate finance history. Investors were shown instead to value distinct, narrowly focused companies more highly than broadly based diversified conglomerates. Not only were investors seeking identifiable opportunities that could be readily appraised, but they saw that the scope of managerial discretion and ability to make unprofitable investments was considerably reduced with focused single-sector companies, and might even merit the unbundling of companies within an industry segment so that each major function was separately incorporated. Recall that asset-specificity indicates the level to which an asset can be deployed in alternative uses, and we have pointed out the high level of specificity of project company assets. A power station can hardly be turned into a shopping mall overnight, as we say again in a later chapter.[13] This is an extreme form of narrowly focused company – in fact it is a single asset company, which has only one possible function. This imposes direct discipline on the project company and its managers in several important ways, for the nature of the asset allows no true managerial discretion in operations. It can be managed only as a power station, further focused by the necessity to honour both supplier and customer agreements. In turn this allows a narrow managerial and cost accounting discipline to be observed, as does the tendency for the technologies employed in project facilities to involve low risks and be well understood. In general, it may be noted that physical capital intensive industries such as public utilities or steel producers are often heavily debt financed, and project companies are a further development of the notion that relatively mature, stable industries have significantly higher debt capacities than volatile emerging sectors undergoing high growth.

2.6.3 *High leverage in capital structure*

Third, we pointed out in Chapter 1 that project companies typically have a high proportion of debt in their capital structure. Leverage concentrates managers' efforts in several respects. First, borrowing costs will tend to reduce free cash flow, making debt an important mechanism to restrain managers by lessening their control over free cash flow. Further, bank lenders will set extensive rules for a project company that allow for cash flow to be applied in a prioritised way; these often include conditions requiring debt to be wholly or partly settled before dividends may be distributed to shareholders. It is fair to say that during the early operational phases of most projects, servicing and repaying debt is the primary focus of typical project company directors. Major lenders may in some cases be represented on the project company's board to reinforce the monitoring mechanisms of their contractual covenants and together allow little managerial discretion over use of cash flow. As far as the problems of asset substitution and underinvestment are concerned, there is no managerial discretion over investment decisions, because the project vehicle is a single-asset company with the investment already chosen and its sole reason for existence. Asset substitution and underinvestment are problems seldom seen in project companies, and if either does occur then it will usually represent a contractual or institutional failure

---

[13] The example of the long-decommissioned Battersea power station in London shows how difficult it can be to find alternative uses for even the most iconic facility.

on the part of the shareholder-sponsors or financiers. If a project company faces financial distress, its lenders will be motivated either to renegotiate their loans or exercise their right to take over and liquidate the facility, or the equity providers may be strategically motivated to provide an infusion of further resources. High leverage commits and compels project companies to meet their debt service requirements or risk losing control of their facility. Important empirical evidence suggests that in certain cases increases in a company's debt tend to increase the share price, and vice versa – it is fair to suggest that the discipline it imposes upon managers does not go unnoticed (Jensen, 2000).

2.6.4   *The nexus of contracts*

Fourth, in corporate finance terms project financed companies are typically large, legally separate (standalone) single asset companies that are extensively controlled by contracts. The notion that the project company acts as a nexus of contracts has particular meaning for our purposes, which in a sense encapsulates each of the influences described here. The essential features of the project company influence the behaviour of shareholders, managers, lenders, suppliers, customers, and all operating personnel. This important function is that in project companies, the culture enforced by interaction of all the contracts is that of *rules over discretion* – there are very few significant incentive conflicts in the project company structure that can undermine corporate objectives, and thus little leeway for discretionary behaviour by any party to any contract (excluding simply criminal or fraudulent behaviour, of course). Further, it is not possible to single out a predominant feature of a project company in this arrangement, for the main features appear to be mutually reinforcing. High equity concentration facilitates direct monitoring by equity holders and lenders; high asset specificity facilitates narrow managerial objectives and removes discretion over investment decisions; high debt concentrates management effort. An important further reason for intensive contracting is to manage the risks associated with the company's ability to service and repay its debt, to which we give attention in Chapter 4. This imposes further direct discipline on the project company and its managers in several important ways, because the nature of the asset allows no managerial discretion – it can be managed only as a power station, further focused by the necessity to honour both supplier and customer agreements. As with particular contracts that need rules to govern the transaction, we view the internal consistency of the project finance model as more akin to a transaction governance arrangement than an example of corporate governance.

Key concepts

The following concepts are considered sufficiently important to memorise as key vocabulary for use in subsequent chapters.

Agency costs of finance
Agency theory
Anatomy of transactions
Asset specificity
Behavioural assumptions in transacting
Bonding and monitoring
Bounded rationality
Complex transactions
Corporate finance instruments
Corporate governance
Dimensions of transactions
Fundamental transformation
Imperfect contracts
Incentive conflicts
Information asymmetry
Opportunism
Relational contracts
Transaction governance

Appendix 2.1   Comparison of long-term debt instruments

Loans and bonds are generic terms for financial instruments representing current or future claims against a debtor.[14] They are taken here to be long term in duration, which by convention means an original contractual life of more than 12 months. The practices described here draw on market conventions, and on formal traditions from several sources, notably English, German and US law. It is unusual for more than one debtor to share liability under a loan or bond, but both instruments are commonly guaranteed by an entity affiliated with the borrower.[15] International bonds and syndicated loans are transactions widely used in major project funding.

Traditionally, loans and bonds were easily distinguished transactions, separate in purpose, application and often in law. Bonds were simple contracts largely free of conditional clauses known in loans as financial or operational covenants, and were

---

[14] Not considered here are borrowings made by individuals or instruments containing equity conversion rights for holders, which are treated as debt instruments until conversion. Financial instrument is used here in a generic commercial sense and not in relation to any formal system of law.

[15] Guarantees are a form of credit support provided by a third party, without regard to any particular legal system. Several jurisdictions may not recognise debt transactions where a borrower and guarantor are closely affiliated, as with parent companies and their wholly-owned subsidiaries. Bonds of lesser credit risk may often be guaranteed or 'wrapped' by specialist monoline insurers.

valued for their ease of sale by one investor to another. Bond issuers tended to be governments or companies of relatively high standing; by contrast, loans were made by banks to a wide range of governments, state organisations or companies, and contained terms restricting their transfer. This helped create an interactive relationship between debtor and lender, for example, in the disclosure of commercial information or compliance with operating covenants. As a result, the parties to a loan would remain largely unchanged until the debt was discharged. The transferability or potential liquidity of a debt instrument was, until the mid-1980s, a good indication of whether the claim was a bond or a loan.

These distinctions have grown increasingly obscured. Loans may trade as freely as bonds, bond issuers may be weaker risks, and both can be subject to equally intricate terms. For some lenders or debtors, these instruments are near substitutes, and while market practice dictates that different contractual structures are used to give effect to loans and bonds, the true differences are becoming increasingly minor. This trend results from several factors, including greater focus on the cost of capital to support bank lending, the entry as lenders of many insurers, hedge funds and other non-bank intermediaries, and the wish of investors of all kinds to freely manage their portfolios by buying or selling financial assets, however categorised.

Project finance is one of few global market segments in which loan transfer is comparatively modest, and where loans and bonds remain generally distinct. This division is likely to fade, even though freely transferable loans are disfavoured by project sponsors who may be constrained from sharing commercial confidences with loan buyers with which they have no working relationship. For example, this may result from loan covenants requiring disclosure of private supply or sale contracts and this also concerns transaction arrangers expected to provide for adequate disclosure to potential investors. International practice can be expected to produce a working protocol to give comfort on this matter to sponsors and lenders.

To date, most loan transfer associated with non-recourse finance has occurred well after project completion, or in defaulted transactions, where lending banks may choose to sell their claims to specialist investors.

A2.1.1   Syndicated loans

Concept and origins

The international market in syndicated loans is a contemporary phenomenon that has grown since the 1970s into a prolific source of funding for borrowers and projects of all kinds. Such transactions are collaborative institutions by which largely self-selected groups of financial intermediaries extend credit to a borrower on common terms. Each lender obtains a separate claim against the debtor,[16] but the loan's

---

[16] This is contractual practice under English and New York law, which taken together govern most international syndicated loans.

legal documents provide for coordination among lenders or their successors in exercising contractual or statutory rights against the borrower, for example, after a breach of commercial terms or payment default. They will also stipulate that payments received from the issuer be shared among lenders in proportion to the scale of their claims.

International syndicated loans first became prominent in large-scale capital asset and energy finance, especially for ships or deep water offshore oil production. These fundraisings were substantial, complex and beyond the prudential capacity or financial resources of any single lender. The complexity of terms of similar financings gave rise to transaction scale economies, so that costs could be reduced by coordination among sponsors, lenders and their advisors. The uses of loans made in this way quickly broadened as banks sought ways of deploying soaring Eurocurrency deposits. With no conspicuous official encouragement, this collaborative practice became established in the international financial markets even for modest loans, bringing greater choice to banks of all kinds. Provincial banks from many countries became able to lend more widely than ever; all banks were induced to manage the spread of credit risks to which they were exposed. Large banks competing as putative loan arrangers could raise transaction returns by exploiting the tiered structure of arrangement fees, passing on to small participants only a share of the upfront facility fee paid by the borrower.

Syndicated loans grew structurally elaborate, with flexible advances, repayments and re-borrowing, provisions for borrowers to switch the currency of the loan or interest payments, or terms that adjusted with the borrower's credit standing. They became a staple of 1980s acquisition and energy finance, 1990s telecommunications funding, and all current forms of construction, infrastructure and leveraged acquisition finance. Only from the early 1980s, when national regulators began to consider problems associated with extravagant lending to emerging market borrowers, did it become clear that the syndicated loan was as likely to encourage risk accumulation as its dispersal. Loan losses resulting from 1980s debt crises led to the creation of common international capital standards for banks in the 1992 Basel Capital Accord.

Arrangement

All syndicated loans have two phases. The first involves negotiation and agreement to terms, syndication of risk, and closing or execution of documents. The second phase maps the loan's life after closing to final repayment, including the making and administration of a loan, monitoring the condition of the issuer or its project, and the transfer of claims among banks and other lenders, often known as secondary trading. If a borrower becomes financially distressed, the second phase will include negotiations for a change of terms, either alone (as with the financing of the Channel Tunnel project), or in a wave induced by a deterioration in general conditions, for example after foreign payments moratoria initiated by certain Latin American states in 1982,

1989 and 1995, or in South Korea and Southeast Asia following the 1997–98 Asian financial crisis.

Prominent banks vie to arrange loans, competing on cost, reputation, or the closeness of their relationship with a potential issuer or project sponsor. This process can be protracted, with rival banks offering advice on financing strategy in the hope of being asked to arrange a deal. Experienced issuers begin the process with requests for proposals (RFP), calling for banks to indicate the terms upon which they will arrange an outlined transaction, and their willingness to commit to provide all or part of the intended proceeds. When the issuer is satisfied with a proposal, it awards a mandate to one or more banks to arrange the deal, usually setting a period for completion. These mandated lead arrangers (MLAs) often underwrite the loan, making a commitment in principle to provide the amount to be raised and its interest and expense cost, or in less favourable conditions a best effort undertaking expressed more loosely as to amount and timing. Underwriting commitments are rarely withdrawn.

The arranger then markets the deal to investors using an information memorandum based on material gathered from the issuer, and in many cases through roadshow meetings attended by senior officials of the issuer. If successful, this syndication process will result in a group willing to join the deal, reducing MLA underwriting risks. The lead banks also profit by inducing others to join the transaction on marginally less favourable terms, since lenders receive upfront fees (usually deducted from the proceeds advanced to the borrower) scaled in proportion to their final lending commitment and initial underwriting, if any. The process relies on the strength of network relationships and reciprocity, and the arranger's skill in balancing a borrower's cost objectives with the lenders' wish for adequate returns on capital. International loan and bond transactions display path dependence in relying on issuer and arranger reputations, so that successful arrangers or coveted borrowers command loyalty from lenders and often obtain favourable borrowing terms.

During syndication the arranger will instruct lawyers to prepare a credit agreement and supporting documents to give effect to the deal, ensure that it will be legally enforceable against the debtor, and as far as possible verify the commercial assumptions on which the loan relies, said to be part of due diligence. In many cases these elements can be simple to put into effect, but complex project financing will entail many stages, several tiers of lending and other risk support, and interlocking credit agreements and security documents. The outcome of a simple transaction is shown in Appendix 2.1 Figure 1, illustrating the contractual links between parties.

Only the borrower, lenders and the lenders' agent are party to the credit agreement. The arranger takes no formal part in the loan but is indemnified against the risk of actions taken before signing, especially in distributing information to potential lenders.

Banks join large transactions in tiers depending upon the scale of their underwriting or lending commitments and whether or not they contribute to arrangements before signing. This often produces an array of deal titles, for example, co-arranger, lead manager or manager, but these are contractually superficial, part of market practice

Credit agreement

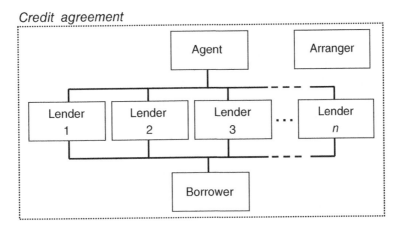

Appendix 2.1 Figure 1 Syndicated loan structures.

due to the importance of league tables of bank transaction rankings. Only MLAs have duties before completion but to avoid unwanted liability all syndicate members participate in the transaction purely as lenders, with other titles expressly dissolved. The credit agreement will appoint a facility agent to administer the loan, usually the arranger or a professional fiduciary. Agents act for lenders in collecting and distributing payments and commercial or financial information, monitor the debtor's compliance with financial covenants and in project finance supervise the maintaining of collateral, insurances, supply and delivery contracts or assignments of revenue.

Syndicated loans are usually floating rate transactions, so that interest is calculated using a money market benchmark that is reset from time to time. This lessens mismatches between the duration and cost of loans and deposits, two important risks inherent in bank intermediation, and offers the borrower a degree of cost-effective flexibility by allowing early prepayment without penalty on any interest payment date. Most loans can be drawn in single amounts or a series of tranches according to need. Lenders receive interest on their outstanding claims accrued at the sum of a benchmark rate such as 3 or 6 month US\$ London interbank offered rate (Libor) which is the most widely used market-determined rate of interest on wholesale deposits for defined periods in one of many currencies. Added to this is a spread or credit margin based on the market's risk perception of the issuer at syndication, including the credit rating of other debts. Lenders may also receive a small periodic commitment fee as compensation for any commitment that is contractually firm but undrawn, to mitigate the opportunity cost of allocating capital yet receiving no revenue.

During the life of the loan, its applicable interest rate is fixed for periods corresponding to that of the benchmark. Interest is paid as each period ends and the loan rolled over for a further discrete period, with interest accruing at a new rate given by prevailing conditions. However, this neglects any interest rate management conducted separately by the borrower or lenders, for example using options or swaps. Lenders are also usually indemnified by the borrower against unexpected funding costs, for example, if a disrupted market makes normal deposits unobtainable at the start of an interest period. Syndicated loans are commonly of 3–10 year maturities; project loans are often longer in tenure. Many long-term loans typically provide for scheduled amortisation of principal by instalments for credit risk or cash flow reasons, beginning after a grace period, or use a composite structure whereby each tranche is separately repaid. These devices can substantially reduce the loan's average life, a comparative measure of the period for which the average principal is outstanding. All these terms will vary with the credit risk of the issuer.

Using transparent benchmark interest rates was crucial to the growth of the syndicated loan market, especially before financial derivatives such as swaps and options made interest rate risks more easily manageable. Banks could assume that they would match fund their loans, using deposits raised in the wholesale interbank markets for the same period as each loan fixing, so that the credit margin (adjusted for the discounted value of any fees) would be identified as the lender's net interest margin on any transaction.[17] This also allowed loans to be easily compared in terms of actual or risk-adjusted returns on capital. Project and corporate loans increasingly use grid pricing, setting the credit margin with a matrix of variables including issuer credit ratings, leverage, or remaining loan life, which rewards the borrower for an improvement in credit risk and compensates lenders for any deterioration in the quality of their claims.

The particular value of syndicated loans is a capacity to achieve the complex relatively quickly. This relies on experienced intermediaries making collaborative commitment decisions and meeting aggressive deadlines, and on the efficiency of standardised procedures and legal documents. Peer pressure and the need for transferability have made credit agreements increasingly uniform in content and language, drawing on generic models developed by banking industry associations. Structural standardisation has impacted even project finance deals that involve distinct phases of funding and layers of collateral. A generic non-recourse loan is shown in Appendix 2.1 Figure 2, representing the financing of a major capital asset such as a ship, a basic form used in syndicated lending since the 1970s.

A loan is made to a special purpose vehicle (SPV) controlled by a project sponsor, the beneficial owner, enabling it to build or acquire its sole asset, the revenue generating ship. The terms of the loan will require that all other elements shown in Figure 2 are kept in place to the satisfaction of the lenders. Thus the owner or project manager must ensure that until the loan is discharged the SPV engages in no other activity, and the vessel is properly mortgaged, employed, insured, and maintained so as to meet regulatory requirements. Structural mismatches that make the loan riskier will require compensating credit support; for example, if the term of a charterparty

---

[17] Interbank dealing includes credit margins on deposits, so that only banks of high credit quality are always able to borrow at Libor.

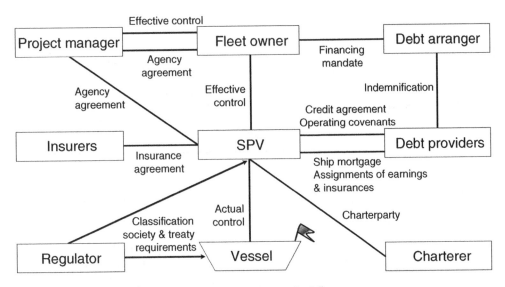

Appendix 2.1 Figure 2   Simplified non-recourse project finance.

is less than the loan's remaining maturity, the sponsor may be required to grant partial recourse in the form of a liquidity reserve or part guarantee. The transaction is an institutional attempt to devise a complete contract over the vessel's commercial operations, so that the loan's terms and covenants match as closely as possible the financial performance of the project for which it is made.

Banks were once reluctant to use the most transparent syndication techniques in project finance unless the scale of funding necessitated otherwise, and confined syndication and secondary loan sales to lenders sharing experience of specific industries. The dilemma reflects the competing objectives of loan market liquidity and lender–debtor relationships. Lenders holding client sensitive information might be restrained from making a corresponding disclosure to potential loan buyers, and be thought to misrepresent the nature of risks associated with a loan. The concern affects both prominent banks and issuers: banks concerned at the risk of improperly inducing a loan purchase, issuers fearing the loss of commercial information. These questions were addressed in court decisions in England, the US and elsewhere,[18] but today's extensive lending by non-bank investors is controversial in this respect and not fully resolved in regulation or market practice. Problems of disclosure traditionally constrained the use of bonds in project finance but now affect loans and bonds alike.

In complex projects, the single loan shown in Appendix 2.1 Figure 2 would represent a succession of transactions, each predicated on the completion of identified stages and providing a repayment platform for its predecessor. The sequential concept is shown in Appendix 2.1 Figure 3.

---

[18] Notably involving syndicated loans owing upon the mid-1970s collapse of Colocotronis Shipping Group.

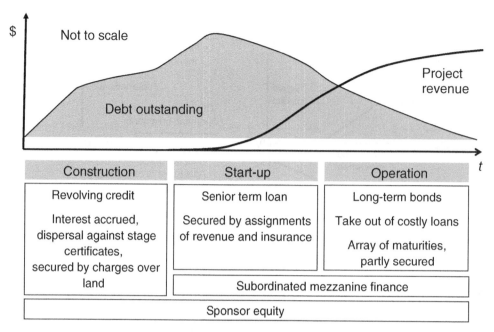

Appendix 2.1 Figure 3 Project lending sequence.

A greenfield project might have five financing components, each with different priorities of claim. Banks provide a revolving credit throughout construction, with the amount owing at any time adjusting to take account of progress, dispersals, and receipts and linking advances to verifiable milestones, so that the facility combines control of cash and debt management, and defines the lenders' risk. With construction completed, the project is commissioned and accumulated debt and any accrued but unpaid interest refinanced with a senior loan on terms sympathetic to cash flow potential but without the need to micromanage cash. This may combine with a longer-term mezzanine facility ranking junior to the senior debt, especially in highly leveraged cases. Finally, when the debt burden peaks and the project is fully operational, the outstanding senior loan is replaced by bonds of long duration, with redemption geared to the project's accumulating reserves. All or part of this element may be provided by securitisation, a form of structured finance which isolates specified project revenues to meet the claims of bondholders. Throughout this timeline the sponsors will be expected to provide and maintain an element of equity capital.

A2.1.2 International bonds

Concepts

Bonds are tradable long-term debt instruments bought and sold by investors of many kinds, including non-professional retail investors. Most are simple in structure: an outstanding principal sum remains constant until repaid in one amount at maturity, until when the investor receives a periodic interest coupon. Unless unusually small, the bond may be traded by banks and investors until some time before its final year. Many types of coupon have been devised to meet the preferences of investors, but most have either a fixed percentage rate paid once or twice annually, or a floating rate calculated as with syndicated loans, usually paid more frequently. Thus the markets can be divided into fixed rate bonds, floating rate notes (FRNs), and more complex issues known as structured notes. Fixed rate bonds form the largest component of the international bond markets and are the focus of these remarks.

Many aspects of the arrangement of new loans and bonds are similar, although all but complex bonds tend to be more quickly initiated, syndicated and closed, a corollary of loans giving more potential flexibility to the borrower. Issues for well-known or frequent borrowers may take only hours to arrange and execute and no more than several days need lapse between an arranger receiving a new issue mandate and the payment date, when the borrower receives its proceeds. Two simple factors distinguish the commercial aspects of loans and bonds: bonds tend not to amortise but are repaid in a single amount, and traditionally have longer feasible maturities than loans. Unlike loan participants, investors in bonds that are subject to early redemption require compensation for the resulting reinvestment risk. In such cases, the investor effectively writes a call option for the borrower, and so expects payment in the event of its use in the form of a premium to the nominal value of the bonds.

One further difference is systemic rather than transactional. The prevailing price of liquid fixed rate bonds forms in advanced economies a transparent link between the way that interest rates are determined, aggregate flows of savings and investment, and the valuation of capital assets and commercial enterprises.

The market in international bonds is an example of a financial innovation developed organically by participants without direct or significant official input. Today's bond markets grew from the Eurobond market of the 1960s and 1970s, once a permissive alternative to national markets dominated by government issues and cross-border capital restrictions. Issuance and trading were largely untied to any national market but after the 1970s became based predominantly in London. Since the late 1980s, the international bond markets have developed practices and protocols for issuance, trading, settlement and disputes that have become highly influential in the domestic markets that they once sought to circumvent, so that international market practice now infects all major domestic bond markets, reflecting a high degree of integration between domestic interest rate determination, all debt markets and markets for interest rate and currency swaps. International bond markets are now virtually global, as with the markets in foreign exchange or interest rate swaps. Only retail investor activity or those minor markets still subject to capital restrictions (such as many in Asia or Latin America) can be considered national in focus.

Bonds are debt claims that in certain jurisdictions are defined as a form of security in order to invoke laws protecting relatively unsophisticated investors. US securities law is the most complex, setting procedures for public issuance that are often demanding and costly. By comparison international bonds are more lightly regulated and traded mainly among banks and other investors. Unlike most shares, international bonds are dealt and settled directly between counterparties, although by convention they are often formally listed on an exchange since many institutional investors are restricted in purchasing non-listed securities. Reliability of settlement may have been the Eurobond market's most powerful innovation, as this was the first over the counter (OTC) market to provide for generally quick and error-free transfer, payment and custody of securities, using systems intended to minimise counterparty risks. The bulk of trades settle through one of two international central settlement and depositaries (ICSDs): the Euroclear system in Belgium and Clearstream in Luxembourg, part of Deutsche Börse.

New issues and pricing

The new issue process for prominent borrowers is similar to that for loans, although the contractual structures of the two instruments differ and most bond issuance is completed far more quickly. As with loans, issuers use RFPs to solicit proposals but banks bidding for mandates from frequent issuers typically submit outline proposals of different types each day or as often as market conditions change. In each case an outline of terms will describe the currency of issue, market segment and target investors, transaction size, the basis of pricing, and the net overall cost of funds to the issuer, particularly when cross-market arbitrage is used to generate a competitive cost. In a transaction that takes advantage of favourable prevailing terms in one bond market segment, but uses interest rate or currency derivatives to achieve cost savings for the borrower expressed in another currency, the overall borrowing cost and the terms of the new issue may be all that the issuer knows, with derivative pricing kept confidential.

Competitive bidding usually leads to the award of a mandate to one bank or a very small group to lead a transaction. A fully underwritten issue will entail delivering funds in a set minimum amount and maximum cost to the issuer, and such bought deals are common practice for well-regarded borrowers. However, a borrower concerned only with its cost of funds and indifferent to the deal's reception among investors will find that such tactics rarely bear repetition. Capable or well-advised issuers try to balance their optimum costs against winning a favourable market reaction to new issues, using investor regard to assist future transactions.

Market practice prefers relatively small bond syndicates in order to enhance the arranger's control of pricing, launch and sale process. Syndicates for frequent borrowers are especially small, with the arranging bookrunner taking by far the largest allocation of bonds. This increases the arranger's potential profit if the issue is successful, especially given that it must manage the risk of the commitment to deliver funds to the borrower at a fixed overall cost. The need for control stems from the effect of active grey market trading, which begins when a new issue is disclosed electronically and before its final terms are known. Grey market trading, which presupposes the terms of bonds as and when issued, means that banks outside a syndicate may influence the bond's initial performance.

For new or infrequent borrowers, bookrunners often conduct open price formation, discussing a range of pricing with potential syndicate members and leading investors, and hope that a favourable response will allow final terms to be fixed tightly within that range. Other banks will do likewise, so that selling effectively begins at the same time as syndicate formation. More generally, the US practice of bookbuilding is increasingly used internationally, with investor interest being canvassed and pooled by syndicate members, leading to final pricing and the initial allocation of bonds. Bookbuilding helps establish a market-clearing price for new issues, and avoids mispriced transactions being left unsold with underwriters. Critics suggest it lessens the value of the grey market, is anti-competitive and oppressive to investors.

Fixed rate bonds are priced using real-time comparisons, most relating to the yield to maturity of the issue. At any time, yield to maturity is an estimate of the present value of cash flows emanating from a fixed rate bond, that is, the discounted sum of coupon payments and net principal, adjusted for any accrued interest, using that yield as the reinvestment rate. Since the coupon and maturity are fixed and bonds are assumed to be redeemed at their nominal or par value, the bond's price and yield will be inversely related. The traditional pricing mechanism is shown in Appendix 2.1 Figure 4. The reoffer yield available to investors at launch is the sum of a credit spread and the rate for comparable maturities given by a benchmark yield curve. Reoffer derives from bonds being bought by the bookrunner and then reoffered to investors at a distinct price.

The benchmark curve is given by the risk-free rate in the currency of issue on outstanding government bonds, or on interest swap rates. In reality, the bookrunner makes all comparisons it believes appropriate, for example, prevailing yields of existing bonds of a similar credit rating, domicile, or borrower type, including those of the issuer. Until the 1990s, the benchmark was always a government bond yield curve, such as for US Treasuries, French *Obligations Assimilables du Trésor* (OATs) or Japanese Government Bonds (JGBs), but the markets in interest rate swaps are now often more reliable for new issue pricing, especially for bonds denominated in euros. This underpins the close links between bond market prices and prevailing interest rates.

The issuer's all-in cost will be marginally greater than the reoffer yield as a result of upfront fees, conceptually similar to fees on loans. Bond fees tend to conform to standard levels varying with issuer credit rating, currency of issue and bond tenure, but are seldom separately collected. Instead the bookrunner and syndicate members will seek to profit from the difference in the cost and sale price of the issue, within the tolerance of the market and issuer. Fees may be passed on to leading investors when conditions are poor.

For bookrunners, the new issue timeline consists of negotiations with the issuer, price formation, syndicate building, bookbuilding or selling, and some form of market making after closing. In international practice, market making is a disclosed

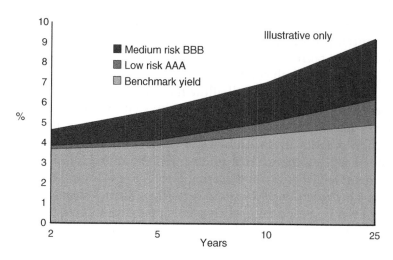

Appendix 2.1 Figure 4 Pricing a fixed rate bond.

willingness to buy or sell bonds in a given minimum amount, either at large or only to investors that are not themselves market makers.

First-time issuers are likely to need advice in obtaining a credit rating from at least one major rating agency.[19] Certain of these functions involve risks that need hedging, including cases where a bond is considered overpriced or cheap to the investor, and the bookrunner may become short of a rising issue, that is, to sell more bonds than it holds. To compensate, successful issues can be increased by prior agreement, although the use of such greenshoe options must avoid offending investors by dampening the issue's price performance. Very large issues can command a yield liquidity premium that represents a marginal reduction in costs for the borrower.

Structures

The simplified structure of an international bond issue for non-frequent borrowers involves three main agreements, shown in Appendix 2.1 Figure 5. The terms used may vary according to the arranger's domicile.

The bond represents an obligation of the issuer to bondholders, and a trust deed or agency agreement governs the administration of payments and information. Before payment, a subscription agreement with the potential issuer governs the creation of bonds and their initial purchase by the arranger and syndicate members. In contrast to the syndicated loan, this structure provides no representation for investors. The contractual structure of bond issues differs from loans for the traditional motive of providing for unrestricted transfer, and the ongoing need for the law to provide protection for unsophisticated retail investors. Bond contracts are usually free of extensive

---

[19] Credit ratings are customary for bonds and may become common for major creditor loans.

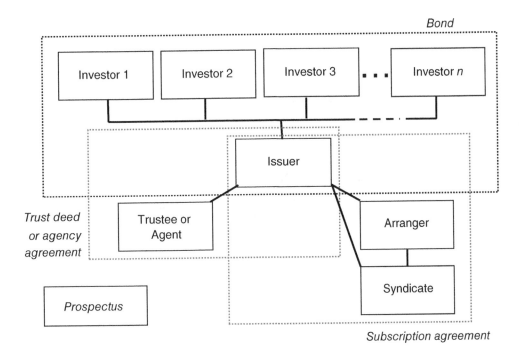

Appendix 2.1 Figure 5 Simple international bond structure.

terms and covenants for the same reason, and may include few other provisions, notably to protect the relative status of the bonds against subsequent issues. Their main function is thus to specify the commercial terms of the issue, notably the coupon and redemption provisions.

Frequent issuers usually establish a medium-term note (MTN) or debt programme to act as an umbrella facility for many kinds of individual issues. The concept is shown in Appendix 2.1 Figure 6.

The contractual terms common to all possible issues are gathered into a programme memorandum, together with an outline of procedures for issuance. A core of named dealers will be expected to bid to manage individual issues and others will be admitted as one-off bookrunners for their command of particular markets or arbitrage opportunities. The programme arranger will periodically update the memorandum with information, so that it serves as both a prospectus and an assembly of terms. Whenever a new issue is made, a short pricing supplement will set out its commercial terms and form a more detailed obligation of the issuer (by incorporating by reference the programme document).

Convergence

These descriptions show that while long-standing distinctions are fading, only a minority of bonds and loans are simple substitutes. Both instruments can be complex

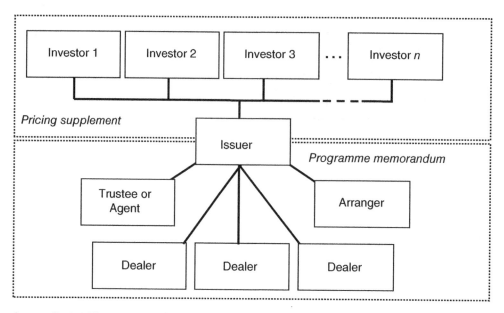

Appendix 2.1 Figure 6 Debt programme structure.

in structure. Most fixed rate bonds and FRNs are inherently straightforward and thus give less flexibility of use but this is barely relevant to borrowers whose paramount concern is their marginal costs of new funds. Historically, bonds were governed by simpler terms, because investors were concerned primarily with ease of transfer, which complex terms could undermine. Further convergence in documentation is likely, especially as structural and contractual homogeneity has markedly assisted secondary loan sales and dissolved some fundamental differences between the two generic instruments.

3

Financial Evaluation

Chapter 2 set out to introduce transaction cost economic concepts to provide context for analysing project finance transactions. We turn now to entirely conventional corporate finance questions, and also apply conventional techniques. We aim to explain the nature of financial decisions faced by the promoters of a project company, and then outline commonly used methods to evaluate the decisions, while using highly simplified corporate finance and accounting concepts to do so. There are many reasons why valuing a company may be necessary, including when the purchase of a private company is considered either by private investors or a corporation (no share price information), or it may be required if shareholders decide to sell a private company and require an indication of value to commence negotiations. It is thus commonplace in mergers and acquisitions in the corporate world, and it is equally important for governments considering participating in some form of venture with private sector participation – to know both its own financial position and that of its private sector counterparty. In the case of developing a new project company its value (i.e. present value) is an essential input into the capital budgeting decision, plus the way the project value is created (the pattern of cash flows generated over its life) is also important in determining the project company's debt capacity and for promoters (and possibly governments) to understand the scale and structure of necessary equity participation.

In this chapter we consider the basic project finance questions to be what a proposed project company is worth (value), how it is to be acquired (capital structure), including how decisions may be reached about the project company's debt capacity (permanent project company debt), and consequently decisions about equity participation. We consider then the nature of the capital budgeting decision itself, which requires consideration of the development phase and the nature of financial analysis surrounding the construction contract and construction finance. In all, this chapter considers three situations that are typical in corporate finance: a company's opinion about how much it might pay for another company (in this case a project company); a bank's decision about how much it might lend to finance such an acquisition; and, assuming the venture under consideration is *not* a going concern, the financial circumstances surrounding its development.

We have ordered the chapter as follows. In Section 1 we present a pressure-cooked version of business valuation. We do not dwell on detail or debates surrounding these subjects, the intention is to provide insight and vocabulary. In Section 2 we demonstrate these principles by generating a mock valuation for a hypothetical project company, and find that project company circumstances somewhat simplify the application of business valuation principles. In Section 3 we consider more closely the financing decision, in particular those elementary methods used by banks and other lending institutions to make decisions about a project company's debt capacity, which also draws attention to contracts put in place to stabilise project company cash flow (again raised in Chapter 4). In particular, this allows us to indicate also how debt capacity could influence capital structure decisions, and in principle where capital budgeting decisions are considered in the project finance venture. In Section 4 we consider the capital budgeting decision in a more conventional sense. Again using our hypothetical project company, we consider the investment decision, and also the nature of the design and construct partner's capital budgeting decision. We consider also essential fundamentals such as financial structure, capital budgeting, and weighted average cost of capital (WACC).

In all, we use again the various stages of the project lifecycle to provide structure, but in this chapter we are forced to use extensively a further prop to demonstrate the financial concepts we deal with – the fictitious Baguio Power Company, introduced in earlier chapters.

3.1   Valuation and the project company

As with most investments, businesses, and thus project companies, these are assets that may be bought or sold, or used as collateral for debt. Their sale or purchase may be via trading shares in the market if a public company, or by private sale of shares if not listed. Our concern here is with what the business is worth to potential sellers (shareholders) or purchasers, investors who might consider it an attractive asset for numerous reasons – strategic, portfolio, acquisition or for private purposes. If the company's shares are publicly traded, its share price of course represents the best current valuation of the business, and if the securities exchange is functioning reasonably efficiently, it represents current market thinking about company management, every product line and every customer's satisfaction, supply chains, about the effects of its treatment of stakeholders, reputation with the financial community, possible legal disputes, such as for environmental damage. But if it is a private company there is typically no publicly available share price, and buyers or sellers may need external advice on these matters in the form of a range of values to commence negotiations. Where a purchaser aims to secure corporate control, however, the dynamics of competition for corporate control and defensive actions (as discussed in Chapter 2) may cause share price movements that reflect behavioural dynamics underlying bidding processes in auctions not entirely explained by rational valuation principles.

Circumstances that require a business valuation generally would require appraisal of the going concern with plant, equipment, premises, established relationships with suppliers, clients, current market share, and a portfolio of opportunities. With few exceptions, an established business that is fundamentally healthy and reasonably well managed will be more valuable than the sale of its physical assets in liquidation. For present purposes we will not dwell on circumstances surrounding liquidation of failed businesses or work-out arrangements for distressed businesses; these are highly specialised subjects which will not cast much light on our interests. We will also not dwell on business start-ups. When entrepreneurs start a new venture, its growth and development rarely follow the trajectory its founders, or even venture capital partners (if any), may have envisaged in a business plan, if ever there was a business plan prepared to support the original decisions.[1] Further, it anyway requires special insights to evaluate a business plan for feasibility, and venture capitalists that evaluate these in search for investment opportunities are a special, perhaps even strange, breed of risk-taker. Evaluating a business plan is not the same as valuing an existing business.

With project finance companies the sequence of events follows a somewhat different, almost reversed logic. First, recall the nature of a functioning project company: it is generally a one-asset company, highly specific and capital intensive, in a well-established industry, typically with concentrated equity participation, high debt in its capital structure, possibly with supply and demand contracts, and ideally with as many risks as possible managed or mitigated by contract, or manageable in markets. It may take decades for any corporation to evolve to the stage where it is able to negotiate such arrangements, assuming it is a corporation with a portfolio of specific businesses. But the project finance approach requires that these characteristics preferably are in place *before* the decision is made to go ahead with the first stage in the project's physical development, long before the venture becomes operational. This means a well-defined specific purpose venture is largely created on paper before investment commences in any physical assets. In effect both the investment and the financing decisions are supported by a conventional business valuation – it informs the investment decision (a corporate capital budgeting decision of the net present value (NPV) rule variety), while it informs the financing decision (the valuation informs decisions about the project company's debt capacity and thus its ultimate financial structure). It is for this reason that we start by introducing the nature of

---

[1] We may well ask the question: did Bill Gates and Steve Allan have a formal business plan when they started Microsoft? Did Li Ka-shing have a formal business plan when he started Cheung Kong? Did Richard Branson have a plan with Virgin?

business valuation, as a prerequisite to understanding the importance of estimating project company value before the project enters its physical lifecycle.

3.1.1 The present value of a business

In keeping with our approach of limiting assumptions about the level of finance knowledge users of this book have, we commence by agreeing with thousands of finance textbooks in defining the *value* of any business as being equal to the sum of the present value of all expected future cash flows generated by the business, discounted to the present by an appropriate risk-adjusted discount rate (WACC). Expected value is thus present value (*PV*) at time t = 0, i.e. now. Discounted cash flow (DCF) is the conventional methodology for business valuation, and so we have the value of the business represented by the generalised present value expression:

$$PV_{BUSINESS} = \sum_{t=1}^{n} \frac{Cf_t}{(1+k)^t} \tag{1}$$

where

Cf_t = cash flow generated in time period t

k = the risk-adjusted discount rate (WACC) in period t, appropriate to the type of business under consideration (we will elaborate on WACC in further sections).[2]

Often the generalised expression is presented without much concern for users, because in use it becomes obvious that the expression has to be disaggregated and restated in a form that applies more specifically to the type of asset (or business) being valued. Thus it is helpful to rewrite the general case into a form that is directly appropriate to a project company, which is our destination in this chapter. But deriving a specific project company valuation model from the general model requires a few important detours, and these gyrations also provide important vocabulary to understand fundamental financial differences between project companies and general companies in trading, manufacturing, services or other sectors.

First, substitute cash flow (Cf_t) with a more conventional accounting-derived description of business cash flow, earnings (E_t), which refers to the cash flow generated by the business for its owners in a particular period t.[3] Because companies potentially may have infinite lives, we represent PV under such a circumstance by:

---

[2] We are not going to explain the mechanics of discounting. All general finance textbooks have sections that cover this in detail, and spreadsheet application software makes it easy. We recommend Brealey *et al.* (2006) for readers who require a primer on discounting.

[3] Business cash flow is conventionally defined as earnings before interest, tax, depreciation and amortisation (EBITDA), or the operational cash flow generated by the business before liabilities, allowances and other accounting adjustments.

$$PV_{BUSINESS} = \frac{E_1}{(1+k)^1} + \frac{E_1(1+g^1)}{(1+k)^2} + \frac{E_2(1+g^2)}{(1+k)^3} + \frac{E_3(1+g^3)}{(1+k)^4} \cdots \infty \qquad (2)$$

where earnings in each subsequent year (E_{n+1}) grow by some factor g for that year (g can be positive or negative). If we make two critical assumptions, namely that the investment is held forever (∞) and earnings (the periodic cash flow) grow at a constant periodic rate g in every period, then:

$$PV_{BUSINESS} = \frac{E_1}{(1+k)^1} + \frac{E_1(1+g)^1}{(1+k)^2} + \frac{E_2(1+g)^2}{(1+k)^3} + \frac{E_3(1+g)^3}{(1+k)^4} \cdots \infty \qquad (3)$$

At $t = 0$ this expression is mathematically summarised simply as:

$$PV_{BUSINESS(t=0)} = \frac{E_1}{k-g} \qquad (4)$$

As a valuation model this assumes that the business will generate constantly growing periodic earnings forever for its owners, the investors, with valuation taking place now.[4] If we view earnings and dividends for now as synonymous,[5] Equation 4 is the familiar (somewhat infamous) constant growth dividend discount model (CGDDM). We may rewrite Equation 4 also as:

$$\frac{E_1}{PV} = k - g \qquad (5)$$

We have stated that share price is the best estimate of a company's value, so if we observe stock market pricing of companies, we may substitute PV with current price, P_0:

$$\frac{P_0}{E_1} = \frac{1}{k-g} \qquad (6)$$

This provides the logic behind the price/earnings ratio, a rather old-fashioned measure commonly encountered in analyses of shares. With the denominator expressed in decimal form; $1/(k-g)$ becomes the p/e multiple, which is used to compare businesses with similar characteristics by applying the multiplier to expected end-of-year ($t = 1$) earnings.[6] WACC, k, is the same as before.

In turn, Equation 5 is also used often in another form, namely:

$$k = \frac{E_1}{P_0} + g \qquad (7)$$

---

[4] Note that E is at $t = 1$, not $t = 0$. This is part of the mathematical summation of a constantly growing perpetual annuity.
[5] They could be synonymous if earnings = profits and the company pays all profits as dividends.
[6] In principle the p/e multiplier functions in identical fashion to the current yield and its inverse, the capitalisation rate (or rental income multiplier) in real estate.

Frequently g is referred to as suggestive of a company's growth nature, i.e. it may be viewed as a *high* growth company (a company with good business opportunities, often referred to as growth options). For example, if the Global Consilium Corporation's p/e is comparatively high (say 20) relative to observed historical k (say $k = 15\%$, so $g = 10\%$), it may be viewed as a *high* growth company; or it may be viewed as a *low* growth company if p/e is low (say 10) relative to historical k (say $k = 15\%$, so $g = 5\%$). This is a very useful way of interpreting a company's growth options, because g represents that proportion of a company's total observed value that is represented by expected future growth in earnings – which of course must come from successfully taking its growth opportunities. Observant readers will notice that in order to solve the variables we need a value for k. Observing an historical k in the case of public companies requires a few manoeuvres that we will return to below, but for truly private companies these variables are practically not observable for outsiders.

The CGDDM assumes that the investor holds the investment forever. While impractical, this is not a totally impossible assumption, as stated. Nevertheless, a perpetual holding period does prove highly problematic when the distribution of an investment's values over time is of interest to investors. Further, a constant perpetual rate of growth in expected earnings is hardly reasonable under most operational circumstances, as the historical lifecycle trajectories of most industries would show. Industries typically follow a lifecycle which exhibits stages: emergence, growth, maturity and decline, which suggests that different growth assumptions are appropriate to different stages in an industry's development – hence the idea of a *multi-period valuation model*. It may look as follows:

$$\text{PV (Business)} = [\text{PV (Period 1)} + \text{PV (Period 2)} + \ldots \text{PV (Period } n)] \qquad (8)$$

The PVs for different stages (periods) are estimated using different assumptions, and the value of the business is the summation of all different period PVs.

At this stage further practical considerations enter the process. First, in Equation 8 Period n does not rule out a perpetual model, but the underlying mathematics of discounting renders a perpetual model superfluous. At low discount rates (say 2–3%), cash flows occurring after 30–40 years contribute very little to PV, and at high discount rates (15% +) most cash flows that occur after around 12–15 years do not have much influence on total PV. At higher rates the influential cash flows are even closer – after ten years cash flows have little significance on total PV. Overall, these realities place a premium on early cash flows, and in valuation practice this suggests that special care is important when estimating early cash flows. The result is that a practical PV model could have as few as two stages, and may look as follows:

$$\text{PV (Business)} = \text{PV (Period 1)} + \text{PV (Continuing Period)} \qquad (9)$$

Dividing the model components into time periods represents a practical approach to valuing a business which takes into account the reality of the dynamics of industry development and its influence on the pattern of growth that a business' earnings may be expected to follow, plus the practical influence of DCF mathematics on the timing of cash flows. Period 1 may be split into Period 1 and Period 2 according to circumstances. The intention of a multiple-period PV model is to concentrate forecasting effort on the cash flows that weigh most in the composition of value – those that occur in the early years. Copeland, Koller and Murrin (2000) call Period 1 the *Explicit Period Forecast*, which requires explicit views to be expressed about industry environment and a business's strategic position within its industry, which must be consistent with assumptions about earnings forecasts and growth assumptions.[7] Ongoing Value is an estimate of the business' ongoing operations after year n, and reintroduces the potentially perpetual life concept, but dilutes its influence on business value by discounting it from year n to year 0. So, assuming two explicit periods we can rewrite the model as follows:

$$PV_{BUSINESS} = \sum_{t1=1}^{n} \frac{E_{t1}}{(1+k)^{t1}} + \sum_{t1=n+1}^{n2} \frac{E_{t2}}{(1+k)^{t2}} + \frac{PV\,(Ongoing\ Value)_{n2}}{(1+k)^{n2}} \qquad (10)$$

where

E_t = forecast earnings in explicit periods t_1 and t_2
k = WACC
n_2 = the final period in the explicit period t_2
$PV(Ongoing\ Value)_{n_2}$ = the business's Ongoing Value at year n_2.

As is clear, PV (Ongoing Value) does not actually solve the problem of valuing as vague a notion as '*a perpetual stream of earnings growing at a constant rate*', but simply moves it into the future where it has a smaller influence, while forcing the constant growth assumption to be replaced with explicit assumptions about the behaviour of earnings and its growth in the near term where it matters most to PV. In most cases (with project companies a notable exception, as we see later), the PV(Ongoing Value) may be represented simply by a constant growth perpetuity, as the p/e model assumes. But it also forces attention to assumptions about long-term industry growth/decline and cost of capital, because PV(Ongoing Value) is much more sensitive to changes in the denominator $(k - g)$ than the numerator (E). Thus for PV(Ongoing Value) we write:

---

[7] The relative importance in valuation of the model's first phases arises from several related factors. First, the time value of money makes us rate more highly the cash we collect today than cash we expect to receive later, which may be subject to erosion by inflation or in some cases by exchange rate changes. In the same way, cash flows received in the future will always be more sensitive to adverse changes in real interest rates than those expected sooner. Second, there is inherently less certainty to the outcome of almost all commercial ventures when looking into the future, largely due to the impossibility of predicting unknown events.

$$PV(\text{Ongoing Value})_n = \frac{\dfrac{Earnings_{n+1}}{(k-g)}}{(1+k)^n}$$

and the final valuation model becomes:

$$PV_{BUSINESS} = \sum_{t1=1}^{n} \frac{E_{t1}}{(1+k)^{t1}} + \sum_{t2=n+1}^{n2} \frac{E_{t2}}{(1+k)^{t2}} + \frac{\dfrac{Earnings_{n2+1}}{(k-g)}}{(1+k)^{n2}} \qquad (11)$$

This is the general model we will operationalise to demonstrate valuation of a project company. Periods $(1 \ldots n)$ and $(n + 1 \ldots n2)$ represent the explicit period forecast of company earnings, and Ongoing Value is estimated for the balance of the business' assumed life – in this model it is assumed to be perpetual. The explicit period t_1 may be anything between 1–10 years, less if an explicit period t_2 is assumed to be appropriate for the purposes of analysis. When we turn to valuing project finance ventures in Section 3.2, things simplify significantly as we drop altogether the assumption of perpetual life from the valuation model.

Before we proceed to the project company valuation model, it is necessary to consider the importance of key variables and their influence on valuation outcomes, because these are important also in project company variables, for reasons that will become clear. We have introduced the fact that timing of cash flows matters as a consequence of DCF mathematics. Given the outline of the valuation model presented above, let us ask a more probing question: which variables influence a business valuation most? An interesting question, following the supposition underlying Copeland *et al's* (2000) approach: management could put in place strategies to influence those variables that may influence value, either as measured by share price or the price that a private company may fetch in an acquisition. This also requires us to revisit the way we view PV(Ongoing Value) in Equation 11.

In order to keep the explanation compact, let us draw again on the CGDDM, the perpetual growth model. We assume that earnings are discounted to give present value, but what causes earnings growth? In the technology bubble years (1999–2001), shares in companies with no net earnings were selling at fantastic prices – try and work out the mathematics for that notion according to the conventions outlined above.[8] The end of economics as we knew it was greatly exaggerated, and the technology bubble duly deflated. We follow traditional thinking, in principle as outlined by Copeland *et al.* (2000), in their very accessible explanation of what drives business value. Drawing in large measure on principles underlying the DuPont system (Brealey

---

[8] Without putting too fine a point on it, it may be argued that irrational share prices for Internet companies represented nothing more than no-costs barred bidding for a limited number of opportunities (available shares) to place bets on which companies would be the next Microsoft (similar to deeply out-of-the-money options on extremely risky shares). It seems that rational valuation had little to do with share prices during this period, while risk-taking had everything to do with it.

et al., 2006), they isolate two fundamental concepts that have to be addressed explicitly for corporate valuation purposes. All being equal, these two variables are return on assets (ROA) and, you guessed it, expected earnings growth (g). ROA recognises that companies that use their assets effectively are valuable.[9] Expected earnings growth (g) in turn draws in concept on what is often referred to as the corporate sustainable growth rate. This in turn recognises that companies which reinvest in assets help sustain the circumstances for effective use of their assets. This principle is demonstrated extremely well by using the CGDDM – the perpetual assumption actually dramatises the effect of ROA and earnings growth, as we demonstrate below.

The theoretical perpetual growth assumption is not supported by evidence, as economic history shows. Historic values for ROA and industry earnings growth rates (g) for all industries are high at the emergent stages of an industry, but growth rates decline as the industry progresses to maturity and eventually to decline, so assumptions about long-term earnings growth for any company are fundamentally constrained. Industry dynamics demand that new entrants and competition will eventually force down ROA, and high earnings growth assumptions over long periods are therefore unrealistic. Management talent requires further close analysis, because it underlies expected success in capturing successfully growth opportunities. Management talent, as we also understand from corporate and transaction governance matters raised in Chapter 2, is not simple to analyse, but in this chapter we will assume management and governance favour shareholders' interests. So, assume for now that perpetual growth assumption is not important, because we are trying to obtain some insight into the influence of ROA and earnings growth on value. In the process we will also obtain insight into the nature of free cash flow, the essential variable in valuation of project companies. It is best to start the explanation with an example.

Assume we have two comparable companies making electric micro-motors of a certain size range for use by original equipment manufacturers (OEMs) in the automotive and power tool industries. The industry is mature, and is not expected to grow much in the future. Global Consilium (GC) implemented a creative marketing strategy that resulted in GC securing several new supply contracts and maintained a steady investment programme in human resources and new facilities, so continuously renewing the company and creating additional output capacity. Global Barnium (GB), on the other hand, has relied on long established relationships with OEMs. It has not strived to create new opportunities, its manufacturing facilities are old although still functional, and consequently GB experiences tough price competition and eroding margins. GB understands it has to reinvest in its plant if it wishes to retain its present market position in the industry, but has postponed investment for some years now. GC, on the other hand, has steadily maintained an appropriate reinvestment rate (often referred to in corporate finance literature as the required reinvestment rate for

---

[9] Note that ROA is often seen by analysts as an easily observed and calculated proxy for return on equity, which is usually their preferred performance indicator but is often harder to assess quickly and accurately.

Table 3.1 Earnings growth ($ thousands) (following Copeland *et al.*, 2000).

Earnings growth	Year 1	Year 2	Year 3	Year 4	Year 5
Global Consilium	20,000	20,600	21,218	21,855	22,510
Global Barnium	20,000	21,200	22,472	23,820	25,250

Table 3.2 Reinvestment assumptions ($ thousands).

		Year 1	Year 2	Year 3	Year 4	Year 5
Global Consilium						
Earnings		20,000	20,600	21,218	21,855	22,510
Reinvestment		(5,000)	(5,150)	(5,305)	(5,464)	(5,628)
Net cash flow		15,000	15,450	15,914	16,391	16,883
Present value at	10.00%	60,039				
Global Barnium						
Earnings		20,000	21,200	22,472	23,820	25,250
Reinvestment		(10,000)	(10,600)	(11,236)	(11,910)	(12,625)
Net cash flow		10,000	10,600	11,236	11,910	12,625
Present value at	10.00%	42,267				

sustainable growth in earnings). In the industry in which GC and GB find themselves we assume for illustration that the reinvestment rate to generate sustainable corporate growth in earnings is 25%. Assume now the two companies generated historical earnings as shown in Table 3.1.

For Global Consilium this reflects an earnings growth rate of 3% annually, while it shows an annual earnings growth rate of 6% for Global Barnium – twice that of GC.

Yet, as a person who is considering investing in either company now, and with a required rate of return of 10%, which company would you think is the better long-run investment prospect? The strategic position of the two companies makes a big difference, and we see that GC is valued at $60 m while GB is valued at $42.3 m. How is this so? This is the result of differential reinvestment rates and consequent expectations as to future rates of earnings growth – GB immediately needs to reinvest a large proportion of future earnings to renew its physical and human productive base, simply to maintain its present state (recall our assumption that GB requires to reinvest significantly in its operations in the near future). If we now assume that the same set of figures is actually a forecast of earnings, we may illustrate the effect of a differential reinvestment requirement by the figures presented in Table 3.2.

Table 3.2 shows that earnings for GB grow at 6% p.a., which results from a reinvestment rate of 50% of earnings. This follows from an implied ROA of 12% ($12.00 additional earnings for every additional reinvestment of $100). For interest, note that GC's reinvestment rate is far less at 25% and it thus generates more cash in the near term – and we also see the influence of early cash flows on PV illustrated. We explain this set of circumstances as follows.

For GB, expected growth g is explained by:

$$Growth = ROA \times Reinvestment\ Rate\ (IR)$$
$$6\% = 12\% \times 50\%$$

Rearranged we have:

$$IR = g/ROA$$

So how do expected earnings growth and return on assets influence business valuation? Using our two hypothetical companies, we refer again to the Constant Growth Model (CGM), where we have:

$$PV_{BUSINESS(t=0)} = \frac{E_1}{k-g}$$

With constant growth we have:

$$Earnings_n(E_n) = E_{n-1}(1 + g)$$

Substitute net cash flow (NCF) for earnings, so recognising required reinvestment, and for Global Barnium this reduces net cash flow to the following expression:

$$NCF_n = E_{n-1}(1 - g/ROA)$$

Remember the Constant Growth Model uses NCF_1 as numerator. Thus we may revise the CGM model as follows:

$$Value\,(PV) = \frac{NCF_1\,(1 - g/ROA)}{k-g}$$

We may demonstrate the relevance of different assumptions about earnings growth and ROA by presenting the following example.

Assume starting earnings $= \$20\,000$; WACC $= 5\%$ and a 10-year explicit period $(n = 10)$, after which ROA $=$ WACC. This requires us to forecast explicitly NCF for 10 years, estimate a continuing value based on expected year 11 NCF into perpetuity, and discount it all back to year 0. This scenario is represented by the following expression:

$$PV_{BUSINESS} = \sum_{t=1}^{n} \frac{NCF_t}{(1+k)^t} + \frac{PV\,(Ongoing\ Value)_n}{(1+k)^n} \tag{12}$$

Each net cash flow for $n = 1 \dots 10$ is simply:

$$NCF_n = E_{n-1}(1 + (1 - g/ROA))$$

and Ongoing Value is represented by:

$$PV(\text{Ongoing Value})_n = \frac{\dfrac{E_n(1 - g/\text{ROA})}{(k - g)}}{(1 + k)^n}$$

Given a range of assumptions for earnings growth (g) and ROA, and 10% WACC we generate Table 3.2 which shows the results as PVs. Table 3.3 further demonstrates the influence of assumptions about earnings growth and ROA, and indicates one critical principle clearly: if ROA is below WACC, the company value reduces in each period and must inevitably collapse, but the company's value increases with any assumed increase in growth. Similarly, at any assumed ROA above WACC, the company value increases (see Copeland *et al.*, 2000).

The above analysis serves to outline in principle the important variables in a business valuation and their relationships. It is essentially academic, and because we are interested in the value of the equity (this is what is bought or sold), we rewrite the valuation expression in Equation 12 in ordinary language in order to operationalise it for project companies and specify its data requirements. Using common industry terminology Equation 12 can be rewritten for practical purpose as the Enterprise Discounted Cash Flow model (ECDF), where:

Value of company equity of a company = Value of a company's operations (enterprise value) available to all its investors [i.e. *PV explicit period* plus *PV Ongoing Value*], **minus** Value of debt and other claims superior to common shareholders (e.g. preferred shareholders).

The square bracketed expression is identical to Equation 12. The EDCF model simply recognises that if sold, the proceeds of any business have to be distributed in order of seniority of claims, and it also reflects the fact that equity is always the residual claim. Of course, it also recognises that the business can change hands while some claims are still outstanding, unless covenants in debt contracts restrict it without permission (common in small and medium-sized enterprises where equity owners may still be personal guarantors for debt.

In the general case of an existing and operational company, say for acquisition purposes, a valuation must draw on all and any existing information about the business itself – which means it has to be clearly disentangled from all its associated companies, holding company, and further corporate ties that may function to make financial information opaque. Information on three variables is important. First, we need insight into the historical revenue and cost structures of the company, in order to explain the nature of earnings and earnings growth. Second, information on the performance of the business' capital investment is required, in order to obtain insight

Table 3.3 The importance of assumptions about earnings growth and return on assets (invested capital).

Earnings growth		Return on assets in $ (invested capital)									
		8.00%	9.00%	10.00%	11.00%	12.00%	13.00%	14.00%	15.00%	16.00%	17.00%
Earnings growth	2.00%	187,500	194,444	200,000	204,545	208,333	211,538	214,286	216,667	218,750	220,588
	2.50%	183,333	192,593	200,000	206,061	211,111	215,385	219,048	222,222	225,000	227,451
	3.00%	178,571	190,476	200,000	207,792	214,286	219,780	224,490	228,571	232,143	235,294
	3.50%	173,077	188,034	200,000	209,790	217,949	224,852	230,769	235,897	240,385	244,344
	4.00%	166,667	185,185	200,000	212,121	222,222	230,769	238,095	244,444	250,000	254,902
	4.50%	159,091	181,818	200,000	214,876	227,273	237,762	246,753	254,545	261,364	267,380
	5.00%	150,000	177,778	200,000	218,182	233,333	246,154	257,143	266,667	275,000	282,353
	5.50%	138,889	172,840	200,000	222,222	240,741	256,410	269,841	281,481	291,667	300,654
	6.00%	125,000	166,667	200,000	227,273	250,000	269,231	285,714	300,000	312,500	323,529
	6.50%	107,143	158,730	200,000	233,766	261,905	285,714	306,122	323,810	339,286	352,941
	7.00%	83,333	148,148	200,000	242,424	277,778	307,692	333,333	355,556	375,000	392,157
	7.50%	50,000	133,333	200,000	254,545	300,000	338,462	371,429	400,000	425,000	447,059
	8.00%	NF	111,111	200,000	272,727	333,333	384,615	428,571	466,667	500,000	529,412
	8.50%	NF	74,074	200,000	303,030	388,889	461,538	523,810	577,778	625,000	666,667

NF denotes not feasible.
Following Copeland *et al.*, 2000.

into historically achieved ROA and capital efficiency, and to form an opinion about expected future performance. Whoever is tasked with valuing an existing project company, the principal source of these data will most likely be the historical accounting data of the company. The third important information requirement is the company's cost of capital, which is unfortunately somewhat more complex and is not directly observable. This is dealt with in Section 3.3.1, and will be considered in Box 3.2. As outlined above, we need information on at least three important variables to conduct a business valuation, each in itself complex, plus an informed opinion of the industry environment, before we can commence valuation.

3.1.2   The project company

All fine – but what about project companies? As alluded to above, project companies differ in important ways from a general business and this affects the approach to their valuation. There are at least two distinguishing project company characteristics that are also interrelated:

- a single, highly specific asset
- project companies typically have limited lives, either determined by the nature of the asset (such as a mineral deposit), or by concession agreement (as with public/ private infrastructure).

These two factors significantly simplify the valuation framework. The high asset specificity very harshly introduces to promoters, developers or owners the economic principle of irreversible investment – 'sunk costs'. If the single asset is fixed and specific, highly capital intensive, with well-known and low-risk technology, short of retrofitting the asset with more efficient technology, there is very little that can be done through use of the asset itself to manage ROA to improve efficiency, or grow earnings beyond plant capacity. Improvements in ROA will have to be achieved from managing costs and/or improving revenue. With long-term supply and distribution contracts, and agreements about price/volume and changes, it means that ROA itself becomes less influential in the valuation exercise, and earnings growth become more important in the valuation exercise. And when the second factor, limited project company life, is considered simultaneously with irreversibility, it essentially determines that there is no expected ongoing value component in valuation of the project company, because it is planned to have a finite life and then decommissioned. Essentially the valuation methodology simplifies only to the present value of future expected cash flows, and forecasting of cash flows over the expected project life becomes the key valuation variable. This input does not mean that there are no project companies that continue to produce beyond their planned lives, there are many, but the point is that there is typically a planned term for the company's existence, and this has to be a lead assumption in the valuation exercise.

If we assume no complicating classes of liabilities, for example preferred shareholders, we can thus simplify the project company EDCF model as follows:

Value of the equity of a company = Value of a company's operations [i.e. PV explicit period]

minus

Value of debt and other claims superior to common shareholders.

This indicates that there is simply one explicit period cash flow forecast, from years 1 to *n*, and there is no assumed ongoing value. In DCF terms, this is a simple model, and for credibility in use it will rely on in-depth research into the project company's earnings characteristics and cost structure. And be aware of an observation that is obvious to some – the value of an operational project company does not consider at all what might have been the cost of developing the project company asset, it depends solely on periodic cash flows generated by the project company. The typical case with a planned term for the project life further restricts assumptions about terminal cash flows at decommissioning or disposal with the end of the planned project period. In some cases the physical plant may be recycled with some associated income, while there may be value in the land if it is freehold or held by a long land lease. In other cases the asset must be handed over in good working order to a public authority following original concession agreements, as with BOT-type facilities. In modern times it is not unreasonable to assume that any terminal income from disposal of physical projects assets is incidental and minor, although circumstances of course differ from case to case. Recall also the financial responsibilities concerning the costs associated with contamination of land and pollution that were flagged in Chapter 1 – it is responsible to assume no net income from disposal/recycling, and apply managerial best practice to contain it absolutely during the operational term. In the rest of this chapter we will interpret reinvestment as including retrofitting of efficiency enhancement and pollution and contamination-controlling technology as the project company life progresses, even if the asset itself is planned to be functionally depreciated for economic purposes at the end of the project life.

3.1.3 *Towards free cash flow*

We turn now to practical matters and data requirements. In the EDCF model, we still need to define what is meant by cash flow and extract it from somewhere before we can discount it. While we are all familiar (to a greater or lesser extent) with the nature of accrual accounting, and are aware that its principles are roughly similar in most jurisdictions, we still need to explain reverse engineering of accounting information to arrive at cash flows as required by the EDCF model. We also understand that profit and cash flow do not mean the same thing – accounting information requires reinterpretation (if historical) and restatement to be useful in DCF valuations based on forecasts of expected cash flows. Without inviting explanation, one such oddity is that a company can make a profit without having actually been paid in sales revenue by any customers. When valuing a business, accounting information inevitably has to be re-engineered to present an economic model of business performance, as opposed to the historical performance represented by the accounting model. For example, in its normally used form, the earnings per share and price/earnings ratio model rely on accounting information and dates from the days when financial market participants had more respect for published accounting information than at present.

In principle, information drawn from two financial statements is essential to forecast cash flows in order to prepare a business valuation. Unfortunately the terminology used in financial statements is often not consistent, and can be particularly confusing for inexperienced analysts attempting to assess performance or compare company performances. We are going to use the most common and simplest meanings we know for all the terms we will introduce. In principle we can say that the main purpose of analysing historical financial information is to develop cash flow information that gives insight into how net cash flow is generated and how it is used before considering the influence of financial structure and taxation on the claims to the cash flow. We are aiming to derive an expression for what was termed free cash flow in Chapter 2. Typically the point of departure is the profit and loss statement, and later the balance sheet. The first variable to be derived from the profit and loss account is earnings before interest, tax, depreciation and amortisation (EBITDA). This is intended to reflect cash flow generated by operations after costs of generating earnings, but before considering how the business is financed. However, EBITDA is not the final stage in what is defined as profit in accounting terms for taxation purposes since there are two further considerations. Consider the typical summarised profit and loss account for Baguio Power Company (BPC) presented in Figure 3.1, assuming a corporate tax rate of 20%.

The profit and loss account is effectively divided into two parts. First, we have accounting data for that part of the business which is concerned with actual operations – actual cash earned, actual costs incurred – so arriving at EBITDA. For those readers who have not had exposure to accounting, cost of sales groups those items in the normal conduct of business that may be expended in the ongoing effort to generate earnings over an accounting period (quarterly, annually). These include direct and indirect expenses incurred, such as cost of inputs, labour, marketing (advertising, public relations, promotion), selling, general and administration expenses, it may include rent paid for offices and factory space and transportation of goods.

The second part of the profit and loss account is concerned with how the business is financed, and the influence of its institutional environment, such as taxation regulation. BPC costs might include interest paid on debt, if the company had borrowed. If the company has no debt, which is where we will start our financial analysis, there would be no interest charges recorded – for now we would simply record $0 against it. In later cases we see how it is relevant. There is usually also one further non-

	$
Total sales	**1000**
Less: cost of sales	400
EBITDA	**600**
Less:	
Depreciation	100
Interest	0
Other interest	
Expense/income	0
Profit before tax	**100**
Less: tax	20
Profit after tax	**80**

Figure 3.1 BPC profit and loss account: 1 July 2010 to 30 June 2011.

operational deduction, a depreciation allowance. But, as all finance textbooks immediately point out, *depreciation is not an actual expense*. The business does not actually pay a depreciation charge to anybody, it is an allowance – a gift from the taxation authorities. Think of it this way: taxation regulators are fully aware of the mathematics above which indicates the importance of reinvestment in order to keep the business viable. Depreciation allowances could be thought of as an incentive to business to continue reinvesting, effectively by allowing the business to exclude a sum of money from the calculation of taxable profit.[10] In most jurisdictions taxation law allows businesses to depreciate capital equipment over an agreed period of time which depends on the nature of the asset, using an agreed method of depreciation (mostly straight-line or reducing balance). For example, in Baguio Power Company, it may have been agreed that the actual initial cost of the whole process facility would be depreciated over ten years, using the straight-line method (i.e. 10% per annum of the cost of the facility). Only after interest and depreciation charges have been deducted do we arrive at profit before tax, upon which 20% tax is paid, to arrive at profit after tax. In order to arrive at free cash flow, the profit and loss account has to be re-engineered, and actual reinvestment considered.

The form of the profit and loss account presented in Figure 3.1 is the subject of re-engineering to arrive at the commonly used interpretation of free cash flow – the actual cash flow generated by the business after real expenses and allowances have

---

[10] Depreciation policies may be expressly covered by law, but in the corporate sector details are often agreed between individual corporations and taxation authorities before capital budgeting decisions are made. Taxation and tax law is complex in all jurisdictions, and we recommend seeking expert advice before making any important decisions.

been adjusted for. Technically at least two items have to be added, firstly investment in fixed assets (which may reflect the amortisation component in EBITDA, but is intended to reflect actual amounts reinvested) and net investment in working capital, which reflects, when simplified, the net position of current assets and current liabilities at the end of an accounting period, and is thus obtained from the end of period balance sheet. It may be negative, for example reflecting that the business had to extend trade credit and so invest more funds in receivables; or it may be positive, indicating the opposite. As will be observed below, we do not consider this particular item as crucial to our analyses, and consequently assume it to be neutral from one period to the next, but please note, if defensible (to some extent) with a project company, this will not be a defensible assumption with a trading or manufacturing company. The format for the simplest re-engineered profit and loss account is presented in Figure 3.2. Depreciation, as a non-cash flow item, is added back, and net capital investment is deducted – the logic being that the business is not sustainable without reinvestment and it is thus a necessary expenditure.

Free cash flow has to be forecast over the project's planned life and discounted to arrive at project company value. With an assumed Ongoing Value of **zero** at the end of the project's planned life, the company valuation depends entirely on assumptions made in forecasting free cash flow and assumptions about the discount rate (WACC). We deal with assumptions about free cash flow first and deal with WACC in later sections. Assumptions about two variables are obviously critical (namely cost of sales and earnings growth), and have to be based on in-depth analyses of the company's environment. The company's internal environment determines its production cost function (cost of sales), and assumptions about the cost structure are thus very important. These cost functions are the subject of much analysis in the project finance departments of project lenders, investment banks, and fund managers. For present purposes we may assume that the cost structure of industries where use of project finance is well established, such as the electricity generating industry and road transport infrastructure, has been thoroughly researched and some accepted cost structure heuristics have developed within markets, even if these may not be directly compa-

	$
Profit after tax	**80**
Add: Depreciation	100
Less:	
Net capital investment	50
Net changes in working capital	0
Free cash flow	**130**

Figure 3.2　BPC free cash flow 1 July 2010 to 30 June 2011.

rable across markets. We are not in a position to comment on such cost structure assumptions, but will simply assume that total cost of sales is a variable that may be expressed for different project industries as a typical percentage which can be managed, but only at the margin. In order to conduct the forecasts and the valuation we thus base assumptions about the cost structure of project company production on assumed typical industry production functions expressed as a set of cost and financial structure relationships relative to the level of total sales. In this respect, we follow the well established tradition of avoiding the complexity of these typical production cost functions (Benninga, 2006; Brealey *et al.*, 2006). We assume for simplicity that these relationships are linear.[11]

Earnings growth expectations are only that – expectations – but from Equation 12 the importance of assumptions about earnings growth is clear. Forecasting cash flows for valuation purposes requires analysis of the nature of the industry, its stage of development, its competitive environment, all to support an opinion of the company's future prospects. A number of excellent models exist to assist analysts in structuring industry analyses and environmental scanning, including the venerable extended rivalry industry analysis framework popularised by Michael Porter (1980; 1985) and possibly still the most internally coherent environmental analysis model of all. While industry analyses and business strategy are very exciting subjects in themselves, our limited scope for this book does not allow us to digress into these (also see Copeland *et al.*, 2000 for an introduction). However, in what is presented below, we will frequently make assumptions that would have been the outcome of using such models. Information about endogenous industry growth and development, and the competitive environment, could come from a variety of well-respected industry and economic sources, usually not free (specialist research publications, periodic investment bank research reports), and from lesser respected sources (some print and electronic media) – as they say, you get what you pay for.

3.2 Valuation and the project company as a single-asset business

The importance of working through this process is fundamental, because at the centre of every capital budgeting decision is the expected value of the asset that is being analysed. If the asset is non-specific and freely available such as a computer or a light commercial vehicle, it will have an externally determined cost to the company (the market price of the asset), and the company is left to estimate what it can produce with the asset to determine what its present value is to the company. However, with project companies where the project facility still has to be created, the corporate

---

[11] However, when there are returns to scale, as is expected with process-based projects, the relationships may of course show lower unit production costs. It requires in-depth industry knowledge and knowledge of operational cost structures.

capital budgeting decision is significantly more problematic, introduced by uncertainty about *both* the value *and* the price of the asset (the cost of having the facility built), and if it requires a long development period additional uncertainty is introduced about the effect on the decision to go ahead at all (the influence of the development period is considered in the next section). Further, with investment decisions that require equity participation from several sources, where some participants may have multiple roles, such as design and construct contractors, the level of complexity increases further. Capital budgeting decisions in capital goods industries are notoriously risky.

There is very little chance that this complexity can be adequately conveyed without referring throughout to an example. This section therefore considers the process of capital budgeting decisions where a project facility is to be created involving several equity investors and a design and construct contractor, rather than an isolated and abstract capital budgeting decision in a confined corporate finance environment as is often encountered in corporate finance textbooks. In order to reflect the decision-making process with project companies, we assume that the decision to develop the main facility of the fictitious Baguio Power Company is being evaluated by Global Consilium Corporation.[12] There are several stages to this evaluation for GC. Moreover, each potential equity investor in BPC will conduct its own evaluation from its own perspective, but if the project is attractive to one it is likely to be attractive to other participants, unless the sharing of risks and returns in the development phase has not been negotiated with symmetry (for example, equity participations do not reflect relative risk-taking). Further, not only is there typically more than one equity investor with project companies, there is also the principal provider of project debt – in our case, Fibonacci Bank.

With project evaluations of this kind, the point of departure is always the estimated enterprise value of the company to be created, and it is here that we begin our analysis. The process from here follows a number of steps, some of which will not be required in all cases. We will deliberately present some viewpoints that are not practical, only because they help to disaggregate the process into all the relevant steps and help to make the process more tractable. For example, while we present an evaluation of BPC as if creation of the company with no debt is an actual consideration, we do so only in order to illustrate where Fibonacci Bank will start its evaluation to determine BPC's debt capacity. But we are proceeding too rapidly – let us return to basics, namely estimating the enterprise value of BPC. As BPC's enterprise value is ultimately dependent on details of the concession agreement it enters into with the Manila Electricity Distribution Board (MEDB), we present a summary of the hypothetical concession agreement and power purchasing agreement in Box 3.1.

---

[12] Please note again: assumptions surrounding BPC's financial and physical circumstances are made for their convenience to demonstrate evaluation principles, and not for technical accuracy.

Box 3.1   Baguio Power Company – concession and power purchasing agreement.

The details of the concession agreement between the Baguio Power Company (BPC) and the Manilla Electricity Distribution Board (MEDB) is assumed to contain the following conditions, all of which have a bearing on the enterprise value of BPC. First, the concession agreement determines that BPC will be regulated by the Philippines Central Government's Department of Energy. The concession period is for 25 years, after which the whole facility is to be handed over to MEDB in fully maintained and operational form, subject to international expert inspection – the concession agreement can thus be summarised as a build-operate-transfer (BOT) arrangement. In addition, as incentive the Manila Metropolitan Government has leased the site to BPC at a negligible rent of US$1 per annum also for 25 years. Further, a power purchasing agreement with a 25 year term starting 1 January 2011 stipulates that MEDB will buy, from BPC 75% of the facility's design output for 25 years, i.e. 13,140,000 MW-hr (13.140 GW-hr) per annum at US$55 per MW-hr for the first 10 years, payable in Philippines pesos, on a 'take or pay' basis (more about 'take or pay' contracts in Chapter 4). The take or pay agreement is guaranteed by the Asian Development Bank, and the Philippines government assures the availability of foreign exchange to convert pesos into US$. The power price is fixed for the first ten years, and is then reviewed every three years to reflect a market price less 5%, based on a weighted average of a basket of Southeast Asian wholesale electricity prices, in US$. Increases in the market price are capped at a maximum of 5% over the price in the preceding three year term. BPC is free to sell its spare generating capacity to industrial and commercial users, but at the same price as it charges MEDB using MEDB's grid for transmission.

Global Consilium has also been active in other matters whilst negotiating the power purchasing agreement (PPA) with MEDB. It has simultaneously negotiated a coal supply agreement with WA Mines, a large listed coal mine in the Western Australian outback, located some 500 km from the nearest port on the Australian west coast, Port Mathilda. WA Mines has been operational for five years and has reserves to operate for another 100 years at current output (including serving BPC), and has its own dedicated railway line to Port Mathilda. BPC, through GC, negotiated a supply contract with WA Mines to purchase a minimum of 6.57 million tonnes and a maximum of 8.76 million tonnes of coal per annum of the specified quality for 25 years on a 'put or pay' basis (more about 'put or pay' contracts in Chapter 4), for delivery at BPC's coal terminal at Subic Bay, so risks in the supply chain between

Port Mathilda and Subic Bay will be borne by WA Mines. The price is fixed at US$45 per tonne for the first 10 years, thereafter prices will be adjusted annually based on the average annual price of the relevant grade of coal in the spot market for coal based on daily quotes on the Australian commodities markets in Sydney, also less 5%. The combined effect of the PPA and the coal supply agreement is that BPC has partly matched supply and distribution contracts, and provided the facility functions effectively in accordance with its designed capacity, it has tentatively secured a gross income stream for BPC.

3.2.1   Valuation of the project company: all equity

As stated, Global Consilium's decision to go ahead with Baguio will depend in the first instance on its estimate of BPC's enterprise value. Using the enterprise valuation model, it requires estimates of operational earnings, or sales less cost of sales (EBITDA). Recall the enterprise valuation model:

> Value of company equity = Value of company's operations
> ***minus***
> Value of debt and other liabilities.

In this framework we commence by observing that BPC is not a going concern, and thus that there is still no certainty of what its debt and other liabilities might be. Thus estimating its potential free cash flow (FCF) is the starting point for analysis of both the value of the enterprise and particularly the amount it may potentially borrow. This requires an estimate of what the operating structure of BPC might be, in order to estimate its possible EBITDA, and thereafter FCF. The starting point is estimated sales and the expected cost structure of the company (this is the practical equivalent of the economists' production function). We present in Figure 3.3 assumptions about the cost structure of BPC in operation[13] and our first analysis is of the project company *as if it has no debt*. We assume no debt because we need to estimate FCF in order to estimate cash available for debt service (the principal determinant of project company debt capacity, which we explain in the next section).

The fact that BPC is designed to be a single-asset project company simplifies matters significantly. As two essential variables, electricity sales and the cost of fuel (coal), are largely fixed for a ten year period, it allows a fairly uncomplicated estimate

---

[13] Under no circumstances do we pretend that these figures are representative of the actual economics of coal-fired power stations. These assumptions are made entirely for the sake of demonstrating the process of financial evaluation of a power generating project company, as an illustration of one example of a typical project company. Each project company will exhibit a unique cost structure.

of both sales and cost of sales for at least the first ten years. In addition to sales generated through the PPA, we assume that BPC succeeds in generating additional sales of $25 million in 2011, growing by 2% per annum up to year 2030. Also, because such a large proportion of output is captive, we assume a very modest cost of sales, current assets and current liabilities as a proportion of sales. Recall BPC's operating generating design capacity is 2000 MW-hr, which allows an estimate of its fuel consumption rate and thus an estimate of its annual fuel costs. Further, because it is a highly automated facility producing a single product with 75% of its capacity accounted for by one PPA, its cost of sales is expected to be dominated by the cost of fuel. For the sake of simplicity, we assume the cost of sales to account for approximately 55% of operating costs (dominated by fuel costs), but in any event that the plant's fuel requirement per annum is based on it operating at an average of around 80% of capacity over the evaluation period.

We need to make a special note about depreciation allowances. We assume the original rate of depreciation is 10% per annum (straight line) of the original cost of the facility, which is a somewhat accelerated rate. Further, we assume that BPC reinvests 5% of the original cost of the facility per annum, which amounts to replacing the plant over 20 years. Reinvested capital is depreciated, however, over five years (i.e. 20% per annum straight line). The last reinvestment occurs in year 2030, so that it can be fully depreciated when ownership of the facility is transferred to MEDB. This creates a 'rolling' depreciation account as shown in Table 3.4, which ends with *zero* net invested capital in year 2035.

One further matter requires attention, and that is the discount rate applied to BPC's free cash flow to estimate enterprise present value. For now we postpone any deliberation about the WACC (i.e. the discount rate), until we present a discussion in Box 3.2 below. For now, therefore, we simply assume that a WACC of 9.5% fairly represents BPC's required financial performance relative to its risk to its equity investors. In Table 3.5 we present an estimated pro-forma income statement and free cash flow statement for BPC over its 25 year concession period as a privately owned project company, based on the assumptions outlined above. Our analysis is further based on an assumed corporate tax rate of 20%, and on the assumption that if BPC makes a loss in any year, the loss is not set off against future profits – every tax year is treated independently from the previous year. BPC is also not allowed to declare dividends in years 1–3, coinciding with an assumed contractors' liability for defects in the construction for a period of 3 years after completion.

It can be seen that with this set of assumptions, BPC's enterprise value is estimated to be around $2.5 bn, if it was a going concern with Global Consilium paying fully for the development of the whole facility without any equity partners or debt. However, while Table 3.5 does not present a serious attempt to determine BPC's enterprise value, it does help in estimating BPC's actual cash flow from which debt service and debt principal can be repaid, because any interest payments will alter BPC's tax liability and thus also its FCF. We refer to this as cash available for debt service (CADS), which broadly interpreted means cash flow to service all contracted liabilities

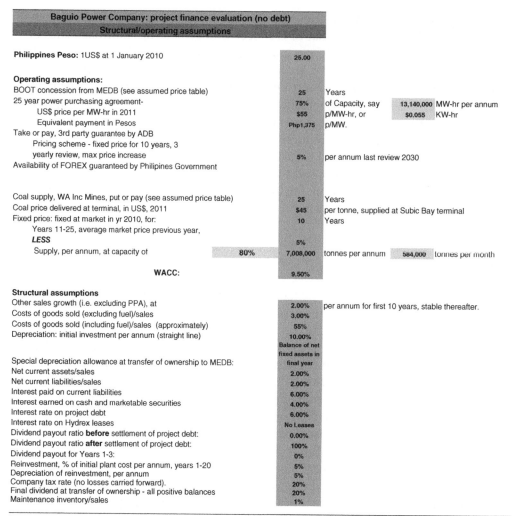

Shaded columns are the result of calculations or have been taken into further figures.

Figure 3.3 Assumptions – Baguio Power Company's cost structure in operation.

(including leases). Cash available for debt service is an important variable in financial evaluation of project companies, because it forms the starting point of departure for any prospective project lender to assess the potential debt capacity of a project company (or any company, for that matter).

Table 3.5 reflects a single estimate of BPC's FCF based on assumptions about variables and their behaviour, and also does not consider the additional influence that debt will have on BPC's FCF. The estimate presented above could possibly be viewed as the *expected scenario*, but relying on only this one scenario is not prudent, it also requires some alternative assumptions. An important assumption here is that project

	Basic facts	
Energy density of coal, approx:	6.67	KW-hr/kg
At 30% efficiency in coal powered generation, approx:	2	KW-hr/kg
i.e.	500	kg/MW-hr
At approximately 2.0kW-hr/kg,	2,000	MW/hr uses
i.e.	8,760,000	Tonnes per annum
Minimum	6,570,000	Tonnes per annum @
and	7,008,000	Tonnes per annum @

1,000,000 kg coal/hr

75% capacity
80% capacity

	MEDB PPA electricity price assumptions/MW-hr: years 2011–2035				P.o.P. coal price assumptions: years 2011–2035		
	Year	$	Effective price $		Year	$	Effective price $
	2011	55	55		2011	45	45
	2012	55	55		2012	45	45
	2013	55	55		2013	45	45
	2014	55	55		2014	45	45
	2015	55	55		2015	45	45
	2016	55	55		2016	45	45
	2017	55	55		2017	45	45
	2018	55	55		2018	45	45
	2019	55	55		2019	45	45
	2020	55	55		2020	45	45
1st Review:	2021	60	58	**1st Review:**	2021	50	48
(+ 5% Max)	2022	60	58	(Market less 5%)	2022	50	48
	2023	60	58		2023	50	48
2nd Review:	2024	60	61		2024	50	48
(+ 5% Max)	2025	60	61		2025	50	48
	2026	60	61		2026	50	48
3rd Review:	2027	70	64		2027	50	48
(+ 5% Max)	2028	70	64		2028	50	48
	2029	70	64		2029	50	48
4th Review:	2030	75	67		2030	50	48
(+ 5% Max)	2031	75	67		2031	50	48
	2032	75	67		2032	50	48
	2033	75	67		2033	50	48
	2034	75	67		2034	50	48
	2035	75	67		2035	50	48

Figure 3.3   *Continued*

promoters and prospective project company lenders must be comfortable with the cost structure of the project company's earnings characteristics (measured as the ratio of total costs : total earnings). Because the project's economic and financial performance is heavily influenced by two variables, i.e. the price of electricity and the cost of coal, we present in Figure 3.4 a best case scenario and a worst case scenario, so that we can imagine a space within which we may expect project free cash flows to occur – we may term it the feasible project FCF space. This is an example of what is termed scenario analysis in Chapter 4, but it also represents a curtailed version of more complex simulation exercises. We say more about risk analysis in Box 4.1.

As stated, CADS is the starting point for any prospective lender to assess how much it may lend to a project company, if it considers the company's basic economics to be sound. In ideal circumstances the decision to lend to any company, and the amount to lend, are separate decisions – the first requiring a decision to assume the credit

Table 3.4   Calculation of BPC depreciation allowances, years 2011–2035.

Depreciation account Year ($ thousands)	2010	2011	2012	2013	2014	2015	2016	2017	2018	2019	2020	2021
Fixed asset at cost	2,500,000	2,500,000	2,500,000	2,500,000	2,500,000	2,500,000	2,500,000	2,500,000	2,500,000	2,500,000	2,500,000	2,500,000
Depreciation		250,000	250,000	250,000	250,000	250,000	250,000	250,000	250,000	250,000	250,000	
Less: accumulated depreciation		(250,000)	(500,000)	(750,000)	(1,000,000)	(1,250,000)	(1,500,000)	(1,750,000)	(2,000,000)	(2,250,000)	(2,250,000)	(2,500,000)
Net fixed assets (excl reinvestment)	2,500,000	2,250,000	2,000,000	1,750,000	1,500,000	1,250,000	1,000,000	750,000	500,000	250,000	0	
Add reinvestment		125,000	125,000	125,000	125,000	125,000	125,000	125,000	125,000	125,000	125,000	125,000
Depreciation reinvestment Yr 1		6,250	6,250	6,250	6,250	6,250						
Depreciation reinvestment Yr 2			6,250	6,250	6,250	6,250	6,250					
Depreciation reinvestment Yr 3				6,250	6,250	6,250	6,250	6,250				
Depreciation reinvestment Yr 4					6,250	6,250	6,250	6,250	6,250			
Depreciation reinvestment Yr 5						6,250	6,250	6,250	6,250	6,250		
Depreciation reinvestment Yr 6							6,250	6,250	6,250	6,250	6,250	
Depreciation reinvestment Yr 7								6,250	6,250	6,250	6,250	6,250
Depreciation reinvestment Yr 8									6,250	6,250	6,250	6,250
Depreciation reinvestment Yr 9										6,250	6,250	6,250
Depreciation reinvestment Yr 10											6,250	6,250
Depreciation reinvestment Yr 11												6,250
Depreciation reinvestment Yr 12												
Depreciation reinvestment Yr 13												
Depreciation reinvestment Yr 14												
Depreciation reinvestment Yr 15												
Depreciation reinvestment Yr 16												
Depreciation reinvestment Yr 17												
Depreciation reinvestment Yr 18												
Depreciation reinvestment Yr 19												
Depreciation reinvestment Yr 20												
Fixed reinvestment at cost		125,000	250,000	375,000	500,000	625,000	750,000	875,000	1,000,000	1,125,000	1,250,000	1,375,000
Less: accumulated depreciation on reinvestment		(6,250)	(18,750)	(37,500)	(62,500)	(93,750)	(125,000)	(156,250)	(187,500)	(218,750)	(250,000)	(281,250)
Less: special writeoff at transfer of ownership to MEDB												
Net fixed reinvestment		118,750	231,250	337,500	437,500	531,250	625,000	718,750	812,500	906,250	1,000,000	1,093,750
Total depreciation allowance for the year		256,250	262,500	268,750	275,000	281,250	281,250	281,250	281,250	281,250	281,250	31,250
Fixed assets at cost		2,625,000	2,750,000	2,875,000	3,000,000	3,125,000	3,250,000	3,375,000	3,500,000	3,625,000	3,750,000	3,875,000
Less: accumulated depreciation		(256,250)	(518,750)	(787,500)	(1,062,500)	(1,343,750)	(1,625,000)	(1,906,250)	(2,187,500)	(2,468,750)	(2,750,000)	(2,781,250)
Net fixed assets		2,368,750	2,231,250	2,087,500	1,937,500	1,781,250	1,625,000	1,468,750	1,312,500	1,156,250	1,000,000	1,093,750

Table 3.4 *Continued*

2022	2023	2024	2025	2026	2027	2028	2029	2030	2031	2032	2033	2034	2035
2,500,000	2,500,000	2,500,000	2,500,000	2,500,000	2,500,000	2,500,000	2,500,000	2,500,000	2,500,000	2,500,000	2,500,000	2,500,000	2,500,000
(2,500,000)	(2,500,000)	(2,500,000)	(2,500,000)	(2,500,000)	(2,500,000)	(2,500,000)	(2,500,000)	(2,500,000)	(2,500,000)	(2,500,000)	(2,500,000)	(2,500,000)	(2,500,000)
125,000	125,000	125,000	125,000	125,000	125,000	125,000	125,000	125,000					
6,250													
6,250	6,250												
6,250	6,250	6,250											
6,250	6,250	6,250	6,250										
6,250	6,250	6,250	6,250	6,250									
	6,250	6,250	6,250	6,250	6,250								
		6,250	6,250	6,250	6,250	6,250							
			6,250	6,250	6,250	6,250	6,250						
				6,250	6,250	6,250	6,250	6,250					
					6,250	6,250	6,250	6,250	6,250				
						6,250	6,250	6,250	6,250	6,250			
							6,250	6,250	6,250	6,250	6,250		
								6,250	6,250	6,250	6,250	6,250	
									6,250	6,250	6,250	6,250	6,250
1,500,000	1,625,000	1,750,000	1,875,000	2,000,000	2,125,000	2,250,000	2,375,000	2,500,000	2,500,000	2,500,000	2,500,000	2,500,000	2,500,000
(312,500)	(343,750)	(375,000)	(406,250)	(437,500)	(468,750)	(500,000)	(531,250)	(562,500)	(587,500)	(606,250)	(618,750)	(625,000)	(625,000)
1,187,500	1,281,250	1,375,000	1,468,750	1,562,500	1,656,250	1,750,000	1,843,750	1,937,500	1,912,500	1,893,750	1,881,250	1,875,000	1,875,000
31,250	31,250	31,250	31,250	31,250	31,250	31,250	31,250	31,250	25,000	18,750	12,500	6,250	0
4,000,000	4,125,000	4,250,000	4,375,000	4,500,000	4,625,000	4,750,000	4,875,000	5,000,000	5,000,000	5,000,000	5,000,000	5,000,000	5,000,000
(2,812,500)	(2,843,750)	(2,875,000)	(2,906,250)	(2,937,500)	(2,968,750)	(3,000,000)	(3,031,250)	(3,062,500)	(3,087,500)	(3,106,250)	(3,118,750)	(3,125,000)	(3,125,000)
1,187,500	1,281,250	1,375,000	1,468,750	1,562,500	1,656,250	1,750,000	1,843,750	1,937,500	1,912,500	1,893,750	1,881,250	1,875,000	1,875,000

Table 3.5 Pro-forma income statement and free cash flow for BPC – years 2011–2035.

Pro-forma income statement ($)	Year	2011	2012	2013	2014	2015	2016	2017	2018	2019	2020	2021
MEDB power purchasing agreement		722,700,000	722,700,000	722,700,000	722,700,000	722,700,000	722,700,000	722,700,000	722,700,000	722,700,000	722,700,000	758,835,000
Other sales		25,000,000	25,500,000	26,010,000	26,530,200	27,060,804	27,602,020	28,154,060	28,717,142	29,291,485	29,877,314	29,877,314
Total sales		747,700,000	748,200,000	748,710,000	749,230,200	749,760,804	750,302,020	750,854,060	751,417,142	751,991,485	752,577,314	788,712,314
Less: Coat of Sales, including fuel		(337,791,000)	(337,806,000)	(337,821,300)	(337,836,906)	(337,852,824)	(337,869,061)	(337,885,622)	(337,902,514)	(337,919,745)	(337,937,319)	(356,541,369)
EBITDA		409,909,000	410,394,000	410,888,700	411,393,294	411,907,980	412,432,959	412,968,439	413,514,627	414,071,740	414,639,995	432,170,945
Less:												
Depreciation		(256,250,000)	(262,500,000)	(268,750,000)	(275,000,000)	(281,250,000)	(281,250,000)	(281,250,000)	(281,250,000)	(281,250,000)	(281,250,000)	(31,250,000)
Interest (project company debt)												
Hydrex Turbines lease payments												
Profit before tax		153,659,000	147,894,000	142,138,700	136,393,294	130,657,980	131,182,959	131,718,439	132,264,627	132,821,740	133,389,995	400,920,945
Less: tax		(30,731,800)	(29,578,800)	(28,427,740)	(27,278,659)	(26,131,596)	(26,236,592)	(26,343,688)	(26,452,925)	(26,564,348)	(26,677,999)	(80,184,189)
Profit after tax		122,927,200	118,315,200	113,710,960	109,114,635	104,526,384	104,946,368	105,374,751	105,811,702	106,257,392	106,711,996	320,736,756
Less:												
Amotisation												
Dividends		0	0	0	(109,114,635)	(104,526,384)	(104,946,368)	(105,374,751)	(105,811,702)	(106,257,392)	(106,711,996)	(320,736,756)
Retained earnings		122,927,200	118,315,200	113,710,960	0	0	0	0	0	0	0	0
Accumulated retained earnings		122,927,200	241,242,400	354,953,360	354,953,360	354,953,360	354,953,360	354,953,360	354,953,360	354,953,360	354,953,360	354,953,360

Free cash flow ($)	Year	2011	2012	2013	2014	2015	2016	2017	2018	2019	2020	2021
Profit after tax		122,927,200	118,315,200	113,710,960	109,114,635	104,526,384	104,946,368	105,374,751	105,811,702	106,257,392	106,711,996	320,736,756
Add back depreciation		256,250,000	262,500,000	268,750,000	275,000,000	281,250,000	281,250,000	281,250,000	281,250,000	281,250,000	281,250,000	31,250,000
Add back interest on debt												
Add back lease payments												
Less: reinvestment in fixed assets		(125,000,000)	(125,000,000)	(125,000,000)	(125,000,000)	(125,000,000)	(125,000,000)	(125,000,000)	(125,000,000)	(125,000,000)	(125,000,000)	(125,000,000)
Add/deduct net change in working capital		0	0	0	0	0	0	0	0	0	0	0
Free cash flow:		254,177,200	255,815,200	257,460,960	259,114,635	260,776,384	261,196,368	261,624,751	262,061,702	262,507,392	262,961,996	226,986,756
ENTERPRISE VALUE: present value of free cash flow:	2,550,719,550											
Less: outstanding debt at 2011	0											
Value of equity:	2,550,719,550											
Outstanding shares:	2,500,000,000											
Value per share:	1.02											
Cash available for debt service (FCF + Tax):		284,909,000	285,394,000	285,888,700	286,393,294	286,907,980	287,432,959	287,968,439	288,514,627	289,071,740	289,639,995	307,170,945

Table 3.5   *Continued*

2022	2023	2024	2025	2026	2027	2028	2029	2030	2031	2032	2033	2034	2035
758,835,000	758,835,000	796,776,750	796,776,750	796,776,750	836,615,588	836,615,588	836,615,588	878,446,367	878,446,367	878,446,367	878,446,367	878,446,367	878,446,367
29,877,314	29,877,314	29,877,314	29,877,314	29,877,314	29,877,314	29,877,314	29,877,314	29,877,314	29,877,314	29,877,314	29,877,314	29,877,314	29,877,314
788,712,314	788,712,314	826,654,064	826,654,064	826,654,064	866,492,902	866,492,902	866,492,902	908,323,681	908,323,681	908,323,681	908,323,681	908,323,681	908,323,681
(356,541,369)	(356,541,369)	(357,679,622)	(357,679,622)	(357,679,622)	(358,874,787)	(358,874,787)	(358,874,787)	(360,129,710)	(360,129,710)	(360,129,710)	(360,129,710)	(360,129,710)	(360,129,710)
432,170,945	432,170,945	468,974,442	468,974,442	468,974,442	507,618,115	507,618,115	507,618,115	548,193,971	548,193,971	548,193,971	548,193,971	548,193,971	548,193,971
(31,250,000)	(31,250,000)	(31,250,000)	(31,250,000)	(31,250,000)	(31,250,000)	(31,250,000)	(31,250,000)	(31,250,000)	(25,000,000)	(18,750,000)	(12,500,000)	(6,250,000)	0
400,920,945	400,920,945	437,724,442	437,724,442	437,724,442	476,368,115	476,368,115	476,368,115	516,943,971	523,193,971	529,443,971	535,693,971	541,943,971	548,193,971
(80,184,189)	(80,184,189)	(87,544,888)	(87,544,888)	(87,544,888)	(95,273,623)	(95,273,623)	(95,273,623)	(103,388,794)	(104,638,794)	(105,888,794)	(107,138,794)	(108,388,794)	(109,638,794)
320,736,756	320,736,756	350,179,554	350,179,554	350,179,554	381,094,492	381,094,492	381,094,492	413,555,177	418,555,177	423,555,177	428,555,177	433,555,177	438,555,177
(320,736,756)	(320,736,756)	(350,179,554)	(350,179,554)	(350,179,554)	(381,094,492)	(381,094,492)	(381,094,492)	(413,555,177)	(418,555,177)	(423,555,177)	(428,555,177)	(433,555,177)	(438,555,177)
0	0	0	0	0	0	0	0	0	0	0	0	0	0
354,953,360	354,953,360	354,953,360	354,953,360	354,953,360	354,953,360	354,953,360	354,953,360	354,953,360	354,953,360	354,953,360	354,953,360	354,953,360	354,953,360

2022	2023	2024	2025	2026	2027	2028	2029	2030	2031	2032	2033	2034	2035
320,736,756	320,736,756	350,179,554	350,179,554	350,179,554	381,094,492	381,094,492	381,094,492	413,555,177	418,555,177	423,555,177	428,555,177	433,555,177	438,555,177
31,250,000	31,250,000	31,250,000	31,250,000	31,250,000	31,250,000	31,250,000	31,250,000	31,250,000	25,000,000	18,750,000	12,500,000	6,250,000	0
(125,000,000)	(125,000,000)	–125,000,000	–125,000,000	–125,000,000	–125,000,000	–125,000,000	–125,000,000	–125,000,000					
0	0	0	0	0	0	0	0	0	0	0	0	0	0
226,986,756	226,986,756	256,429,554	256,429,554	256,429,554	287,344,492	287,344,492	287,344,492	319,805,177	443,555,177	442,305,177	441,055,177	439,805,177	438,555,177
307,170,945	307,170,945	343,974,442	343,974,442	343,974,442	382,618,115	382,618,115	382,618,115	423,193,971	548,193,971	548,193,971	548,193,971	548,193,971	548,193,971

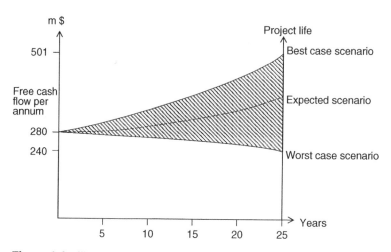

Figure 3.4   Illustration of a potential feasible project free cash flow space.

risk, and thereafter how much to lend and at what price. Frequently these steps are blurred both in corporate lending and lending to project companies because of the competitiveness in funding what are perceived to be good projects. In all, the process of making lending decisions relies on credit risk analysis, which for project companies includes the complete process of assessing the project company's viability, and all the financial risk measures to be discussed in Chapter 4. We assume further that input cost and supply chain risks, sales volume and price risks have been given acceptable risk management attention by GC and the promoter/equity investors (discussed in Chapter 4, and all likely to be preconditions to lending to the project company). Once a bank has decided that a project is bankable, and that it is happy with the cost structures, the parties to the venture and the constructural agreements, its most pressing concern becomes what amount to lend to the company.

We are interested in the use of ratio analyses with project companies for their application in credit risk analysis generally, but in particular for their application in estimating a project's company debt capacity by a lender to a project company. Financial ratio analysis is a cornerstone of credit risk analysis in bank lending decisions to project companies, as with businesses of all kinds. But there are important differences. The first observation about project loan characteristics is that the overall nature of the loan is that of a term loan to a project company. Term lenders are primarily interested in the stabilised internal cash flow generating ability of the project company, as a principal indicator of ability to service debt over the long run – i.e. the company's CADS, which is also influenced by the capital structure of the firm and projections of future profitability. As has been pointed out, lenders to project companies look initially to the cash flows of the project company as the principal source of funds from which any loan will be repaid, and, in the case of non-recourse loans, secured only by the assets of the project company.

This suggests that for credit risk analysis the principal concern of potential lenders is the project's cash flow generating ability over its expected life, not the value of the project assets in circumstances of a distressed sale. An assumption from which we proceed is that uncertainty in the net expected project company cash flows as measured by EBITDA have been mitigated by hedging, long-term contractual arrangements, insurances, guarantees and co-financing arrangements to the fullest extent possible (as further discussed in Chapter 4). If this assumption is accepted and expected cash flows are assumed to be stabilised, at least three particular aspects of cash flow are immediately important to prospective lenders:

- the expected level (or scale) of the net cash flow in absolute dollar terms
- the expected profile over time of the cash flow generated
- the length of the term over which the project is expected to generate cash flow.

These variables, and those interest rates at which the debt will be contracted, provide the essential variables to estimate a project's debt capacity – that amount in money terms that the project is able to bear without stress over its expected life. There is an important further observation to be repeated – how much the project costs to construct does not enter the picture with initial assessment of debt capacity. CADS forms the point of departure to estimate the size of the loan that the project lender accepts.

Expected CADS is thus the point of departure for assessment of the project company debt capacity, while the contractual conditions of the loan form the second component of credit risk assessment and management. Another important measure is project company enterprise value. Our first assumption is that the lender is comfortable with the cost structure of the project company's earnings characteristics exhibited in Figure 3.3, but it is also likely that lenders to a project company will contractually monitor net margins anyway. We are thus assuming that measures to manage earnings volatility have been put in place to the extent that the lender's project finance credit analysts are comfortable. Based on the assumption of acceptable net margins and supply and demand risk management such as the PPA and CPA outlined above, project company debt capacity can be estimated beginning with project ratio analysis. A combination of at least three measures of projected company performance is important for initial analysis of debt capacity. These are the projected debt service coverage rules, the capital ratio, and the profile over time of projected CADS and its debt service coverage ability. An important further point here is that the analysis is simultaneous – an assumed loan amount is entered into the model in order to test its effect on FCF, CADS and debt service capacity, and its effect is then used to adjust CADS and test the outcome through observed debt coverage ratios.

We commence with what have become known as debt service coverage rules. These are not 'rules' because they apply to all lenders and projects (unless there is an unlikely circumstance where regulatory authorities prescribe it), it merely reflects that lenders typically do apply some version of comparable rules based on debt service ability as internal guidelines when assessing loans to project companies. The point of departure here is the estimated value of the project company, dealt with at some length in above sections. We present two such approaches, known as loan life coverage ratio (LLCR), and project life coverage ratio (PLCR). Both in fact result in measures of debt capacity in stock terms, and do not consider how cash flows are expected to occur over project life. We represent PLCR as follows:

$$PLCR = PV \ (CADS \ over \ project \ life)/PV \ Total \ Debt$$

This ratio measures the project company's ability to cover its total loan obligation over the total period of the project's expected life. It is a measure that indicates the project company's overall long term capacity to cover the project debt as a total capital amount, but does not consider how the cash flows may occur during a project's life. Obviously the larger the ratio, the less risky the loan is considered to be. Figure 3.5 illustrates how a PLCR of 2:1 for BPC might be represented graphically, assuming a loan life of say ten years and a project life of 25 years.

The LLCR is in principle comparable to PLCR, but it concentrates on assessing the project company's ability to cover project debt over a shorter period than the project life – i.e. the term of the loan only, which will always be for a term less than the project's expected life. It also does not consider how the cash flows may occur during a project's life. We represent LLCR as follows:

$$LLCR = PV \ (CADS \ during \ loan \ life)/PV \ Total \ Debt$$

This ratio measures the project company's ability to cover its total loan obligation over the term of the loan only. It is thus a shorter term measure of how comfortably

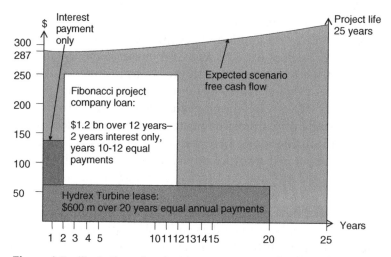

Figure 3.5   Illustration of project loan coverage ratio of 2:1 for BPC.

CADS can cover the total debt than PLCR. While it does not take into account the pattern of project cash flows explicitly, it does provide a measure of how risky the earlier stages of a project's life might be, and what might be thus an appropriate project loan structure. Similarly, the larger the ratio, the less risky the loan is considered to be. The total life of projects depends on the nature of the project. In infrastructure projects developed under BOT principles, for example, the total period of the concession determines project life, 25 years in BPC's case. Mining and energy project life depends on the size of the deposit and how long it may take to extract – in these project companies an important component of their assets is represented by the reserves not yet extracted. Figure 3.6 further illustrates how a LLCR of 1.5 : 1 might be represented graphically for BPC, assuming a loan life of ten years.

The next matter for analysis concerns the profile of the project company's projected cash flow over both its life and the loan term. This is of course important as a matter of time value of money, because earlier cash flows have a relatively higher present value than later cash flows.[14] It is further important because from a project lender's risk management perspective it is desirable to match the debt service and repayment profile with the project company's expected cash flow profile. Given that there is fierce international competition to arrange major project loans, flexibility in structuring debt service and repayment schedules to suit expected project cash flow makes

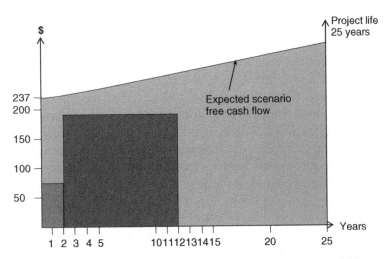

Figure 3.6 Illustration of loan life coverage ratio of 1.5 : 1 for BPC.

---

[14] Duration analysis reduces the cash flow profiles of different investments to a single numerical metric for comparative purposes, and is used extensively by financial institutions in fixed income portfolio management and asset-liability management. Interested readers may refer to Saunders (1997) for a review of duration analysis and its application in banking. Value at risk analysis is relatively widely used in stress-testing of proposed project financial ventures.

competitive sense. This observation helps explain many project loan features, such as grace periods, interest only periods and progressively growing loan repayments. For example, project lenders may allow the first two years of a ten-year loan to a mine to accrue but not demand interest, based on the expectation that the richest seams of ore may be reached after say three years of operation. The BPC cash flow profile in Figure 3.6 suggests no periods of cash flow stress under the best and expected case scenarios, but with the worst case scenario it is clear that debt service problems may occur. This would be a matter for risk management attention (Chapter 4).

While expected cash flow profile over project life is important for indicative purposes in structuring the profile of project debt service and repayment, the project company's operational cash flow management also matters and is likely to be the subject of loan conditions as part of monitoring arrangements. Debt service coverage requirements for ongoing operational management monitoring may be prescribed as part of the loan contract, and may inform loan structuring considerations in the same way as overall project cash flow profile does. A typical measure is the debt service coverage ratio (DSCR):

$$DSCR = CADS/Total\ DS$$

The important matter is that debt service (DS) includes all debt service obligations, including lease payments. Times interest earned, for example often encountered in trading or manufacturing company ratio and trend analyses, is not a sufficient measure. We suppose a DSCR of 1.5× for BPC with a PPA to be reasonable for purposes of illustration. It is however clear that ongoing debt service ability of the project company in operation forms part of the consideration of project cash flow profile and its influence on loan structuring.

We continue with capital ratios. Recall that one aim of project finance companies is to minimise equity and maximise project company debt, and we have loosely suggested that a debt:equity ratio of 60–80% might be typical for project companies. There is one very important caveat. In this respect the estimated project company value (a prospective project company lender's valuation of the proposal) is much more important than the estimated cost of the project, although it clearly will also be concerned about project cost. From a debt capacity perspective a lender will be more concerned with a capital ratio of 60% of debt:total value, leaving the risk of project costs and its management to the promoters – in Chapter 4 we explain that the risk of project cost and time overruns is best borne by the contractors, but is also dependent on planning and design. You guessed it, as a precondition to lending to a project company yet to construct its facility, project lenders are likely to require agreed measures to be put in place to ensure that the project cost, time to completion and quality are managed according to best international practice so as to prevent distortion of incentives. In practice, therefore, we are more concerned with debt:total value and this is how we will interpret the capital ratio.

While we have been discussing debt capacity at length, we have not yet considered any capital structure decisions for BPC; these will be considered in the next section. We have only been considering in general how much debt BPC might be able to sustain, not what form this might take (for example whether in the form of a large term loan or combined term loan and capital leases). For the purpose of evaluating BPC's overall debt capacity and valuing the company with debt, we assume that LLCR of 1.5× is acceptable, while a PLCR of 2.0× is desirable. For the purpose of structuring BPC's balance sheet, however, we assume its debt:total capital ratio will reflect what it costs to construct the facility and how it will be paid for, while project company debt itself will reflect BPC's debt capacity based on a set of debt service coverage rules.

In order to complete the requirements for estimating BPC's debt capacity, we commence by assuming BPC is offered a term loan of twelve years, with no repayment the first two years, and with the loan amortised over the remaining ten years with ten equal payments. We assume that BPC considers entering into a loan with Fibonacci Bank reflecting these conditions. Also, we assume that Hydrex Turbines (HT), a prominent supplier of combined cycle electricity generating equipment, has agreed to lease electricity generating equipment to BPC over twenty years, at annual lease payments reflecting a capital value of $600 million for the turbines. Under these circumstances, Table 3.6 demonstrates the estimate of BPC's debt capacity based on these conditions, and an assumed floating interest rate of 6.0% for the debt and 6.0% for the leases. Under these conditions the debt capacity estimate suggests that at a LLCR of 1.5× BPC has a debt capacity of around $1.3 bn, while a LLCR of 2.0× suggests some $1.8 bn. However, it also transpires that total DSCR might fall below 1.5× during years while the debt is being amortised and taking leases into account, which suggests that the profile of cash flows generated by BPC might require attention (for example, recall that coal price increases are not capped after Year 10). Taking all these factors into account, with the $600 million in capital equipment from HT, it seems reasonable to suggest that BPC could service and repay a $1.4 bn project debt over twelve years, because it exhibits a reasonably healthy LLCR of almost 2.0×. However, as stated above, this is not the end of the analysis, because we have to consider also how BPC's debt service capacity has to be distributed, and this will alter the analysis somewhat as is demonstrated below where we return to the proposed project term loan as outlined here.

The above process to determine a project company's debt capacity is versatile. Obviously it can be used by financial managers in the same way as lenders may use it, for example to develop their own views of debt capacity prior to approaching prospective lenders. It can also be used when considering purchasing a distressed project company that requires refinancing, keeping in mind that under these circumstances the original cost of the distressed project company's facilities is entirely irrelevant to determining its new value and its debt capacity – ask the successive owners and debt financiers of the Channel Tunnel about this unfortunate fact.

Table 3.6   An estimate of BPC's debt capacity.

Estimate of debt capacity ($)	Year	2010	2011	2012	2013	2014	2015	2016	2017	2018	2019	2020
Hydrex Turbines equipment value:	600,000,000											
Cash available for debt service			284,909,000	285,394,000	285,888,700	286,393,294	286,907,980	287,432,959	287,968,439	288,514,627	289,071,740	289,639,995
Less: equal annual lease payments for 20 years			(52,310,734)	(52,310,734)	(52,310,734)	(52,310,734)	(52,310,734)	(52,310,734)	(52,310,734)	(52,310,734)	(52,310,734)	(52,310,734)
Net cash available for debt service			232,598,266	233,083,266	233,577,966	234,082,560	234,597,246	235,122,225	235,657,704	236,203,893	236,761,006	237,329,261
Present value (total CADS):	4,256,185,606											
12 year loan: years 1 and 2 interest only, 10 equal amortisation payments at year-end, years 3–12:	(190,215,142)	project loan:	1,400,000,000									
At LLCR = 1.5, project debt capacity for 12 year loan:	1.5	1,325,309,299										
At PLCR = 2.0, project debt capacity:	2	1,828,092,803										
Interest coverage years 1–2:			2.77	2.77								
Interest coverage years 3–12:					2.78	3.02	3.31	3.69	4.20	4.91	5.99	7.78
Total debt service coverage ratio years 1–20, combined lease & debt:			5.45	5.46	1.18	1.18	1.18	1.19	1.19	1.19	1.19	1.19

	Year	2010	2011	2012	2013	2014	2015	2016	2017	2018	2019	2020
Outstanding balance (at beginning of year):			1,400,000,000	1,400,000,000	1,400,000,000	1,293,784,858	1,181,196,808	1,061,853,475	935,349,543	801,255,374	659,115,554	508,447,346
Equal payments					190,215,142	190,215,142	190,215,142	190,215,142	190,215,142	190,215,142	190,215,142	190,215,142
COMPONENTS:												
(1) Interest			84,000,000	84,000,000	84,000,000	77,627,092	70,871,809	63,711,209	56,120,973	48,075,322	39,546,933	30,506,841
(2) Capital					106,215,142	112,588,050	119,343,333	126,503,933	134,094,169	142,139,819	150,668,208	159,708,301
Outstanding balance (at end of year):			1,400,000,000	1,400,000,000	1,293,784,858	1,181,196,808	1,061,853,475	935,349,543	801,255,374	659,115,554	508,447,346	348,739,045

3.2.2   *Valuation of the project company: equity and debt*

However, as stated above, this is not the end of the analysis of debt capacity. We also have to consider in more detail the financial structure of BPC. Recall that our first assumption was that development of BPC's power generating facility was financed in its entirety by Global Consilium – GC paid for constructing the facility and all the costs associated with incorporation, and so on, and owns 100% of BPC's equity. This

ble 3.6 *Continued*

021	2022	2023	2024	2025	2026	2027	2028	2029	2030	2031	2032	2033	2034	2035
170,945	307,170,945	307,170,945	343,974,442	343,974,442	343,974,442	382,618,115	382,618,115	382,618,115	423,193,971	548,193,971	548,193,971	548,193,971	548,193,971	548,193,971
310,734)	(52,310,734)	(52,310,734)	(52,310,734)	(52,310,734)	(52,310,734)	(52,310,734)	(52,310,734)	(52,310,734)	(52,310,734)	0	0	0	0	0
860,211	254,860,211	254,860,211	291,663,708	291,663,708	291,663,708	330,307,380	330,307,380	330,307,380	370,883,236	548,193,971	548,193,971	548,193,971	548,193,971	548,193,971
12.18	23.67													
1.27	1.27	5.87	6.58	6.58	6.58	7.31	7.31	7.31	8.09					

2021	2022	2023	2024	2025	2026	2027	2028	2029	2030	2031	2032	2033	2034	2035
3,739,045	179,448,247													
0,215,142	190,215,142													
924,343	10,766,895													
9,290,799	179,448,247													
9,448,247	0													

is unrealistic, as one of the reasons for using project finance is often that companies are actually balance sheet constrained – they are interested in minimising equity investment and maximising debt in project companies' financial structure. Further, term loans are not the only form of credit transaction; capital (financial) leases as referred to above are also considered to be debt for the purposes of debt capacity. Thus technically the lease over HT's equipment categorises as a credit – and is treated as a loan alongside the project loan as BPC total debt. We also thus have to consider how BPC's debt service capacity is distributed, because this also matters when potential equity partners are considered for BPC. Further, in an imperfect financial system with taxes and behavioural concerns, capital structure choices also affect enterprise value, which reflects in adjustments to WACC, as we explain in Box 3.2. Figure 3.7 reflects all these financing decisions on BPC's balance sheet.

Box 3.2 Weighted average cost of capital (WACC).

It has been suggested that finance, and indeed corporate finance, joined the modern era with two critical developments in the 1950s, namely modern portfolio theory (the brainchild of Harry Markowitz with some important additions by James Tobin), and the Modigliani and Miller propositions about corporate financial structure. In the early 1960s William Sharpe and Jack Treynor separately developed modern portfolio theory (MPT) further into the capital asset pricing model (CAPM). These developments some 50–60 years ago still provide us with the framework to understand the nature of thinking behind choosing a discount rate for project evaluation.

To commence, Modigliani and Miller's Proposition 1 shows that company success depends on investment decisions, and NOT how it is financed. The return on capital invested depends on the projects it chooses, not how it funds them (i.e. with equity only, or with equity and debt). They argued that the required return on the asset is critical for success in business, not attempts to engineer financial benefits by manipulating balance sheet structures. Modigliani and Miller's Proposition 2 recognised that companies do borrow, but also showed that this implied that while the return on a particular investment depends on the nature of the investment, payments to the two different liabilities (equity and debt) depend on their proportions of overall company value, and return on the asset and interest rate paid on debt. This becomes the weighted average cost of capital (WACC).

$$r_{assets} = WACC = \frac{D}{V} r_{debt} + \frac{E}{V} r_{equity}$$

In capital structure D means debt, E refers to equity, and V refers to company value (market values of D, E and V). In all cases **r** refers to the cost of that financing instrument, either expressed as the *rate of interest for debt*, or as a *required rate of return for equity*. If we keep WACC constant (recall it is always equivalent to r_{assets}) then we see that as the level of debt and equity changes, so the required return on equity changes. In practice r_{assets} is conveniently dropped, for reasons that will become clear. So:

$$r_E = r_A + \frac{D}{V}(r_A - r_D)$$

In short, the more debt a company has, the higher is the required rate of return on equity. Investors demand higher returns to hold equity of highly geared companies. However, while Modigliani and Miller's propositions were developed for a perfect capital market, capital markets have important imperfections, people and taxes. Where interest payments are tax deductible for taxation purposes, and including and representing the corporate tax rate with t, the expression for WACC becomes:

$$WACC = (1-t)\frac{D}{V}r_{debt} + \frac{E}{V}r_{equity}$$

In fact, the WACC expression changes with the addition of every different financing instrument together with the tax implication of using that instrument (for example, lease payments also have tax implications). For illustration we add leases to the WACC expression, as we would have to for Baguio Power Company. So for a company with financial leases in its capital structure we would have:

$$WACC = (1-t)\frac{D}{V}r_{debt}(1-t)\frac{L}{V}r_{Leases} + \frac{E}{V}r_{equity}$$

While r_{Debt} and r_{Leases} are known when these contracts are entered into, the required rate of return on a company's equity is more tricky. We come back to the notion of a typical return on any asset, r_{assets} in the original WACC expression, and note that this would be the return on an investment had it no debt at all, and as if the company was 100% equity financed. This suggests that we have to determine that unique required r_{equity} for the kind of investment we are analysing. Unfortunately we cannot observe this as with r_{Debt} and r_{Leases}, we have to extract this from price information generated in capital markets. Price information generated in capital markets trading allows us to operationalise MPT, and it allows us to operationalise (sort of) the Sharpe/Treynor CAPM, which allows us to derive a value for the required rate of return for holding equity of a company that carries on a particular kind of activity. The CAPM has it that:

$$R_{equity} = r_f + \beta_i(r_m - r_f)$$

where r_f indicates the return on risk-free investments (taken to be government debt instruments) and r_m is the return earned on a fully diversified market portfolio derived from MPT – in practice assumed to be the broad index of the stock exchange in a particular jurisdiction. The term $(r_m - r_f)$ is referred to as the market premium, i.e. the reward for taking the risk of being exposed to the market at all. **Beta**$_i$, or β_i, refers to asset i's risk relative to that of the market portfolio – it may be more risky, in which case it has to earn a higher return than the market, or less risky, in which case it is expected to earn a lower return. The market portfolio's Beta is conveniently measured as 1, so all other assets have a Beta relative to 1. A Beta >1 is riskier than the market, and should earn a higher return, while a Beta <1 should earn less than the market return. Beta is measured by:

$$\beta_i = \frac{\sigma_{im}}{\sigma_m^2}$$

where σ_{im} is the covariance of returns between the overall market and *asset i*, measured as $p_{im}\ \sigma_i\ \sigma_m$, while σ_m^2 is the variance of returns on the market portfolio.

As an example, assume we are interested in developing a power company like BPC. Assume we can borrow at 5%, lease finance costs us 6%, and the tax rate is 20%. Assume r_m is 12.5%, and assume r_f is 2.5%. Assume the value of the company is $2.5 bn, and it is financed with $0.5 bn in equity, $1.4 bn in debt, and $0.6 bn in leases. Assume research shows that a sample of power companies similar in nature to BPC (with comparable PPAs and CPAs and debt profiles) yields an average Beta of 0.7 after adjusting for leverage (set r_E in the equation on page 130 equal to observed r_E and solve for r_A).[15]

We are now in a position to estimate WACC for BPC. First, required r_{equity}:

$$R_{equity} = r_f + \beta_i(r_m - r_f)$$

$$R_{equity} = 2.5 + 0.7(12.5 - 2.5) = \textbf{9.5\%.}$$

We can now complete the WACC calculation:

$$WACC = (1-t)\frac{D}{V} r_{debt}\ (1-t)\frac{L}{V} r_{Leases} + \frac{E}{V} r_{equity}$$

$$WACC = (1-0.2)\frac{1.2}{2.5} 5.0\% + (1-0.2)\frac{0.6}{2.5} 6.0 + \frac{0.5}{2.5} 9.5$$

$$WACC = 1.92 + 1.15 + 1.9 = 4.97,\ \textbf{\textit{say 5\%.}}$$

This is used as the discount rate to estimate the enterprise value of BPC. In practice, as the capital structure changes every year with debt being reduced, so the WACC should change. For our purposes we simply assume WACC for the evaluation throughout BPC's life, which will result in somewhat of an overvaluation of later cash flows. We are interested in demonstrating the process only. We assume Global Consilium and Bagatelle Design and Construct to have higher WACCs (see Tables 3.8 A, B, C).

---

[15] For those interested, current (typically leveraged) Beta's for various companies are widely reported in the financial press and specialist websites.

In Box 3.2 we summarised a commonly followed approach to estimate WACC. In order to apply this to BPC we first have to consider its capital structure. We have estimated above that BPC has a debt capacity of around $1.4 bn[16] which we assumed reasonable, but that was after also including HT's leases of $600 million. How did this come about? We may assume that Global Consilium interested two other very big companies in participating in developing the facility. The first is Hydrex Turbines, a large USA company which produces electricity generating equipment and control systems worldwide, and the second is Bagatelle Design and Construct Inc. (BD&C), a prominent global total project design and construct and facilities management company. BD&C and HT also agreed to contribute equity to the project, with BD&C agreeing to design, engineer, construct, and maintain the plant after completion, and HT agreeing to supply and maintain the electricity generating equipment and control systems for the facility. This agreement results in a pro-forma balance sheet for BPC at commencement of operations in 2011 as presented in Figure 3.7.

As indicated, the price at which HT's turbines and control systems is supplied to BPC is assumed to be $600 million. As reflected in Figure 3.7, HT also invested $150 million in a 30% equity stake in BPC, which of course allows it the right to participate in dividends and governance. The equipment will be leased to BPC over a twenty year period at equal annual payments, reflecting a rate of interest of 6.0%. In turn, Fibonacci Bank has agreed to provide a twelve year term loan for $1.4 billion to BPC, interest only for years 1–2 and with ten equal annual payments of interest and principal in years 3–12. A final note: as debt is repaid, BPC's financial structure changes every year, also if refinancing occurs. This means that BPC, and indeed most companies, will have WACCs that change every year according to the relative weights of the various forms of capital. However, overall WACC, the return required on the asset, will **not** change – this is the fundamental message of Modigliani and Miller's first proposition. For present purposes, however, we assume that WACC stays the same over the life of the project.

We now have the revised set of information to estimate the enterprise value of BPC with its revised financial structure. The revised BPC enterprise valuation is presented in Table 3.7.

The purpose of the revised estimate is to value BPC's equity, and thus what each equity holder's share in the company is worth. The reason for valuing each equity holder's share is important – it is this figure that will guide each equity partner's capital budgeting decision. While the nature of GC and HT's capital budgeting decision will be very similar, we are particularly interested in BD&C's capital budgeting decision, because it has to base its decision on an interesting combination of two factors: its return on the design and construct contract over a four year construction

---

[16] Note that for simplicity we have ignored complex loan transaction details, as explained in Appendix 1, Chapter 2. These would directly influence the pattern of cash flows over the loan life period, and influence all financial aspects of the evaluation.

The Baguio Power Company (incorporated in the Republic of the Philippines 1999)

Balance Sheet at June 30, 2007

Liabilities	($ thousands)			Assets	($ thousands)	
Equity				**Fixed assets**		
Issued ordinary shares:				Power generating facility at Subic Bay (at cost)[a]	2,600,000	
Global Consilium Corporation	175,000	35.0%		Furniture, fittings, equipment, vehicles, etc.	5,000	
Bagatelle Projects D&C Inc	175,000	35.0%		LESS:		
Hydrex Turbines Inc	150,000	30.0%		Accumulated depreciation	(1,000)	
Shareholders' equity		500,000		**Fixed assets**		2,604,000
Debt						
Long-term debt				**Current assets**		
Secured bank debt, due 06/2022	1,400,000			Incidentals, maintenance inventory	5,000	
Capital leases: generating equipment	600,000			Other current assets	310,000	
Long-term office lease	10,000			Cash and marketable securities	5,000	
Long-term debt		2,010,000		**Current assets**		320,000
Current liabilities						
Commercial paper: 3 months	114,000					
Short-term bank loan: 1 month	300,000					
Current liabilities		414,000				
Total debt		2,424,000				
Total		2,924,000		**Total**		2,924,000

a The land at Subic Bay is leased from The Philippines Government at a neglible annual rent of US$1 per annum.

Figure 3.7   Pro-forma balance sheet for Baguio Power Corporation at commencement of operations, 1 January 2011.

period combined with the return from investing in BPC's equity. From Table 3.7 we summarise BPC's enterprise valuation below:

Enterprise value of BPC = PV of BPC's operations **minus** PV of debt and other
liabilities
= $2,731,082,634 – $1,400,000,000
Value of BPC's equity = $1,331,082,634

This translates into a value of $0.53 per share for BPC, a value of around $465 million for GC's equity holding, $465 million for BD&C's holdings, and some $399 million for HT's holding. These values form the principal inputs into the capital budgeting decisions outlined below.

3.3 Capital budgeting decisions

As alluded to above, every equity participant in BPC will conduct its own capital budgeting decision to determine the attractiveness of participating in the development of BPC. It matters little which party was responsible for the original idea or for approaching MEDB and the Philippines Government, GC, BD&C or HT, or indeed if MEDB approached GC or any of the other participants initially. GC, BD&C and HT will each conduct its own capital budgeting analysis and decision from its own perspective of how its interests may be served by BPC. In this section we outline typical corporate capital budgeting decisions for GC and BD&C, to provide an example of how these decisions may be approached. We assume GC and HT's decisions are to some extent comparable, so we present only a suggested analysis for GC. However, because BD&C's decision involves also the construction of the plant it is a bit more complex, so we will analyse its decision also.

An additional point is important when the development phase is factored into GC, HT and BD&C's capital budgeting decisions. We went through the process of evaluating BPC's debt capacity as if it was an operational project company, but the power generating facility still has to be constructed. In Chapter 4 we discuss in some detail reasons for separating the construction phase from the operational phase of project companies when financing is considered, for the simple reason that we have stressed over and over – there is no operational project company until the facility functions according to plan. Hence most construction finance and permanent project company finance are split into two separate and independent transactions: a *construction loan* which has to be repaid in full when the construction period is over (whether the facility is delivered on time, budget and to the required standard or not); and a *project loan* to the operating project company once construction and commissioning have been successfully completed. Our estimate of debt capacity above concentrated on the second loan, but decisions about the potential size of the construction loan are directly influenced by the values determined in the debt capacity estimates. Often

Table 3.7   BPC enterprise value with final capital structure assumptions.

Free cash flow	Year	2011	2012	2013	2014	2015	2016	2017	2018	2019	2020	2021
Profit after tax		13,878,613	9,266,613	4,662,373	5,164,375	5,980,350	12,128,813	18,629,386	25,502,857	32,771,258	40,457,936	262,148,694
Add back depreciation		256,250,000	262,500,000	268,750,000	275,000,000	281,250,000	281,250,000	281,250,000	281,250,000	281,250,000	281,250,000	31,250,000
Add back interest on debt		84,000,000	84,000,000	84,000,000	77,627,092	70,871,809	63,711,209	56,120,973	48,075,322	39,546,933	30,506,841	20,924,343
Add back lease payments to Hydrex		52,310,734	52,310,734	52,310,734	52,310,734	52,310,734	52,310,734	52,310,734	52,310,734	52,310,734	52,310,734	52,310,734
Less: reinvestment in fixed assets		(125,000,000)	(125,000,000)	(125,000,000)	(125,000,000)	(125,000,000)	(125,000,000)	(125,000,000)	(125,000,000)	(125,000,000)	(125,000,000)	(125,000,000)
Add/deduct net change in working capital												
Free cash flow:		281,439,347	283,077,347	284,723,107	285,102,200	285,412,892	284,400,756	283,311,092	282,138,913	280,878,925	279,525,511	241,633,771
ENTERPRISE VALUE: present value of free cash flow	2,731,082,634											
Less: outstanding debt at 2011	1,400,000,000											
Value of equity:	1,331,082,634											
Outstanding shares:	2,500,000,000											
Value per share:	0.53											

the loans are initially made by the same lenders, with the permanent project company loan conditional upon successful completion of an operational plant, but sometimes different lenders are involved in the two separate transactions. We say more about this in Chapter 4.

Construction loans typically also contain several restrictive conditions. First, it is not unusual that the equity investors/promoters are required to invest significant parts of their equity commitment before some tranches of a construction loan are released. This is to ensure that the equity investors demonstrate their commitment to the project by investing their contributions. We will assume for the sake of illustration that the equity investors in BPC are subjected to an extreme version of such a requirement: Fibonacci Bank, which has decided to provide a construction loan to BPC, will only release the first tranche of the construction loan once the equity partners have fully invested their capital. Also, we will assume for the purpose of illustration that the construction loan is to be repaid in full at the end of commissioning (assuming it is successful). Interest on the construction loan is assumed to be 6.5% (more risky than a term loan to an operational BPC), calculated periodically, with principal and accrued but unpaid interest to be settled in one lump sum. Sophisticated interest rate derivatives such as swaps and options give considerable freedom for many lenders to agree to defer interest payments on short and medium-term loans without incurring significant market-related risk. This practice is common in the early phase of many projects, and will usually support project economics, but may also amplify the overall credit risk exposure of lenders.

Table 3.7 *Continued*

2022	2023	2024	2025	2026	2027	2028	2029	2030	2031	2032	2033	2034	2035
270,274,653	278,888,168	308,330,966	308,330,966	308,330,966	339,245,904	339,245,904	339,245,904	371,706,589	418,555,177	423,555,177	428,555,177	433,555,177	438,555,177
31,250,000	31,250,000	31,250,000	31,250,000	31,250,000	31,250,000	31,250,000	31,250,000	31,250,000	25,000,000	18,750,000	12,500,000	6,250,000	0
10,766,895	0	0	0	0	0	0	0	0	0	0	0	0	0
52,310,734	52,310,734	52,310,734	52,310,734	52,310,734	52,310,734	52,310,734	52,310,734	52,310,734	0	0	0	0	0
(125,000,000)	(125,000,000)	(125,000,000)	(125,000,000)	(125,000,000)	(125,000,000)	(125,000,000)	(125,000,000)	(125,000,000)					
239,602,282	237,448,903	266,891,701	266,891,701	266,891,701	297,806,639	297,806,639	297,806,639	330,267,323	443,555,177	442,305,177	441,055,177	439,805,177	438,555,177

We now show how these conditions might affect GC and BD&C's captal budgeting decisions.

3.3.1 *Global Consilium's capital budgeting decision*

Global Consilium's capital budgeting decision appears relatively uncomplicated, albeit a decision that is influenced by a long development period, around four years. In essence its decision depends on two things: first the investment of its equity, and upon completion of the facility and its successful commissioning, it receives the value of its ownership share in BPC. This is why the valuation of BPC's equity as an operating company is important – the payoff for GC for participating in the creation of BPC only emerges once BPC is operational, and this is how it views the capital budgeting decision. In Chapter 4 we discuss the risks associated with this circumstance. Of course, now the expected cost of developing the facility becomes critical, because the NPV rule simply requires the value of GC's equity upon completion and commencement of operations to be worth more than its costs in developing the facility. In Figure 3.8 we present a hypothetical BD&C cost estimate for developing the facility, which includes HT's costs of supplying and installing turbines and control systems. For simplicity we assume BD&C provides HT's contractors with access and minor standby support, for a nominal fee.

We need to explain further the significance of the construction loan before we complete the discounted cash flow analysis. We assume once BPC has invested its equity contribution ($500 million in total), construction loan withdrawals may commence, and are made based on amounts earned by BD&C over interim periods (usually monthly, but we use quarterly periods simply to make the table smaller and more accessible). There is an interesting cash flow phenomenon that occurs with this practice and with construction loans: BPC draws the amount of the interim payment to BD&C from Fibonacci Bank, but the amount drawn technically only passes through BPC and is paid directly to BD&C. As far as BPC is concerned, this results in a *zero* net cash flow. On GC's capital budgeting DCF (Table 3.8B below) we itemise these as an inflow to BPC from Fibonacci with a matching outflow to BD&C. The amounts are paid through Baguio to Bagatelle, which in turns pays subcontractors.

These assumptions yield a very positive result for GC. The cash flow indicates a net present value of $183 763 316 (an IRR exceeding 35%), which suggests that this is a highly attractive project for GC. This is reflected in Table 3.8B. Do not be over-excited by this seemingly very high IRR – many corporations will use several times their calculated WACC as hurdle rates of return.

3.3.2   Cash flow and the construction phase

We turn now to BD&C. As with Hydrex Tubines, BD&C has multiple interests in BPC. First, it has secured the contract to construct the facility, and obviously this is in itself a corporate opportunity that has to be assessed on its merits. Second, it is a significant minority shareholder in BPC, and this must be considered in combination with the construction contract as a corporate opportunity. Third, depending on how matters progress, it may also secure a long-term contract to manage the completed facility it is to construct. But first the facility has to be constructed, to its own budget less HT's supply of generating equipment and control systems (Figure 3.8), in the agreed time and to the required performance specified. It is likely that if BD&C is late in delivering the facility it will be heavily penalised, and if it mismanages the construction phase to a sufficiently disruptive level, its performance guarantors may be called upon to complete the facility – this is a state of affairs that is simply too serious to contemplate, and performance monitoring functions as an early warning system. In order to avoid incentive conflicts, we assume that BD&C is not responsible for commissioning the facility; this is managed by an independent contractor, at BD&C's cost. We present a summary of assumptions governing GC and BD&C's capital budgeting decisions in Box 3.3.

Two observations emerge from this analysis. First, it transpires that the construction contract in isolation shows a positive NPV of some $18 million at a WACC of 20%. This means that despite a low profit margin of 1.5%, BD&C nevertheless earns

Bagatelle Design and Construct Inc.: fixed cost estimate for design and construct of Baguio Power Company facility		
Summary of estimate of cost of construction		
Code **Principal project elements**		**Estimated Total ($)**
1 Planning, design and engineering to approval stage		50,000,000
2 Prepare site and construct plant roads, stormwater sewerage and prepare for all plant civil engineering works		50,000,000
3 Dredge and prepare shipping channel approach to coal terminal		100,000,000
4 Marine engineering construction: collier docking facilities		80,000,000
5 Coal offloading and stockpiling facility: civil engineering works		50,000,000
6 Conveyor systems: civil engineering works		20,000,000
7 Main generating plant, tank farms, cooling towers: civil engineering works		200,000,000
8 Steel structures and cladding: coal terminal, conveyors, main generating plant building		500,000,000
9 Mechanical and electrical: conveyor systems		50,000,000
10 Mechanical and electrical: boilers and ancillary pipework systems		250,000,000
11 Supply: mechanical and electrical, power generating equipment and ancillary equipment (Hydrex Turbines)	450,000,000	
12 Supply: mechanical and electrical, control systems (Hydrex Turbines)	150,000,000	600,000,000
13 Commissioning		250,000,000
14 In loco administration building: Baguio Power Company		5,000,000
Bagatelle only subtotal		**1,605,000,000**
subtotal plus Hydrex Turbines		**2,205,000,000**
A-15 Allow for contingencies		220,500,000
Total		**2,425,500,000**

Figure 3.8   A hypothetical cost estimate for developing BPC's power facility.

Table 3.8A   Baguio Power Company: estimate of interest bearing construction cash flows and total interest on construction loan.

QUARTERS:	1	2	3	4	5	6	7	8	9
	Year 2007:1	Year 2007:2	Year 2007:3	Year 2007:4	Year 2008:1	Year 2008:2	Year 2008:3	Year 2008:4	Year 2009:1
All outflows ($)									
Project construction: interim payments to Bagatelle	10,000,000	30,000,000	50,000,000	80,000,000	150,000,000	180,000,000	138,125,000	138,125,000	138,125,000
Gross quarterly cash outflow:	10,000,000	30,000,000	50,000,000	80,000,000	150,000,000	180,000,000	138,125,000	138,125,000	138,125,000
Previous cumulative total:	0	10,000,000	40,000,000	90,000,000	170,000,000	320,000,000	500,000,000	140,369,531	283,020,067
Cumulative total construction loan outstanding:	10,000,000	40,000,000	90,000,000	170,000,000	320,000,000	500,000,000	138,125,000	278,494,531	421,145,067
Accumulated interest:							2,244,531	4,525,536	6,843,607
Cumulative totals outstanding:	10,000,000	40,000,000	90,000,000	170,000,000	320,000,000	500,000,000	140,369,531	283,020,067	427,988,675
Total interest:	122,897,413								
Total loan:	1,227,897,413								

Table 3.8B   Baguio Power Company: Global Consilium Corporation's capital budgeting decision.

QUARTERS:	1	2	3	4	5	6	7	8	9
	Year 2007:1	Year 2007:2	Year 2007:3	Year 2007:4	Year 2008:1	Year 2008:2	Year 2008:3	Year 2008:4	Year 2009:1
Inflows									
Present value of equity investment in Baguio Power Company in 2011									
Loan drawdowns:									
Construction loan drawings (project construction payments)							138,125,000	138,125,000	138,125,000
Total inflows	0	0	0	0	0	0	138,125,000	138,125,000	138,125,000
Outflows									
Incorporation and pre-incorporation expenses	5,000,000								
Initial loan syndication fee	1,400,000								
Equity investment – 35% ONLY of initial construction payments to Bagatelle	3,500,000	10,500,000	17,500,000	28,000,000	52,500,000	63,000,000			
Project construction: balance of construction payments to Bagatelle							138,125,000	138,125,000	138,125,000
Repayment of construction loan and interest									
Total outflows	9,900,000	10,500,000	17,500,000	28,000,000	52,500,000	63,000,000	138,125,000	138,125,000	138,125,000
Net in/outflow	(9,900,000)	(10,500,000)	(17,500,000)	(28,000,000)	(52,500,000)	(63,000,000)	0	0	0

WACC	10.00%	per annum	2.50%	per quarter
NPV:	183,763,316			
(Equity) IRR:	35.61%			
Decision rule:	If NPV ≥ 0, GO AHEAD			
	If NPV ≤ 0, DO NOT GO AHEAD			

Table 3.8A *Continued*

10	11	12	13	14	15	16	17	18	19	20	
Year 2009:2	Year 2009:3	Year 2009:4	Year 2010:1	Year 2010:2	Year 2010:3	Year 2010:4	Year 2011:1	Year 2011:2	Year 2011:3	Year 2011:4	Totals
138,125,000	138,125,000	138,125,000	138,125,000	138,125,000	0	0					1,605,000,000
138,125,000	138,125,000	138,125,000	138,125,000	138,125,000	0	0					1,605,000,000
427,988,675	575,313,022	725,031,390	877,182,681	1,031,806,431	1,188,942,817	1,208,263,137					
566,113,675	713,438,022	863,156,390	1,015,307,681	1,169,931,431	1,188,942,817	1,208,263,137					
9,199,347	11,593,368	14,026,291	16,498,750	19,011,386	19,320,321	19,634,276					
575,313,022	725,031,390	877,182,681	1,031,806,431	1,188,942,817	1,208,263,137	1,227,897,413					

Table 3.8B *Continued*

10	11	12	13	14	15	16	17	18	19	20	
Year 2009:2	Year 2009:3	Year 2009:4	Year 2010:1	Year 2010:2	Year 2010:3	Year 2010:4	Year 2011:1	Year 2011:2	Year 2011:3	Year 2011:4	Totals ($)
							2,731,082,634				2,731,082,634
											0
138,125,000	138,125,000	138,125,000	138,125,000	138,125,000	0	0	0				1,105,000,000
138,125,000	138,125,000	138,125,000	138,125,000	138,125,000	0	0					1,105,000,000
											0
											0
											5,000,000
											1,400,000
											175,000,000
138,125,000	138,125,000	138,125,000	138,125,000	138,125,000							1,105,000,000
							1,227,897,413				1,227,897,413
138,125,000	138,125,000	138,125,000	138,125,000	138,125,000	0	0	1,503,185,220 (Total equity value less construction loan interest)				
0	0	0	0	0	0	0	526,114,827 (GC's share)				

Table 3.8C   Bagatelle Inc.: constructing the Baguio Power Company facility.

QUARTERS:		1	2	3	4	5	6	7	8
		Year 2007:1	Year 2007:2	Year 2007:3	Year 2007:4	Year 2008:1	Year 2008:2	Year 2008:3	Year 2008:4
Inflows									
Present value of equity investment in Baguio Power Company in 2011									
Interim construction payments		10,000,000	30,000,000	50,000,000	80,000,000	150,000,000	180,000,000	138,125,000	138,125,000
Total inflows:		10,000,000	30,000,000	50,000,000	80,000,000	150,000,000	180,000,000	138,125,000	138,125,000
Outflows									
Insurances, maintenance bond, other setup costs		50,000,000							
Project construction: payments to contractors			9,850,000	29,550,000	49,250,000	78,800,000	147,750,000	177,300,000	136,053,125
Costs of commissioning									
Total outflows:		50,000,000	9,850,000	29,550,000	49,250,000	78,800,000	147,750,000	177,300,000	136,053,125
Net in/outflows:		(40,000,000)	20,150,000	20,450,000	30,750,000	71,200,000	32,250,000	−39,175,000	2,071,875
	WACC	20.00%	per annum	5.00%	per quarter				
	Construction project only NPV:	18,181,242							
	Construction project only IRR:	9.15%	Note: this IRR is an anomaly – it results from the mathematics of IRR estimates when cashflows change signs several times over the timeline of the analysis.						
	Project NPV:	221,442,672							
	Project IRR:	239.72%							
Decision rule:	If NPV ≥ 0, GO AHEAD								
	If NPV ≤ 0, DO NOT GO AHEAD								

Box 3.3   Assumptions governing cash flows presented in Tables 3.8A, B and C.

Summary of assumptions for construction and interest bearing cash flows (Table 3.8A)

In general, the following assumptions are made in order to estimate how the structure and conditions of the construction loan will influence Global Consilium, Baguio Power and Bagatelle's capital budgeting decisions.

1. Start of operations for Baguio Power Company: 1 January 2011.
2. The agreed construction loan facility is for $1 400 000 000. The first tranche of $500 000 000 is payable after equity participants have invested their contributions. The second tranche, $900 000 000, is made available from 1 January 2009. Interest calculation commences from the date of drawing on the facility. (We assume a simplistic loan structure so as to contain Table 3.8.)
3. The initial equity investment required from shareholders is $500 000 000, to be invested over the first four quarters of construction (in proportion to equity-holding).

Table 3.8C *Continued*

9	10	11	12	13	14	15	16	17	18	19	20
Year 2009:1	Year 2009:2	Year 2009:3	Year 2009:4	Year 2010:1	Year 2010:2	Year 2010:3	Year 2010:4	Year 2011:1	Year 2011:2	Year 2011:3	Year 2011:4
								465,878,922			
138,125,000	138,125,000	138,125,000	138,125,000	138,125,000	138,125,000						
138,125,000	138,125,000	138,125,000	138,125,000	138,125,000	138,125,000	0	0				
136,053,125	136,053,125	136,053,125	136,053,125	136,053,125	136,053,125	136,053,125					
						50,000	60,000				
136,053,125	136,053,125	136,053,125	136,053,125	136,053,125	136,053,125	136,103,125	60,000				
2,071,875	2,071,875	2,071,875	2,071,875	2,071,875	2,071,875	(136,103,125)	(60,000)	465,878,922			

4. Interest on the construction loan is initially at 6.5% per annum, i.e. 1.63% per quarter. Interest is calculated quarterly on the cumulative outstanding loan balance, accumulated from the date of first drawing. A syndication fee of 0.1% is payable upon signing of the loan contract (assumed to be Quarter 1). The total loan amount outstanding plus accumulated interest is repayable upon completion of construction of the facility (Quarter 13).

5. Construction is assumed to be completed in 3.5 years from commencement, i.e. over a total of 14 quarters.

6. Interest rate on the construction loan is calculated at a rate over Libor (as the loan contract benchmark). For simplicity, we assume there are no changes in interest rates over the construction period. (This assumption is purely for convenience – it is highly unlikely that this will occur in practice.)

7. For simplicity, construction costs are assumed to be end-loaded, as a consequence of large-scale but relatively low-value civil and marine engineering work required before high-value structural steel frames are constructed and cladded, and mechanical and electrical systems are installed. Thus it is

assumed that 70% of total construction expense is incurred in Quarters 7 -14, with Quarters 15 and 16 set aside for commissioning.

8. Costs in excess of $1 605 000 000 will be financed by equity partners in proportion to their shareholding, by way of a standby equity agreement. These will be categorized as construction cost contingencies.

9. The total accumulated interest on the construction loan will be capitalised and treated as part of fixed investment on Baguio's balance sheet.

Summary of assumptions for Baguio Power Company: Global Consilium Corporation's capital budgeting decision (Table 3.8B).

In general, the following assumptions are made in order to estimate how incorporating the structure and conditions of the construction phase will influence Global Consilium's (GC) capital budgeting decision.

1. GC's capital budgeting decision depends on it investing $175 m (35% of the equity required), and receiving around $465 m (35% of Baguio's equity present value) in Quarter 1, 2011.

2. Construction loan and total accumulated interest on loan over the loan period is repaid in one lump sum at end of Quarter 16, when permanent project finance is put in place.

3. Loan withdrawals are made based on amounts of interim payments to the contractor after Baguio invested its equity ($500 000 000). The amounts are paid directly to Bagatelle, who pays the subcontractors. This results in an effective net **zero** cash outflow to Baguio for the balance of the cost of construction. In practice there will be nominal sums involved, too little to affect the capital budgeting decision materially.

Summary of assumptions for Bagatelle Inc.: constructing the Baguio Power Company facility (Table 3.8C).

In general, the following assumptions are made in order to illustrate how the nature of the project and the structure and conditions of the construction phase will influence Bagatelle's capital budgeting decision.

1. Recall that Bagatelle owns 35% of Baguio. Thus Bagatelle's capital budgeting decision depends on it investing $175 m (35% of Baguio's initial equity required), and receiving $465 m (35% of Baguio's equity present value) in Quarter 1, 2011.

2. Bagatelle contracts with BPC to build and commission the facility, and in turn subcontracts 100% of the overall construction contract.

3. Bagatelle's margin over the cost of construction is 1.5%, for all construction.
4. Bagatelle pays subcontractors in the quarter after it received payment.
5. For simplicity, interim payments reflect accurately the value of work completed in the relevant quarters.
6. Assume all commissioning expenses are for Bagatelle's account, based on the nature of an ideal design and construct transaction.
7. Assume Bagatelle WACC is 20%.
8. Assume both a performance guarantee **and** a maintenance period bond are provided by a banking syndicate led by Bagatelle's principal corporate bankers, HSBC, for a one-off fee at commencement of construction.

In principle, Hydrex Turbines' capital budgeting decision will be a much simplified version of Bagatelle's, because it effectively invests its equity share early, and will receive payment for delivering and installing the turbines, but not their capital cost. The construction period cash flows then have to be combined in a timeline with the capital cost of the turbines and the lease payments received over the life of the project.

a return in excess of 20% on the working capital it invests in the BPC contract.[17,18] Further, a combined analysis shows that BPC could earn an amazing return on its participation in BPC – a return exceeding 200%. This explains quite clearly why so many design and construct firms and general contractors are so active in project initiations of all kinds all over the world. If readers are tempted to think that a return exceeding 200% is absolute nonsense, we have some news – for example, in real estate development projects many established and well-capitalised companies would not give projects suggesting returns below 50% a second glance (assuming normal interest rate scenarios). Such attitudes to risk reflect in large measure the difficulty companies have in managing cash flows during the construction phase. For some insight into operational cash flow management during projects see Box 3.4.

---

[17] One often hears how poor the profit margins are in contracting. While this may be true, it is really return that matters – in this hypothetical case BPC earns a construction project only return in excess of 20%. We on purpose made it somewhat rosy (but not entirely unbelievable). This outcome is the result of the cash flow profile of the project. In effect profits are less important – to earn good returns in contracting requires excellent operational project management and financial management skills. It can easily be demonstrated by entering BDC's cash flows in Table 3.8 in a spreadsheet and moving them around a bit – very quickly returns evaporate, but profit remains. Which of the two is preferable? This highlights a further point: managing construction project cash flows for return can also be done opportunistically, and we say more about that in Chapter 4.

[18] We ignore IRR in this analysis, because the nature of this particular problem yields multiple IRRs. This results from the mathematical properties of IRR when cash flows switch sign from negative to positive more than once – it means that there are multiple solutions to the IRR expression. This is why NPV is the preferred metric – view IRR as additional information.

Box 3.4 Interim payments and construction contracts.

As an extreme simplification, when a client places an order with a construction company to build a facility of any kind (irrespective of scale, complexity or cost), suppose the nature of the transaction is such that the client pays the construction company upon delivery of the facility. Without external construction loan financing, the construction company would complete the facility (a small industrial building, say) using its own finance, and receive full payment by the client upon handing over the facility, subject to quality and other specified requirements. This set of circumstances is highly simplified, and will place stress on the financial resources of both contracting parties. The client must have a large amount of capital to pay on completion and handover, and the construction company must have adequate internal financial resources to incur and carry all the costs of construction until successful completion and handover, when it will be fully reimbursed.

In all, this is a very inefficient economic arrangement, for the simplest reason that it would not only curtail construction industry capacity severely, but also severely curtail those client companies' operations that purchase construction company services. In any economy with reasonably advanced corporate and financial sectors, reputable client companies or reputable construction companies may borrow from a range of financial intermediaries to fund project costs. Such loan facilities are made for projects ranging from building a family residence to the largest project facility that may be imagined. In all these cases the ultimate effect of such lending (whether it is a loan to a client company or a construction company) functions to ease the demand on client or construction company financial resources by allowing the construction company to finance (usually in part) the costs of constructing the commissioned facility before it is reimbursed. Typically thus, a construction company invests some of its own financial resources in the starting up of a construction project, and receives interim payments based on progress during construction, and final payment based on completion.[19] For construction companies this model may be described as initial investment of corporate resources in the startup of the project, subsequently leveraged by receipt of interim progress payments as the project is executed. In financial terms project success for the construction company invariably depends on how well it manages its outflows (project costs) relative to inflows (interim progress payments) – the better its operational financial management during the project, the better will be the returns it earns on the corporate financial resources it invests in the project startup. But operational cash flow management cannot dominate the necessity to construct the facility, and often is the source of tension between project companies concerned with progress and project managers attempting to manage construction cash flows to optimise returns.

---

[19] Many standard forms of building contract exist, and many recognise and formalise the principle of interim payments for progress during a project.

3.3.3   *Capital budgeting and the value of flexibility*

We complete our discussion of capital budgeting and project company valuation by mentioning that developments in the theory and practice of capital budgeting have been extremely exciting over the last two–three decades, particularly around a field that has become known as 'real options analysis'. Even a cursory treatment of real options analysis will require the introduction of options pricing theory and methodology, which we do not attempt in this book. But it is nevertheless important to explain why these developments are important.

It is fair to say that discounted cash flow methodology, the standard for valuing investments, businesses, and project opportunities that businesses may have, has undergone somewhat of a change over the last two decades with what is often referred to as the real options analysis (ROA) revolution, but also with developments in decision theory and real options analysis of computing power. ROA is outside the scope of this book, but it is necessary to explain briefly why it is critical to be aware of its nature. Practical criticism of DCF methodology centres around the fact that it is an equilibrium model, and does not place a value on choices that management may face that allows strategic flexibility – the theory behind the DCF model relies on *now or never* type analyses. For example, the choice to delay investment while investing in more information, the choice to mothball a plant temporarily if market conditions militate against continued production (for example where keeping the facilities in operation incurs high fixed costs), scaling and timing investments, etc. all have strategic value that are typically not accounted for in DCF. For example, to bring the analogy within the sphere inhabited by Baguio Power Company, it would have meant the option that BPC had in designing and engineering the plant to equip it with multiple fuel capabilities (coal, oil or gas) rather than the cheapest one, coal. This would mean that BPC could switch between alternative input fuels at any time when comparative fuel prices make it desirable, rather than being locked in to one technology only. This obviously changes the production function of BPC significantly, and this potential flexibility is seen to have a value beyond a proposal's static NPV. These additional strategic choices available to companies are viewed as real options, in that they are discretionary – like financial options, companies have the discretion to exploit the flexibility options but are under no obligation to do so.

In essence, when viewed in a corporate capital budgeting context, the value of flexibility is added to a project's net present value to indicate what is referred to as expanded NPV, where:

$$\textit{Expanded NPV} = \textit{NPV} + \textit{PV Real Options}$$

This introduces the exciting possibility that a negative NPV project might be attractive when the value of real options is added. Although it is qualified, the view most widely held about this relationship is that DCF methodology undervalues investment opportunities systematically. Estimating the value of real options associated with a particular investment opportunity has drawn variously on options pricing theory and on renewed interest in decision tree analyses and similar lattice-type decision-making frameworks (Dixit and Pindyck, 1994; Trigeorgis, 2000). This is an extremely exciting field, and has developed rapidly.

When considered in the context of valuing a business, the opportunities embedded in a company's strategic choices have also generally become referred to as growth options. In essence this is a parallel, at company level, of the real options that a company may have in its capital budgeting decisions. At the business level, this requires a restatement of the expanded NPV expression, without any change in principle to the underlying logic. In changed form we simply have:

PV of the Business = PV of Current Operations + PV Growth Options

Valuing growth options at company level is a more complex concept, and is far less developed that valuing individual real options, for which there is a large theoretical literature and practical reservoir of experience. Nevertheless, it has to be pointed out that the value of a business's growth options is real – and is reflected in a company's share price through mechanisms such as expected earnings growth, changes in a company's market share, and such competitive environmental, operating and managerial circumstances.

While explicit consideration of a business's growth options has brought about interesting debates in academic circles and amongst business valuation professionals, and ROA has contributed much to invigorate the practice of capital budgeting decisions, these are specialist fields we are not able to pursue because its application is somewhat curtailed by the project finance model. Why? Because the project finance model explicitly aims to remove most discretionary flexibility from the realm of managerial decision-making – the project finance venture is supposed to be asset-specific and is intended to have a narrowly defined scope of operation which is likely to be controlled by covenants in the governing contracts. Managerial flexibility and discretionary strategic actions are not luxuries that project finance executives are afforded, although there may be some flexibility options at the project promoters' corporate level before an irreversible commitment is made to proceed with physical development of the facility – for example delay development of the project, or develop facilities for companies such as BPC that could function on multiple fuels, like gas or oil. Further, to speak of growth options at the project finance venture level is also not a concern, because the conservatism inherent in the model will not attract a particularly high proportion of debt in its capital structure if much of the venture's value is assumed to be in growth options – debt financiers are interested in cash to service debt, and not in growth options that may or may not come about.

Key concepts

The following concepts are considered sufficiently important to memorise as key vocabulary for use in subsequent chapters.

Capital leases, finance leases
Cash available for debt service
Debt capacity
Earnings growth
Enterprise value
Expanded NPV
Explicit period forecast
Free cash flow
Loan life coverage ratio
Ongoing value forecast
Operating leases
Project life coverage ratio
Real options
Reinvestment
Weighted average cost of capital

4

Managing Risk in Project Finance Transactions

No business venture operates without a cloak of risk. Because of their nature, many commercial projects have brought risk management to an advanced state, despite the complexity inherent in the transactions that they comprise, mainly because the relatively high asset specificity of project-related activity facilitates the identification of risks to a more developed, intense and usable degree than with conventional business enterprises. There remains some form of residual risk in all such schemes, despite the best efforts of planners or financiers to create project companies for which risk and risk mitigation are closely aligned. For example, supply and output risks may be institutionally matched through contracts, and price risks managed through long-term supply contracts or through the commodities or financial markets. In spite of such arrangements, or extensive efforts to mitigate residual risks through allocated or unallocated risk–benefit sharing, some risk will inevitably remain.

Risk management is inevitably imprecise, however sophisticated it may appear in its use of quantitative methods and modelling, always being constrained by bounded rationality, information and market imperfections. While all projects experience problems with residual risk that thus goes unmanaged, the concern of project sponsors is usually a matter of how material is the result of potential unforeseen risks, to what extent they may derail or make derelict the project, or how they may undermine the spirit of agreements between parties, or indeed interfere with the successful delivery or commercial operation of the venture. The very nature of what constitutes risk may be seen to be central to this bounded rationality, for what was not considered risky a decade ago may now loom as a major source of risk and uncertainty. As one example, consider the potential cost impact of climate change, and what might happen to the many billions of dollars invested throughout the world in fixed irreversible port facilities if the sea level were to rise materially.

The nature and origins of risks common to contemporary project finance ventures are not considered as mysterious, despite the fact that every project will exhibit some unique risks, if only for the reason that each project has a different business model, geographical location, and feasible outcomes. Broad typical categories of risks in project ventures have been identified and described in many texts. We are thus concerned in this chapter with the character of different approaches to managing common financial risks associated with project ventures, rather than extensive lists of risks that may or may not be encountered in any single scheme. Our general concern is also with what may be described simply as financial risk, which we take to include any risk that might threaten a project company's ability to service and repay its financial obligations, but more broadly represents any risk that might lead to a significant deterioration in a project company's financial condition. In turn, that may be distilled to a focus on the effect of risks on the present value of cash flows emanating from the project, or the extent to which all such risks may disturb the company's projections of such cash flows. Our aim is to present a framework to foster an understanding of when and how commonly encountered risks arise in the project finance model, and identify the techniques of risk management that seem to have been used successfully to manage those risks. We must emphasise that every project will exhibit unique risks, and that we are advocating comprehension of common risks, rather than a checklist of project venture risks to be applied indiscriminately to any single project.

Effective risk management in business generally involves the concept of appropriate risk allocation or efficient risk-bearing. This has come to mean transferring or distributing identifiable risk through some form of institutional mechanism, or to those counterparties or agents that are best able to bear those particular risks. These may include insurers and specialist contractors, or markets in financial or commodity derivative contracts. We generally accept the view which suggests that risk management involves the reduction or mitigation of risk, or creating marginally more certain outcomes to events that are subject to risks. This approach also has a broader meaning, and includes the deliberate decision to bear (retain) selected risks, or acquire exposure to selected risks, as explained by Merton and in particular Mason in Crane *et al.* (1995). In addition to accepting selected risks and allocating risks to parties best able or most willing to assume them, Mason's approach simplifies all risk management techniques into four types, namely risk avoidance, retaining or absorbing risk, loss prevention (or control), and risk transfer. Risk transfer implies one of three generic approaches, which are hedging, diversification and insurance. This characterisation should not be confused with the managerial activity which has commonly become referred to as a risk management process as for example described in Brealey *et al.* (2006). The concept of a risk management process relies on a systematic identification and assessment of risks in a specific activity or venture, the selection and implementation of risk management techniques, and a review of ensuing outcomes. Normatively, as a particular managerial activity, we consider a systematic risk management process a widely accepted approach to managing risk in any venture, generally advocated as a preferred response to perceived concerns that face an enterprise in such a way as to disturb its projected performance. It will be clear from this description that risks are associated with potential deviations from predicted or contractual performance, and are not necessarily harmful to any enterprise.

Three general principles support the structure of this chapter. The first principle relates to project risks and the stages in the project life cycle introduced in Chapter 1. No project can generate revenue until it is constructed and commissioned (even if those stages are reached in a way that is less than successful), so development and construction may be viewed as the most risky project phases in technical and technology terms. It is fundamental to understand that the nature and magnitude of risks change with the stages of a project, as do the feasible approaches to the management of risk. The second principle borrows certain concepts in systems theory, first introduced in Chapter 1 and elaborated here, to give order to categorising project venture risks. Thirdly, we borrow a principle that has permeated financial literature for some three decades and found its way into general risk management vocabulary, namely successful risk management requires risks to be allocated to those parties of a transaction, insurers or other specialist risk takers, that in each case are best able and most willing to bear those risks. This is generally referred to as efficient risk bearing. We follow this principle to cover conceptually the fundamentals of risk management, with emphasis on the nature of techniques used to manage risks, rather than any method's technical details. Our point of departure in Sections 4.2 and 4.3 is the risk position of the *project owners*, which includes sponsors that have capital at risk in the project, and consider *project lenders* separately in Section 4.6. In any single project, the nature of associated risks matters, each stage in project life-cycle matters, and which party assumes and bears these risks and how they do so also matters.

4.1 The project cycle revisited

As stated, the first principle used to order thinking about risk relates to the stages in the project lifecycle. As explained in Chapter 1, the life-cycle of project companies could broadly be summarised into five phases, the first three of which are loosely grouped as the development phase.

1. Planning, design and engineering ⎫
2. Construction, or procurement ⎬ Development phase
3. Commissioning ⎭
4. Operation
5. Decommissioning.

Each phase carries specific classes of risk, which have implications for the project finance model. Issues of planning, design and engineering are inherent in the feasibility of each phase of proposed projects, from economic viability to operational and technological feasibility. It is clearly possible that design and technology decisions in these phases may induce operational failures, so one aspect of managing risks during

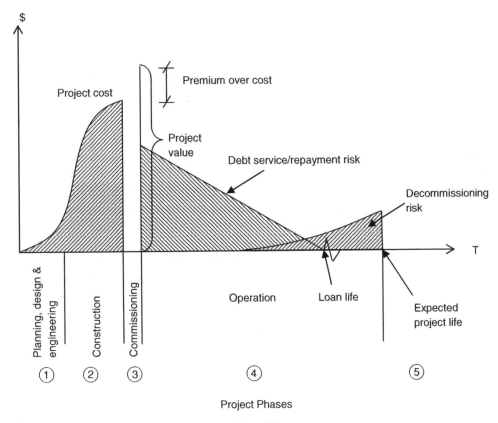

Figure 4.1   Risk phases and project capital at risk.

planning will be to make proven choices in design and technology. Different risks then become evident as the project proceeds from planning to construction, for example the possibility that construction costs exceed a budgeted ceiling, that construction suffers delays or that aspects of the work are of inadequate quality. Each concern has the potential to undermine a project's feasibility. No project can deliver net operating revenue until it is successfully constructed and commissioned, so some sponsors and equity-holders will view the construction phase as the most risky of all phases in the project. We may consider this in terms of capital at risk, with sunk resources accumulating in the project before its successful commissioning. Figure 4.1 illustrates how a project's risk profile changes during its life-cycle. After successful commissioning, project risks change emphasis from developmental to commercial and financial, not least because leverage may be high early in the operational phase. The extent to which total capital at risk is reduced depends on the value created in the project company, given the concerns about irreversibility and asset specificity discussed in Chapter 3.

From Figure 4.1 the logic of two fundamental risk management strategies in the creation of project companies emerges. Given that high leverage is common to project companies, debt financiers consider development (particularly physical construction) and operational risks separately and independently, and thus separate construction finance and permanent finance into at least two sequential transactions to address different needs and risks. It cannot be stated too often that there is no operating project company prior to a successfully completed development phase. If development fails, no operational project company exists at all, with enormous consequences for all those with capital at risk. While our concern is with what we described above simply as financial risk, it becomes clear that the nature of such risks differs in successive phases. Operational risks are any and all risks that may cause the project company financial distress during operations. Recall from Chapter 3 that a narrow definition of financial risk suggests that there is no financial risk without debt in the company's capital structure, but here we consider financial risk far more broadly. However, we remain mindful of the project lenders' position first, and so emphasise those risks facing the operation of project companies that threaten their ability to generate sufficient cash flow to service and repay contracted debt. During the operational phase the project company may face risks from changes in its situational environment, or institutional risks, which we will return to in Section 4.5. The construction phase clearly presents the risk that there may not be a project facility at all, which in the final stages of construction typically represents a large fixed and irreversible capital investment, and the risk that this invested capital may be irrecoverable in the event of non-performance or non-delivery by the construction company. Such risks may also introduce risks of contract enforcement, that is, institutional risks, the subject of Section 4.5.

Over the last two decades decommissioning of facilities has become a matter of substantial concern, even more so as concerns regarding the reality of the planet's environmental state have moved from society's fringes to mainstream social, political and economic agendas in many countries. We would hold that decommissioning risks are most productively considered during planning and design, perhaps decades prior to being adopted, by choosing technologies and processes that prevent contamination and pollution that will be costly to remedy, especially given imperfect information about future impact risks. We do not address decommissioning risks here because our point of departure is that they most critically represent a matter of corporate values and the national or transnational legal setting. Societies are now entering a new era where accountability for environmental impact is expected. It is likely that more risks stand to be identified as the world becomes more aware of the causes of pollution and contamination; think of the case of asbestos insulation, which was in widespread use as recently as thirty years ago.[1] Because these are complex issues, we intend to view

---

[1] The discovery that certain types of asbestos fibre were a hazard to health prompted remedial capital spending in private and public buildings and infrastructural schemes, and led to the collapse of a number of major manufacturers that proved unable to meet the costs of litigation by individuals and governments.

pollution and decommissioning more philosophically in the context of a discussion of institutions in Section 4.5.

4.2   Risk management approaches

It may be appropriate to open discussion of risk management approaches by deciding first what constitutes risk. Two points need to be made in this context: first, the difference between risk and uncertainty, and second, the concept of risk aversion. To commence, few future outcomes in business can be viewed as certain. Indeed, most future events are uncertain in any economy, whether advanced or developing. Following finance convention, we conceptualise risk as that form of uncertainty that is potentially costly, while uncertainty that is not costly does not represent risk. In this manner, there are potentially risky outcomes to many activities in business in general, and so of course in project ventures. The volume of road traffic travelling through a toll tunnel, for example, is uncertain at any time but observations as to traffic numbers may have a stable distribution. If volumes drop because motorists have the option of a new alternative route, the reduced tunnel usage may also settle at a new stable distribution, but it may be risky because project revenue may be insufficient to cover project debt service requirements, and trigger loan agreement default covenants. There is clearly a demand risk in this circumstance. On the other hand, excessive traffic volume may exceed tunnel capacity, lead to congestion, negative user action and a reduction in future volume. Similarly, if a production facility produces above-expected output to meet higher-than-expected demand, it may trigger supply chain disruptions or facility breakdowns, and output quality may suffer. This illustrates that risk is not only a downside phenomenon as is sometimes suggested, but that any deviations from the expected could be costly, and thus risky (Box 4.1).

Box 4.1   Risk and uncertainty.

We distinguish this approach to risk and uncertainty from one that has traditions in neoclassical economics, and which has given a conceptual basis to much of the day-to-day risk management assessments undertaken by banks in providing capital to support their risk generating activities, in accordance with accepted international practice for bank regulation. In the early 1920s, conservative US economist Frank Knight wrote of risk as being associated with hazards or events that were predictable, and uncertainty as those that were unpredictable, or to which no reasonable probability of occurrence could be attached. Modern examples might include the major tsunami that follows some earthquakes, but not all, or the contagion that has been known to follow some financial crises, but not others. Banks are encouraged by regulators to undertake continuing value-at-risk (VaR) assessments of the extent to which their aggregate risk positions – trading books, loans, and all capital at risk – may be affected by changes in key external variables, such as interest or exchange rates, and calculate how much capital would be wiped out by a damaging movement in such influences. The result is a level of capital that the bank is said to be capable of losing in any single period without becoming insolvent: how much of a write-off can the bank stand, and how exposed is its risk portfolio to events over which the bank has imperfect or no control? Knight's concept illustrates the failing in this approach, because however sophisticated are the bank's financial models, they cannot attach a probability to the unpredictable taking place, and it is the unpredictable event that often poses the greatest real risk to the bank or the financial system. This Knightian distinction was given a wider than usual audience by then US Defence Secretary Donald Rumsfeld in February 2002, in stating at a press conference that:

> '*As we know, there are known knowns. There are things we know we know. We also know there are known unknowns. That is to say, we know there are some things we do not know. But there are also unknown unknowns; the ones we don't know we don't know.*'

VaR technology is widely used by major project companies to model the project's net worth.

The second relevant point concerns risk aversion, which may be viewed as willingness to pay in order to reduce exposure to risk. This immediately emphasises the fact that reducing exposure to risk cannot be costless, and is thus best described as a cost–benefit or risk–reward trade-off decision. The cost of reducing risks may be significantly reduced in well-developed markets, but by their nature markets will always be incomplete. So no risks can be managed by market instruments alone, and many must be managed or mitigated by private instruments and contracts. More will be said about market incompleteness in Section 4.5.

Overall, we follow the conventional characterisation of the risk management process as a systematic attempt to analyse and deal with risk, comprising five steps:

1. risk identification
2. risk analysis and assessment
3. selection of risk management techniques
4. implementation
5. review.

The process is iterative, not a single calculation, with frequency decided largely by the degree of risk aversion of the user, the influence of any industry regulator, and the nature of the risks concerned. This is a generic systematic approach that could form a managerial policy response to any problem faced by an organisation. The last two steps address the obvious actions of implementation and reflection as a control mechanism, and we will not deal with these. For the purpose of this chapter risk identification and risk management techniques are the most relevant, while we give only limited attention to risk assessment.

4.2.1   Risk identification, and risk analysis and assessment

Identification of the risks facing a project company would very likely be the subject of a structured risk identification and analysis process as early as possible in the project's planning phase, preferably during a feasibility study, and certainly in planning and design. While the risk management approach described above is labelled systematic, risk identification itself is better described as requiring a systemic approach. It is concerned with identifying the most important risks facing the project company in its entirety, as well as their interrelationships. It makes little sense to consider demand risk independently from price risks if price elasticity is high and more efficient producers threaten to enter the market. Similarly it is nonsensical to consider exchange rate uncertainty separately from earnings risks if major markets for project output are open to international competition. Further, it would be foolish to ignore the influence of interest rate uncertainty on earnings if the project company's debt is primarily floating rate. Similarly production risks, supply chain and demand uncertainties have to be evaluated together with their interrelationships. In general it is to be expected that detailed knowledge of the particular industry, its structure and economics, the project and its likely competitiveness would be critical to identify risks (Bodie and Merton, 1998).

As one method to facilitate risk identification, we draw on systems theory concepts introduced in Chapter 1 to support and structure our approach. We are particularly interested in exploring the notion of systems boundaries and the systems environment as useful mechanisms to assist in risk identification. In turn, these concepts draw on situational complexity ideas introduced in Chapter 2.

System boundaries and the system's environment

In Chapter 1 we introduced a number of systems theory principles, but only alluded to the critical importance of systems boundaries and a system's environment. A system's boundary is intended to indicate where the system or subsystem under analysis ends, and where the subsystem or system's environment begins. It allows us to conceive of and understand the different nature of internal project risks and risks that arise from the project's environment, and why they rely on different risk management approaches. In the systems approach, the project boundary becomes a critical concern, for it is here that risk originates and risk identification can be critically focussed. There are generally comparable business risks in project ventures in all industries, hence the checklists that are often seen in project finance literature. Systems concepts allow us to place checklists to identify and manage risks in perspective – it is not only a matter of which particular risks out of a list may be present in a particular project, it is more about identifying an approach that will allow risks to be comprehended and thus anticipated (Nevitt and Fabozzi, 2000). Understanding and conceiving of systems boundaries is critical in any attempt to understand situational complexity, and identify how, why and where risks originate, and to what extent they are manageable.

Our interest in a business system's environment springs from the fact that we may conceive of most risks facing a project company as originating from the various levels of its environment as an open system. While this may seem self-evident, the usefulness of this concept becomes more clear if we differentiate a business's environment into what may be described as the *internal environment*, possibly containing the production process itself and the normal harder functions such as accounting and administration; the *operating environment*, concerned with the business in its competitive environment; and the *institutional environment*, the regulatory bodies, laws, norms, or customs within the society in which the business is located. This characterisation of a business's environment is further useful, because each of these indicates an increase in the level of complexity in the environment, as well as a decreasing ability to exert influence over the environment – that is, a lessened ability to manage the risks facing the business. Each level is also associated with a higher noise-to-signal ratio – less certainty about information. For example, typically a well-managed business would have good information about its internal environment and possibly less specific information about competitors and other factors in the operating environment. Of necessity there are some overlaps in this categorisation; these are important and we will return to them in the following subsection. With project companies in underdeveloped institutional environments, instability and uncertainties emanating from this level are possibly most difficult to manage effectively. We devote Section 4.5 to the institutional environment.

At some stage in analysing a project company's environment, it becomes clear which risks may adversely affect the project's operation and profitability, and it also becomes clear which risks can be managed or mitigated and which cannot. Finding the boundary (or boundaries) between the project and its various categories of environment is crucial to understanding where risks originate and how they may be approached from a managerial perspective.

We have said very little about risk analysis or assessment. This may be viewed as the quantification of costs associated with risks that have been identified. It may be unfair, but in our view risk identification is a lesser developed activity than risk assessment, and deserves more attention. Risk analysis and assessment has become increasingly sophisticated, particularly with widespread availability of simulation software, to the unfortunate neglect of careful attention to identification of risks. In essence, however, risk assessment continues to depend critically on relevant risks being identified first and the probability of their occurrence estimated (if these risks are not already priced in markets, such as commodity price risks), their interrelationships identified and the severity of their impact on the venture estimated, should the particular risky events materialise. In all, risk assessment aims to differentiate those risks and events that are potentially distressful or calamitous to the venture and its finances from those risks that may have a high probability of occurrence but a low potential to cause financial or other distress to the project company. The principle is that potentially distressful risks and events would be prioritised for risk-management attention. A quick primer on risk analysis and risk assessment is presented in Box 4.2, and see also Section 5.9 in Chapter 5.

Box 4.2   Risk analysis/risk assessment.

The objective of risk assessment is to estimate the financial impact of the various identified risks on the project company. Typically there will be risk identification workshops for detailed aspects of the project, for example the construction phase, the operational phase, decommissioning, and so on. Such structured risk identification sessions will typically be attended by multidisciplinary teams of experienced project planners and executives, design and construction professionals, commissioning professionals, and so forth – in essence a group of specialists in a *Delphi-study* tradition. Risk analysis/assessment typically follows after risks have been identified. Again we propose that the nature of risk assessment/analysis in principle follows different frameworks for operational and development phases, simply because development/construction risks and operational risks are in principle sequential – if the development and commissioning phases are completed successfully and the facility is sound, operational risks are independent of earlier project phases.

Assessing/analysing risks in the project company's operational phase typically follows modelling procedures such as sensitivity analyses and/or scenario analyses, or more complex simulation procedures such as VaR. In sensitivity analyses project value outcomes and debt service/repayment capacity are modelled following a subjectively determined expected range of values for single variables considered to be risky, such as interest rates, exchange rates, input prices, and the like. Generally a minimum sensitivity analysis would consider three values for variables: optimistic, pessimistic and expected. In scenario analyses, recognition is given to conventional wisdom that changes in several variables may occur simultaneously (their movements are often correlated); for example, high input prices, interest rates and exchange rate movements may occur simultaneously, thus leading to periods of financial stress. Several sets of subjectively estimated combinations of variables are then tested and the outcomes analysed, again typically testing optimistic, pessimistic and expected scenarios. While sensitivity and scenario analyses are technically based on simulation methodology, the simulation is often taken to mean a more sophisticated simulation where all risky variables are varied randomly simultaneously taking into account interdependencies between variables, with a project value distribution produced for analysis and interpretation.

Over the last decade risk subjective analysis/assessment of the construction phase of projects has gained acceptance, and much learning has occurred in practice. Following complex early attempts at full simulation of construction programs using operations research methods such as program evaluation and review technique (PERT), design and construction teams have developed more simplified risk analysis/assessment methods based on group workshop-based subjective assessment of risks identified in the construction program. Following Reiss (1996), we may present such risk assessment attempts as a simple process structured on the basis of risks identified in the construction program, their probability of occurrence, potential financial impact and hazard rating. On a scale of one to ten, an identified risk's likelihood is estimated, and its impact is also estimated on a scale of one to ten (scientific facts may be drawn on, but typically subjective probabilities are utilised first to arrive at early indications of potentially damaging events). The hazard rating is simply the product of the risk's probability and its potential impact, thus on a scale with maximum value of 100. Risks with potentially high hazard ratings are then investigated in more depth, and correlations with other risks are also investigated. From Box 4.5 we know the overall importance of correlated risks – if risks are highly correlated and highly probable, there is cause for serious attention. Suppose we are investigating the construction phase of a toll road tunnel project under a busy harbour, where the tunnel is constructed from standardised pre-cast concrete sections towed into the harbour from a remote casting basin, to be aligned and submerged beneath a busy shipping lane without interfering with harbour traffic. Box 4.2 Table 1 demonstrates how the outcome of an initial risk assessment exercise may be represented.

Box 4.2 Table 1 Outcome of an initial risk assessment exercise.

Risk	Probability	Impact	Hazard factor
Harbour traffic accidents to tugs moving tunnel sections	2.5	5	12.5
Time delays with aligning tunnel sections	4.0	5	20.0
Extended program interruptions from rough seas	0.7	7	4.9

And, from a correlation matrix of risks, we may identify that item 2 and item 3 are highly correlated, which increases their potential to cause project delays and thus potentially financial stress. Clearly these two risks require special risk management attention.

4.2.2   Selection of risk management techniques

Managing risk in business generally depends on four techniques, namely risk avoidance, retaining or absorbing the risk, loss prevention (or control), and risk transfer.

Risk avoidance and retaining or absorbing risk are essentially opposite actions. Risk avoidance may follow a conscious corporate decision to avoid a particular business activity because it is considered too risky – a design and construct company may choose not to bid for a contract to design a process engineering project that might employ a perniciously dangerous chemical. On the other hand, risk retention represents a conscious decision to bear the risk, and bidding to design and construct the facility. How it manages the risks it then accepts becomes the subject of two mechanisms, loss prevention and control, and risk transfer. Most prevention and control might simply be regarded as good management, but it first requires potential losses to be guarded against or controlled to be identified, and emphasises the importance of risk identification. To retailers, simple preventable losses may include those resulting from theft, fire, or loss of goods in transit. Management might elect to place operational measures in place to prevent such losses, such as improved supply chain management or physical warehouse security measures. It may also adopt the fourth general risk management technique, which is risk transfer (Mason in Crane *et al.*, 1995).

Risk transfer

Risk transfer suggests a very literal concept, that is finding another party to take on a risk faced by a business, and a naive interpretation might even think that this could be achieved without cost. Risk transfer often involves complex contracting, and is thus expected to be expensive, as one would expect with any form of risk management.

In general, we consider risk transfer to have four dimensions. The first is to contract with a counterparty that will assume the specified risk, thus relying implicitly on that party to manage the risk. Efficient risk bearing would require the counterparty to be better placed to manage the risk than the party facing it. A good example of this would be for a project company to contract with a design and engineering contractor to deliver a project facility on time, within budget and to the required quality and performance specification. After all, this is its business. This transfers all delivery risk for the facility to the contractor company.

After acknowledging the risk transfer function of private contracts, we follow the view of Bodie and Merton (1998), who identify three further mechanisms to accomplish the transfer of risk, namely to hedge, insure or diversify. It is with these three that financial markets play the biggest role. As we shall describe, together with private contracts to manage risks, financial markets provide the opportunity to manage many commodity price risks and financial risks faced by project companies. We outline briefly hedging, insurance and diversification as risk transfer mechanisms, and point out in principle differences in the outcomes from using these three approaches.

Hedging

In financial and commodities markets, hedging is taken to mean acting so as to alter or mitigate a risk exposure, usually by dealing in some form of financial instrument. However, by managing or mitigating losses, hedging usually reduces potential opportunity to gain from risks. Forward contracts are commonly used in this respect, and are contracts for the sale of a commodity or financial instrument at a price agreed today (spot) for future delivery, at a time that is usually a multiple of months ahead. This can help manage and mitigate both quantity and price risks. A forward sale of project output can reduce exposure to risks of price declines, but also eliminates the potential benefit from any price increases that might emerge after the forward transaction is agreed. Thus, while purchasing forward mitigates the risks of input price increases, it also may eliminate the gains associated with possible future price declines. Forward sale and purchase transactions can be arranged for almost anything using purpose designed contracts available through exchanges or arranged as over-the-counter (OTC) transactions by financial intermediaries such as investment banks. Standardised forward contracts available through exchanges are known as futures. Box 4.3 outlines briefly the nature of OTC transactions and exchange-traded contracts.

Box 4.3 Over-the-counter and exchange-traded contracts.

Financial instruments for which there exist organised markets are traded in one of two settings, between parties over-the-counter (OTC) or through an exchange, which implies standardised terms and centralised clearing, settlement and price reporting. Each setting has a long history, and now accommodates trading in claims such as loans and credit risks that were long thought unsuitable for transfer of the kind that trading implies.

The main difference for traders is seen in counterparty risk. OTC markets involve direct contracting while most exchanges act as the trader's settlement counterparty. Exchanges typically limit counterparty risk by requiring traders to deposit centrally a cash or equivalent margin, based on the prevailing value of the traded asset. OTC counterparties assess each other's capacity to complete the contract, especially if the claim is long in duration, as with many swaps.

Markets that are deep and liquid tend to deal in standardised contracts, regardless of the setting. Moreover, in a technologically adept world no market requires a physical location, even though trading in commodities or securities may involve physical settlement. A recent trend is for derivative instruments requiring delivery of goods or securities to be settled in cash, which helps limit market manipulation or price distortion.

Major OTC markets include global dealing in foreign exchange for cash or forward settlement, trading in loans, international bonds and most swaps – a substantial part of the markets in financial derivatives. The most important exchanges trade in equities, base metals, commodities, insurance and ship chartering but OTC markets co-exist in many of these instruments. All futures and most elemental option contracts are exchange-traded, futures being a standardised form of forward contract.

Exchanges typically regulate the markets they host, often under some form of devolved statutory authority but also subject to governmental agency monitoring and over-sight. Both OTC markets and exchanges can be sophisticated and transparent in how they allow price discovery to take place, but OTC markets for illiquid instruments may be opaque to all but a handful of parties.

It is possible to construct limited price hedges using spot markets, for example, by short selling. This involves borrowing a traded commodity today, selling it in the cash spot market for immediate delivery, and simultaneously buying the same amount forward for settlement at some specified future date, often to match the period of borrowing. This locks in the gross revenue from using today's sale price, and protects against a subsequent price fall, but is not costless – the borrowing costs needs to be taken into account in the trade – and the market may move in such a way before the trade is unwound to render it even more costly. The purpose of these kinds of transactions is not necessarily to lower costs or increase revenue, even though this may be pleasing to the financial controller, but to increase the degree of price certainty in the sale of output. Other more complex and widely used hedging instruments include forward and futures contracts, financial options and swaps. The principal characteristics of swaps – which are ubiquitous in long-term project finance – are shown in Box 4.4.

Diversification

Similar to hedging as a risk transfer mechanism, diversification is best characterised by the outcomes it facilitates. The principles which make diversification such an important finance concept are briefly explained in Box 4.5. Diversification similarly applies to risk management. It is an important consideration in the logic employed by insurance companies in managing their liabilities (the policies they sell), as well as with many other financial services. Diversification as a risk transfer mechanism reaches far beyond the management of the insurance industry, the funds management industry and the use of financial market instruments. It influences both the asset and liability side of any business's balance sheet, as well as the supply and demand side of any business. For example, despite the often emphasised attractions of relationship banking, diversification emphasises the desirability of diversifying

Box 4.4   What is a swap?

Swaps are a generic way of identifying contracts for an exchange of obligations between two counterparties, separately or in more complex transactions. Payment on each leg of a swap is contingent upon the other. Beginning in the late 1970s, interest rate and currency swaps[2] were first negotiated case-by-case, but now include many types of obligations and are subject to standardisation of terms and market practice devised by the International Swaps and Derivatives Association (ISDA). Almost all swaps are dealt directly between parties over-the-counter.

Swaps are a means by which asset or liability managers may modify exposure to market risk and return. For example, if a borrower is offered favourable terms for a loan, but prefers to make interest payments in another currency it can contract to exchange interest liabilities over the life of the borrowing, and so pay its preferred costs.

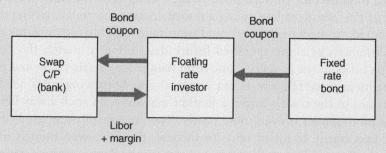

Box 4.4 Figure 1   Swap diagram.

Box 4.4 Figure 1 shows an interest rate swap entered by an investor hoping to effectively alter the coupon paid on a bond. The bank acting as counterparty receives the bond's fixed rate coupon, and returns to the investor a payment based on a floating rate (Libor) preferred by the investor. The transformation made with the swap can be more complex, for example involving changes in currency, timing of payments, or making payments contingent on defined events. There is usually no exchange of principal but only periodic payments, although zero coupon swaps involve payment on one leg only at termination.

Banks dominate the swap markets but investors, borrowers, hedge funds and government organisations all use swaps for reasons associated with portfolio management; that is, to alter marginally the balance of risk and return in an array of assets or liabilities, and bring its expected performance closer to an objective. Any single transaction can be risk-averse or speculative. This applies to the case of a major borrower, that trades frequently in swaps to change its

---

[2] Distinguished from foreign exchange swaps, which is an older-established term for transactions involving simultaneous spot and forward sales and purchases of currency in like amounts.

aggregate interest, foreign exchange or credit risk liabilities, or an investor that decides that its interest receipts no longer match its beliefs about future rate movements, and wishes to collect instead a floating rate linked to an index such as Libor for one or other major currency.

Swaps can be seen as an exchange of liabilities tied to other instruments such as loans or bonds, which in the markets of the 1980s reflected the terms available to borrowers in clearly segmented markets. Thus the pricing of the swap would be taken (derived) from prevailing bond yields or currency rates. Since the mid-1990s, swap markets in many risks have grown such that swap pricing is often more liquid and continuous than underlying markets in securities or risks, so that the terms of a new bond issue will be based mainly on the value of a standardised interest rate swap.

Box 4.5   Portfolio theory.

Portfolio theory explains how the pooling of a large number of different assets into a portfolio allows the emergence of portfolio risk characteristics that will excite any systems theorist; portfolio risk could be lower than the lowest risk of any of the elements of the portfolio, given appropriate underlying conditions and objectives. With risk pooling, the most fundamental precondition that facilitates risk reduction is the same as with portfolio theory, namely that the various risks that are pooled should not be correlated or should have low correlations – thus, they should be diverse. In a diversified share portfolio, for example, combination of a relatively small number of assets (25–30) can eliminate a very large proportion of return/risk uncertainty, and as more assets are added the theoretical limit to reducing risk, 'market risk', is approached. Harry Markowitz won the Nobel Prize for formalising the mathematics of portfolio theory. Milton Friedman famously said that Markowitz's PhD wasn't economics but that he wasn't sure what it was. Portfolio theory neatly confirms the old adage – *'never put all your eggs in one basket'*. An important extension of portfolio theory is the development of the capital asset pricing model (not entirely without controversy), which posits that if rational investors follow the logic of portfolio theory in their investment activities, it has important consequences for the required rate of return on individual investments, and hence provides a mechanism to estimate the required rate of return on individual investments (see Box 3.2).

Modern portfolio theory similarly applies to most fields in risk management, including pension and other fund management activities. It is also an important consideration in the logic employed by insurance companies in managing their liabilities (the policies they sell), as well as with many other financial services.

sources of corporate debt (different banks) and types of debt (bonds versus loans), and similarly further emphasises the desirability of diversifying suppliers to the entity, as well as customers to which it sells. In all cases it reduces the magnitude of risk associated with one large single exposure – irrespective of whether this is an asset or a liability.

Insurance

We follow the characterisation of Mason in Crane *et al.* (1995) of insurance as a precise approach to risk management with the most common forms of insurance being actuarial insurance, guarantees and options. In principle, when a project company insures against some risk, it engages in loss control but retains the possibility of benefiting from upside risk. In exchange for incurring a certain and defined loss (the insurance premium), it contracts with the insurer to receive a defined (certain) amount, dependent upon the occurrence of a defined contingent event (for example, the destruction of a facility in a fire). In essence, one incurs a known loss (the premium) to mitigate a larger loss (the event). In order to bond the insured party to continue to be vigilant in preventing the insured event from occurring (a moral hazard risk), the loss is typically not covered in full (the difference is commonly known as the excess). Most insurance companies function according to the principle of actuarial insurance, which draws in large measure on risk pooling and diversification. Box 4.6 gives some insight into actuarial insurance. The two major sectors of the insurance industry, life and property-casualty (PC) insurance companies, sell many millions of policies worldwide, and cover many classes of risks. For an indicative list of risks covered by PC insurers, where our principal interests lie, see Box 4.7. We are, however, concerned with risk transfer mechanisms that fulfil the *function* of insurance, and not only insurance as it is commonly understood.

When considering two further mechanisms that function as forms of insurance, mainly options and guarantees, the fundamental difference between insuring and hedging becomes clear. Hedging manages and mitigates the risk of loss, but it also eliminates the potential for gain. With insurance, however, the risk of loss is managed or mitigated without eliminating the potential for gain. For example, by purchasing a commodity put option, a project company can eliminate risk of sale price decreases, without sacrificing the possible benefit of price increases. The same project company can also eliminate the risk of input price increases by buying a call option on the input commodity, which does not eliminate the potential gain from input price declines. In both cases the option price is a sure loss (the insurance premium). If the company managed the same risks by executing a hedge through a forward contract, it would not be in a position to benefit from price movements that may be in its favour. Box 4.8 provides a brief primer on options and forwards.

Guarantees often similarly fulfil an insurance function in managing losses that might occur from counterparty performance risks when financial performance or physical delivery of some asset, product or service is required (a typical example is,

Box 4.6   Inside insurance.

There are three basic types of insurance model, actuarial, event and catastrophe. Insurance is a hedge against risk, the assumption of risk by the insurer, and thus the transfer of risk from the insured to the insurer. Risk is usually taken as the chance of loss or damage, but more formally in a project finance setting will imply the possibility of a project outcome that varies from the central predicted or favoured outcome. No aspect of risk is certain: scale, occurrence or frequency, so risks tend to be appraised by the probability of their taking place. Insurers and financiers use statistical forecasting models to estimate those probabilities, and use the result to help quantify the outcome on the undertaking to be insured. The upfront or ongoing fee or premium charged by the insurer to provide compensation against such an occurrence depends on the results of such analysis. For example, in determining premiums for car insurance, the model's input variables would include the car's age, type, place of use, and the driver's claim history, education, gender, occupation and health. This is known as an actuarial model, using a collection of variables for which data histories are known.

Infrequent and distinct natural disasters such as earthquakes, floods or severe weather and the like cannot be predicted in the same way, even though specialist insurers are increasingly able to write contracts to compensate for the effects of such catastrophes. Extreme events that rarely occur but may have serious consequences are assessed using highly iterative models that simulate the effects of any such event in many thousands of ways, informed by inputs from seismologists or meteorologists, for example. This catastrophe model has two aims: to assess the probability of any single event occurring, and to derive a range of estimates for the consequences of that event. This forms the basis for pricing insurance contracts and certain sophisticated types of financial derivatives. At the same time, most simulation models will assume that the most unlikely scenarios, such as the simultaneous occurrence of two or more unlikely shocks, are so unlikely as not to warrant hedging, and may be a problem for both the insurer and insured. One last type of insurance is to hedge against a particular event, for example war, insurrection or revolution. Such risks cannot be forecast by statistical or scientific models. Often they are insured by governments to promote policy objectives. This is the event model.

of course, a project facility). Guarantees are possibly most often encountered in managing credit risks of all kinds. In consumer credit, the personal guarantee of a close relative is often necessary before a first loan is made to young market entrants with no credit history. Although the aim of project finance is often to structure a non-recourse loan to a project company, the shareholder-promoters of the project company

Box 4.7 Property and casualty risks commonly covered by insurance companies.

Aviation insurance	Earthquake insurance	Life insurance
Agricultural all-risks insurance	Fidelity insurance	Maritime insurance (hull and machinery, protection and indemnity, pollution)
Burglary and theft insurance	Financial guarantees and sureties	
	Fire insurance	
Business interruption insurance	Group accident and health insurance	Mortgage protection insurance
Commercial rent insurance	Goods in transit insurance	Motor vehicle insurance
Commercial all-risks insurance	Household all-risks insurance	Political risk insurance
		Product liability insurance
Credit insurance	Industrial equipment insurance	Professional negligence, malpractice insurance
Crop insurance		
Disability insurance	Inland marine insurance	Travel insurance
		Workers' compensation insurance

Following Saunders (1997).

often have to insure repayment of project company debt by formally guaranteeing its obligations. Of course, this does not eliminate the lender's good fortune if the loan is repaid without default. Construction contracts often also call for a surety to guarantee completion, possibly provided by another construction company, but more typically this is achieved by a surety bond provided by the construction company's bank or insurance company (this service is provided for a fee – i.e. an insurance premium). A surety bond functions as a guarantee that sufficient funds will be made available to complete the project by a contractor of the project company's choice, in the event of non-performance by the first contractor. For insight into the nature of guarantees and sureties see Box 4.9.

The general discussion above covered a number of risk management mechanisms, each treated as a discrete concept. Possibly the most exciting (and scary, to some) phenomenon of financial markets and financial contracts is not that risk management often utilises instruments that are not part of the vocabulary of many people but that these instruments can be combined in endless ways to isolate and manage exposures of many further risks. If we combine the risk engineering abilities of banks' over-the-counter activities with risk management products traded in formal exchanges, the possibility to synthesise many more financial positions is imaginable – and does occur every day.

Box 4.8   Using forward or option contracts to manage price risks.

Markets in financial forwards and options allow users to enter many types of price hedging strategies. These may be commonly viewed as risk preferring or risk mitigating but all entail the user altering the balance of risk and return associated with the financial results of a commercial endeavour. For project sponsors, it may be valuable to seek to hedge certain risks associated with changing interest rates, perhaps to guard against an unwanted or unbudgeted rise in financing costs. At the same time, entering into such a hedging strategy will usually involve a cost, either in terms of fees paid in the same way as insurance premiums, or in foregone income. Thus in different market conditions, a project manager may choose to use forward or option contracts singly or in arrays to guard against all or part of interest rate risks. The simplest form of hedge is often based on combinations of call and put options, and might allow a borrower to create a 'cap' on its interest costs, so that the hedge payoff compensates against interest rates remaining above a certain level for the period of the hedge. The borrower might have a view as to the likely path of interest rates during any single phase of project development, so for example if it believes that the project will always be net cash flow positive if interest rates remain at a certain level, it can choose to write an option that gives it a reward if rates should fall below that level. This in turn can be used to lessen the cost of an interest rate cap, and leave the borrower certain that its interest costs will remain within a band, regardless of what happens over the life of the hedge to prevailing rates. In major project finance transactions, longer-term risks of this kind are commonly addressed with various kinds of swaps, especially those involving foreign exchange risks or interest rate risks extending over several years.

4.3   The project company and risk identification

At this stage it is opportune to demonstrate the concepts of a system's boundaries and risks in its environment with an example. When considering risks, risk management and the project company, an appreciation of the project as a *system in its environment* forms a useful way to order thinking about causes of risks – that is, risk identification. Of course, in the final analysis anything that could conceivably disrupt the cash flow and debt service ability of a highly leveraged project company is a potentially serious risk, and so much of what follows focuses on a conceptual risk identification framework, rather than specific risks. We consider separately the development phase (planning, design and engineering, and construction) from the operating phase, in keeping with the different risk characteristics of project companies

Box 4.9   Guarantees and sureties.

Insurance contracts (known colloquially as policies) require the insurer to compensate the insured when a defined event takes place in a specified form and manner. It is usually associated with an insured party wishing to be compensated or protected against losses arising in its commercial activities through some form of uncontrollable exogenous event, but can include occurrences over which the insured has some influence, for example within a manufacturing process. The effective risk is transferred from the insured to the insurer at the cost of the payment of a premium. Similarly, in the growing global market for credit derivatives, instruments such as credit default swaps protect the protection buyer in similar ways to a model insurance contract. Both instruments involve contractual compensation for loss, the main practical difference being that an insured party can be expected to hold or control the subject of the contract, which is not necessarily the case with credit derivatives, and in highly divergent legal structures and regulatory treatment. In some ways, guarantees are similar to both insurance and credit derivatives, and may be regarded by any risk-averse users as close substitutes, but separate laws usually govern guarantees and contracts of insurance, and in advanced economies different regulatory regimes typically exist for insurers and banks and specialist guarantee providers. A guarantee is a contract to meet the financial obligation of a third party. The guarantor will be liable for a payment to the beneficiary in the event that the third party defaults, and will usually obtain subrogation rights against the defaulting party in the hope that it might recover its payment, while the beneficiary of the guarantee is left protected against that loss. By contrast, insurance is not a collateral obligation to pay the defaulted obligations of a third party. There are legal differences between these instruments, especially in the rights that accrue to the insurer or guarantor after it makes payment, but it is often the case that these differences arise from custom or practice, and especially from the convenience of financial regulators.

before and after commissioning. We firstly view the project as a completed and operating facility, and thereafter we return to risks associated with construction of the project.

4.3.1   The project company in the operational phase

When explaining the concept of a project company in its systemic environment as an approach to identify risks, we introduced internal, operating and institutional environments as categories to order thinking about the environment. We now use these concepts to identify risks in a hypothetical project company's environment.

The project's internal systems environment

For now, let us consider the project company narrowly as a business system, developed in order to produce and sell some product or service. We select a power generating plant as an example, because such facilities are good examples of what we termed process projects in Chapter 1. It also allows comparative analysis of project companies that characterise as stock-flow projects, for example where input risks are manageable but off-take risks are complex (for example tolled-road tunnels or oil or gas pipelines without throughput agreements). Process projects allow demonstration of fundamental risk management approaches to supply (input) risks, process risks and demand (off-take) risks, whilst allowing some insight into risks generated at subsystem and system boundaries. To demonstrate our approach, we return to the fictitious Baguio Power Company at Subic Bay in the Philippines, introduced in earlier chapters. A brief description of our imaginary facility is presented in Box 4.10.

Box 4.10   The imaginary Baguio Power Company project.

The Baguio Power Company (BPC) is a US$2.5 billion, 2000 megawatt (MW) coal-fired electricity generating plant built on a 50 hectare site next to a deep sea port where a dedicated coal terminal was developed as part of the power plant's supply chain arrangements, complete with stockpiling, coal-handling and road infrastructure to service the whole plant and integrate it with the nearby metropolitan transportation networks. BPC is a continuous processing system with the function of turning coal into electricity at large scale and low unit cost, for sale to potential purchasers of electricity in the region. As part of this massively capital intensive facility we have numerous process control systems that control the rate of burning, steam distribution, emissions, and the like (the hard systems), including control mechanisms that manage the technical goals of Global Consilium Corporation (GCC) for the project (overall efficiency of factor inputs). Together these describe the business system's *internal environment*. In a narrow sense we can represent the internal environment and its boundary as illustrated in Box 4.10 Figure 1.

For practical purposes Baguio's electricity generating process may be viewed as fully automated, consisting of four large combined cycle turbine-generators with all the associated technical control equipment, driven by steam generated from a continuous coal combustion process. At the Baguio facility transmission exchange facility, the electricity it produces is metered formally and fed into the receiving electricity distribution grid. To manage, operate and maintain this

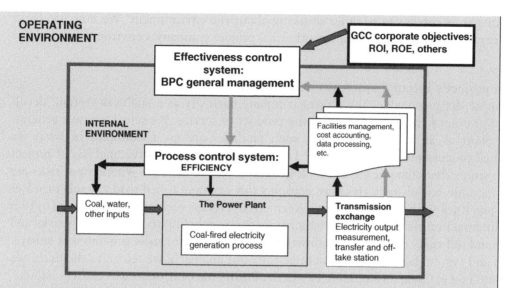

Box 4.10 Figure 1   BPC coal-fired power project: internal system environment.

automated, continuous process facility BPC has in place an experienced and proven team of operations managers and highly trained technical and maintenance professionals from its international pool of facilities operating teams. In Box 4.10 Figure 1, we see that the plant operations management team is supported by the normal business functions that provide efficiency and effectiveness information for control purposes, such as accounting and administration, facility management, data processing, and so on. This information function informs both the effectiveness and the process control systems. Together with the electricity generating system itself, this is presented as the system's internal environment. The effectiveness control system is part of BPC's internal and operating environments, and it reacts to BPC's relations with GCC's corporate objectives for BPC, such as the return on equity or net investment and other corporate objectives. From here it influences the internal environment through the process control system. This is to illustrate that BPC's management take signals from the operating environment and adjust objectives if required, which would likely require effectiveness and efficiency objectives to be altered in accordance. For practical purposes the effectiveness control mechanism and its interface with the internal environment and operating environments facilitates identification of the first risks that need to be managed.

The Baguio Power Company (incorporated in the Republic of the Philippines 1999)						
Balance Sheet at June 30, 2007						
Liabilities		($ thousands)	**Assets**			($ thousands)
Equity			**Fixed assets**			
Issued ordinary shares:						
Global Consilium Corporation	175,000	35.0%	Power generating facility at Subic Bay (at cost)[a]		2,600,000	
Bagatelle Projects D&C Inc	175,000	35.0%	Furniture, fittings, equipment, vehicles, etc.		5,000	
Hydrex Turbines Inc	150,000	30.0%	LESS:			
Shareholders' equity		500,000	Accumulated depreciation		(1,000)	
			Fixed assets			2,604,000
Debt						
Long-term debt			**Current assets**			
Secured bank debt, due 06/2022	1,400,000		Incidentals, maintenance inventory		5,000	
Capital leases: generating equipment	600,000		Other current assets		310,000	
Long-term office lease	10,000		Cash and marketable securities		5,000	
Long-term debt		2,010,000	**Current assets**			320,000
Current liabilities						
Commercial paper: 3 months	114,000					
Short-term bank loan: 1 month	300,000					
Current liabilities		414,000				
Total debt		2,424,000				
Total		2,924,000	**Total**			2,924,000

a The land at Subic Bay is leased from The Philippines Government at a neglible annual rent of US$1 per annum.

Figure 4.2 The Baguio Power Company pro-forma balance sheet.

The Baguio Power Company's controlling shareholder is the Global Consilium Corporation, and its early projected balance sheet at project completion is presented in Figure 4.2. Assume Bagatelle Design and Contruct, a large international whole facility design and construct company, acted as developer of the project and was responsible for planning, design, engineering and construction of the whole facility, in addition to having a significant minority shareholding.

The risks in the internal environment may be described as fairly narrow. They could include, for example, technical malfunction in the burning of coal and steam distribution system, in the generating turbines ending with the transmission exchange facility, or in the control systems. Although daunting for its scale, this is nevertheless a predictable and not inherently risky internal systems environment. The technology is mature and proven; safety and all other control systems have been continuously improved and virtually perfected over more than a century of learning with similar facilities. If anything, the technology is probably threatened with obsolescence for its environmental impact. Information and its flows are accurate, timely and allow the process to be controlled to within a very small efficiency envelope. Once commissioned and maintained properly, these facilities do not commonly break down or fail catastrophically unless poorly constructed. We will thus not dwell on further technological risks;[3] rather we assume the design and construction of the plant are sound, the control mechanisms function properly, and the plant's operations and maintenance management are of international best practice. Risks in the internal environment seem to be small – or are they?

---

[3] Nobody is suggesting that all risks are known, we are boundedly rational, after all. If we talk about technical and technology risks in nuclear power plants, however, it is an entirely different story. There remains much to be learnt about these facilities' risks.

So what can go wrong? Serious things, of course – remember, our first assumption is that this is a continuous process generating power at large scale. It is a highly *asset-specific*, scale-intensive, single-purpose facility that requires continuous burning of coal to keep (say) at least one turbine continuously driven for a minimum generating efficiency. To start with, there may be interruptions for routine maintenance purposes, or following isolated technical malfunctions, both sufficiently inconvenient to cause an interruption in the continuous generating process. So while we may agree that the risk to process interruption is small and in itself not costly to remedy, the consequences may be financially costly because all process disruptions will be by definition costly.

How may such process system risks be managed? To commence, identification of risks in the internal environment would likely be the subject of structured risk analysis procedures as early in the project's planning phase as is possible – preferably at inception and feasibility stages. It is also an accepted technology risk management assumption that there will always be several substitute approaches to manage technical risks that are identified and need to be managed, each substitute with its own costs and benefits – there is *never* only one way to achieve technical risk management objectives. For the sake of illustration, the risk of process interruptions in coal-fired power stations may be mitigated by accepting the risks (self-insuring), or transferring some risks by equipment supplier warranties, equipment leases with maintenance agreements, and a maintenance bond from the design and construct contractor to correct initial facility teething problems. Alternatively, Baguio Power Company (BPC) may pursue a loss prevention approach by designing the plant beyond accepted safety factor standards, for example with dual cooling systems for critical parts of the generating system, or dual systems in steam distribution, with all switching between backups automatically managed by the control systems. Such dual systems are in turn likely to be supported by an on-site mechanical engineering and fabrication workshop with sufficient replacement equipment and staffed by highly skilled and experienced technicians and maintenance personnel able to deal with all but catastrophic failures. Also, a planned maintenance schedule may routinely take one turbine-generator off-line for inspection purposes. Thus we may say losses from unscheduled process interruptions may be practically prevented or controlled by over-design or over-engineering, planned maintenance, and on-site engineering fabrication capacity. This process loss prevention and control mechanism is illustrated by adding a continuous planned maintenance engineering and fabrication subsystem to the internal system environment as represented previously in Box 4.10, which also receives control inputs from the effectiveness control system. This modification is included in Figure 4.3. Much of a planned maintenance unit's work may in fact be prescribed and inspected by regulatory authorities.

At this stage we may summarise the points that emerge from considering risks that are commonly identified from the internal environment of typical project financed ventures. In essence these may be categorised as mainly low technological risks, reflecting the requirements of project debt financiers for low risk, proven technologies

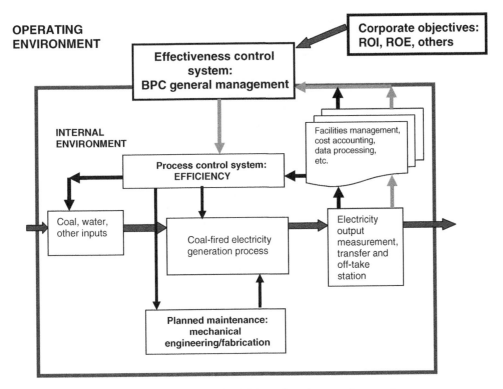

Figure 4.3   BPC's coal-fired power project: internal system environment.

and experienced, proven facilities management. Residual technological risks are effectively manageable through intelligent risk prevention and control, and appropriate insurance (subsystem warranties, post-construction facility and process maintenance contracts, or others). We also may assume that good control system information, both technological and managerial, is generated and used effectively. When compared to a typical business in most service or manufacturing industries, the internal environment of the power project company is well-defined and the risks are well-known. Thus we may assume the facility is constructed properly, and is functioning efficiently and effectively. These are convenient assumptions in a dynamic context, as we shall see.

We summarise these points briefly in Table 4.1, which is an indication of categorising generic project risks within a nested systems environment framework, and is not intended to function as a checklist. The principal risk is technological, with two subcategories, namely technology failure and control systems failure. Every project venture will have its own specific risk subcategories under main risks, and these will be developed and assessed extensively during project planning, design and engineering.

Table 4.1   Risks and risk management: internal project environment.

Risk	Risk management mechanisms
Technological malfunction Efficiency risks	Select proven technology **Loss prevention and control** Proven technology Reliable suppliers Lease key technologies with performance guarantees Make equity investors of key technology providers
Technology malfunction and failure	**Loss prevention and control** Proven technology Over-design and over-engineering Duplication of critical systems Post-construct maintenance periods/bond Planned facilities maintenance Main plant maintenance/replacement agreements Maintenance leases for subsystems Property insurance
Subsystems and control systems malfunction and failure	**Loss prevention and control** Best practice facilities maintenance and management Subsystem suppliers service and replacement guarantees Insurance or continuous operation guarantees Maintenance leases for subsystems Supplier maintenance periods/bond

4.3.2   From internal environment to operating environment

We have simplified BPC's internal environment to a sufficient extent to suggest that most remaining risks to the project company are effectively outside the internal environment. We now illustrate the nature of risks that typically originate in the operating environment (and later in the institutional environment). Let us then proceed from BPC's internal systems environment (Figure 4.3) to its operating environment. It will become clear that risks become more complex, as the systems environment moves from technical and technology risks to risks that are less clearly defined, start assuming softer dimensions, and are less able to be influenced by BPC. In effect, we are building a richer systems perspective of risks – risks in the operating environment are added to the internal risks and often may compound them. Again, we do not consider generating a checklist of risks that may occur generally in operating power plants as productive; instead we wish to concentrate initially on developing an approach to two important categories of operational risks, namely supply and demand risks, and price risks. Often these risks are collectively referred to as market risks, and represent a good example of risks that occur at the boundary in the analytical progression from a subsystem to its host system. We wish to demonstrate also that the approach we follow can be replicated to analyse other risks in the operating environment. What will also become clear in the analysis of supply and demand risks is that every project requires the economics of its operational environment to be very clearly understood before risks management measures can be put in place.

Input risks

What constitutes the operating environment of a project company? Perhaps we should start by stating what are considered to be operating risks. These are generally considered to be any risks that may cause a project venture not to perform at the level expected to achieve its objectives (set by GCC and BPC's boards) and more specifically, anything that will cause it to generate insufficient revenue to service and repay its debt. The first operating environment risk to the continuous electricity generating process may be identified as interruption in the continuous supply of inputs, and risks to the sale of output. We add these as the first operational environment risks in Figure 4.4. Principal physical process inputs are water and coal, but there will be others, some more complex than natural resources (for example, labour,

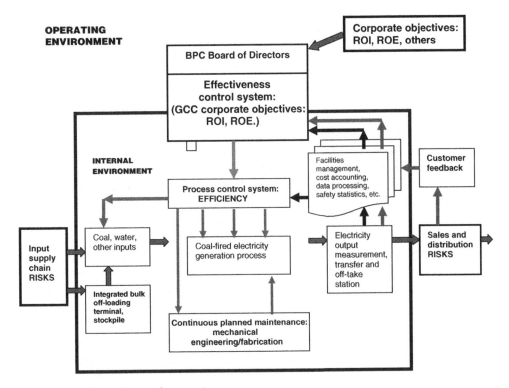

Figure 4.4 Operating environment.

transport, communications and other physical infrastructure, social infrastructure, and more). Let us concentrate on coal, but similar logic will apply to all resource inputs.

To commence, the appropriate quality coal to meet applicable environmental standards will have to be sourced and supplied to BPC at Subic Bay, in sufficiently large volume to fuel a continuous generating process. The market for coal is an intensely competitive international market, with suppliers from quite a few countries including Australia, Brazil, Canada, the United States, South Africa and others with the capacity and reserves and access to bulk mineral carriers to supply BPC for the intended life of the facility. If long-term supply and timely delivery of coal are arranged through long-term supply contracts with reputable suppliers, then we can argue that loss from supply risk is managed, or at least mitigated, by contract. If supply is also contracted at an agreed price or agreed pricing formula for delivery at Subic Bay over the expected project life, then price risks may also be managed, which then addresses market risks in the supply of coal. Such arrangements may even include agreements with the purchaser of the electricity to allow cost increases to be passed through into prices paid – but these are difficult agreements to achieve. Delivery at the terminal generates a whole range of potential losses from further supply chain risks that have to be managed before we can assume delivery and price certainty at the terminal. What if there is an accident at the mine that is contracted to supply the coal and it is closed down? What if there is a derailment on the track that transports the coal from the Western Australian outback to the port from where it is shipped to Subic Bay? All this might convince GCC and BPC's management that it might be prudent to diversify supply sources. It may also consider an alternative fuel, such as oil or gas, or even to use generating technology that can switch between alternative fuels such as oil and gas. Such an approach puts in place diversification of sources and types of inputs as a risk management approach (multiple input technologies are favoured case studies in real options analysis touched on in Chapter 3).

In the first instance BPC will have a reserve stockpile of coal at its Subic Bay terminal to draw on to keep the facility operating for a period, say some weeks, but further supply chain risks nevertheless remain. Such risks may be managed by diversifying suppliers and ensuring that different suppliers are not subject to correlated supply chain risks. This could be achieved by entering into contracts with suppliers from different continents. In the end, BPC can also enter into what are known as *put or pay* arrangements. In effect the supplier guarantees supply, by agreeing to purchase the input on the market and supply it to BPC if there is a disruption in its own supply chain, or alternatively provide BPC with the equivalent financial resources to do so. Such put or pay agreements may in fact require a third party guarantee as precondition before debt financiers will be willing to provide project finance.

Project loans are frequently syndicated among groups of lenders, and usually involve interest payments calculated on a floating rate basis, meaning that the greater interest rate risks are taken by the project company borrower. In many cases this will create a need for debt management, to increase the certainty associated with interest costs, or in some cases lessen the overall interest burden, and the project company will approach this task by entering into swaps, or buying and selling interest rate options, often in the form of interest caps or collars. Swaps are explained in Box 4.4, and aspects of the use of options in Box 4.8. The difference between the way that these types of financial derivatives are traded or made available to users is explained in Box 4.3.

A further common price risk is of course foreign exchange risk – the risk that there will be adverse (or positive) changes in exchange rates during the term of the supply contracts. International commodities markets conduct trading in US dollars. Thus there is a currency mismatch, because BPC will earn income in pesos but pay for its main input in US dollars. As identified above there are essentially three ways of managing this risk besides avoidance, by attempting to fix exchange rates for the supply contract (an unlikely scenario in modern times), namely risk avoidance, hedging or insurance. Hedging currency risks can be achieved in whole or part with forward contracts, currency swap agreements or by trading in currency futures. Insurance may include currency options, currency forward contracts or long-term currency swaps. Swaps may have very long maturities – as much as 20 years or more in many major currencies, and although most options and forward contracts have short-term lives, it is common for risk managers to engage in dynamic hedging, which involves layers or arrays of interlocking or successive contracts, or to defray the cost of any single hedge by receiving a fee for assuming risk in a different aspect where the project company may bear risk more efficiently. Currency risk can be avoided by purchasing coal from a supplier in the same country. Local coal mines with acceptable reputations may be available, but again project bankers may insist further on put or pay agreements, and even third party guarantees for such agreements. As part of institutional development programs, multilateral organisations have provided third party guarantees for supply and demand contracts (we return to this matter in Section 4.5). BPC may also consider integrating backwards by developing a dedicated mine in the Philippines exclusively to supply BPC, but this then raises a whole host of further risks: is the local infrastructure adequate to ensure transport and delivery? Does it require significant further infrastructure development? By now it becomes clear that risks have to be identified and carefully analysed in order to consider the best approach to manage supply risks. We summarise these points in Table 4.2.

Demand and distribution risks

Of course, we have made no mention of demand risks – that is, sale of the electricity that is being generated in a continuous process by Baguio, or the price at which it may be sold. Under the assumption of reasonably functioning markets these risks are always present, but we need to provide general background to the peculiarities of electricity markets before proceeding. The structure of electricity demand and supply

Table 4.2 Risks and risk management: input risks.

Level 2 systems category: operating environment	
Input risks	**Risk management mechanisms**
Coal supply risks, especially inconsistent supply or quality of supply	Co-location and development of plant and mine
	Backward integration – buy or develop mine
	Stockpile supply of coal
	Long-term supply contracts
	Diversify suppliers
	Penalties for non-performance
	Put or pay contracts
	Multi-lateral/third party guarantees
	Agreed access to foreign exchange reserves to manage currency mismatch
	Alternative fuels?
	Diversify fuels: multiple fuel technology?
Coal price risks	Fixed prices
	Agreed pricing formula
	Agreed increased cost pass-through to sale prices
	Hedge price risks short–medium term: forward purchasing contracts.
	Hedge price risks shorter term: commodity futures, options
	Multi-lateral or third party guarantees
Foreign exchange risk with internationally sourced inputs	Co-location and development of plant and mine
	Operational hedge: supplier from same currency jurisdiction
	Hedge medium-term foreign exchange risks, with forward purchase contracts, and long-term currency swaps
	Limit short-term foreign exchange risks, with foreign exchange swaps, currency futures and currency options
Interest rate risks	Hedge medium-term interest rate costs and changes with interest rate swaps, options on swaps (swaptions) and forward swaps
	Insure against short-term interest rate risks with interest rate futures and options

in countries everywhere has undergone changes in the last two decades. In principle many countries, states and cities have moved away from the public sector supply and distribute model that functioned in most countries over much of the twentieth century (in some cases this transition is not complete, but there are a limited number that have not adopted this model). Instead many countries have opted for a private sector influenced model, namely competition in electricity supply with a regulated monopoly in electricity distribution. The pure form of the private sector supply approach forbids the exclusion of any approved electricity producer from the distribution grid and allows consumers to choose freely between competing suppliers over the same grid.[4] Hence the concept of the merchant power plant (MPP) was coined. In all, this scenario presented a fundamental change in the regulatory framework under which the industry previously operated. We return to industry regulation in Section 4.5.

Although this pre-empts the analysis of demand risks and the introduction of institutional risks somewhat, we may assume for the sake of explanation that BPC's promoters intended for it to be an MPP in due course, but that it is presently in a somewhat awkward position. But in the case of BPC we may assume it is an MPP. Further, assume that electricity distribution has not yet been deregulated in the Philippines, and that the distribution grid is still owned by a fictitious public sector body (i.e. it is city owned), the Manila Electricity Distribution Board (MEDB). MEDB technically owns and manages the electricity distribution grid together with a number of other inefficient public sector owned power stations in the Manila metropolitan area, and in turn is answerable to the central government's Ministry of Energy. MEDB sells subsidised electricity to millions of households and businesses at subsidised prices, rather than prices that could allow an MPP to generate and sell electricity at a reasonable return. MEDB power plants are old and poorly maintained for lack of public funding, demand outstrips supply, and the grid frequently breaks down with supply disruptions to many parts of the metropolitan area. MEDB (and its power stations) is regulated through the Ministries of Energy and of the Environment, through vague and loosely enforced emissions and pollution standards, occupational health and safety standards, in addition to periodic inspections for operating licence renewal purposes.

For illustratative purposes, assume the Manila Metropolitan Government has received permission from the Ministry of Energy to attempt to solve this problem by first privatising power supply, and then privatising the distribution grid under a regulated monopoly. BPC is the first experiment with a new, modern MPP supplying electricity to the distribution grid, and it is expected that the other power plants will be sold to private interests as MPPs once the BPC experiment is shown to be successful. In the meantime user subsidies will remain, and will still reach users through lower prices administered by MEDB. The cost of the subsidies will be recovered through the Metropolitan Government's budget, which in turn receives shared tax revenue from the central government. So GCC is effectively in an experimental position with BPC – a newly built MPP selling electricity to a single buyer (the MEDB monopsony). This may require a multilateral co-financing arrangement or guarantee (also referred to in Section 4.5). We summarise this hypothetical regulatory structure in Figure 4.5.

---

[4] This loosely describes the model which has facilitated privatisation of electricity supply in the United Kingdom, some Australian states, some European states, some states in the United States, some Indian states (still in progress), Pakistan (still in progress), some new supply facilities in China, and more.

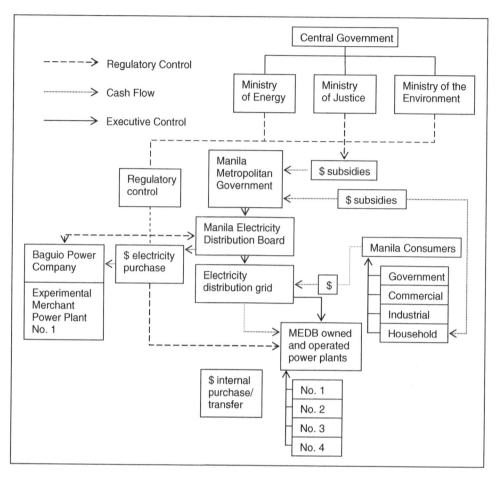

Figure 4.5 A hypothetical regulatory structure for BPC.

This elaborate description was required to place BPC's demand risks in perspective. Remember firstly that science has not yet managed to develop efficient large capacity electricity storage and ultra-long distance transmission systems. Further, fuel-cells technology as one potential technology to overcome the location-specificity and irreversibility of power plants is still in its infancy. While energy commodities such as oil and gas are traded on organised exchanges (notably the Chicago Board of Trade, Chicago Mercantile Exchange and London Metals Exchange), electricity and electricity derivatives are bought and sold between users and suppliers but not traded as commodities, either on an organised exchange or over-the-counter. It is thus immediately evident that without large demand for the output of a plant within a reasonable distance there may be a weak business case for developing the plant, which would most probably not find favour with potential project finance lenders. Also, generating electricity continuously and at scale is necessary for cost efficiency with current technologies, so if there is no continuous buyer for the continuous output, then the process should preferably not be started at all. And remember, a coal-fired electricity generating plant is a highly specific asset – it cannot be turned into a high-end shopping centre, car assembly plant or art gallery overnight. So there has to be a buyer for the electricity, and a buyer preferably also has to commit to purchase electricity continuously and in bulk from BPC for a long term. This means at least long enough to amortise the development costs of the facility and earn GCC's required weighted average cost of capital (WACC) when the investment decision was made, and certainly longer than the reasonable term of a project loan considered by project lenders.

Similar to the coal supply arrangement, demand risk could be controlled if long-term purchase of the facility's output capacity could be contracted, again at an agreed price or price schedule, known as an off-take agreement. But unlike the case of sourcing inputs, BPC's output is not to be sold in a market where sales could be diversified to manage risks. There is only one purchaser allowed by the Philippines power industry regulation for the Manila metropolitan area, namely MEDB. But similar to risks in the input supply chain, there are also residual risks in the relationship with MEDB. As a precondition to providing project debt, BPC's project lenders may insist it enters into what is known as a *take or pay* contract: MEDB would agree to pay for the output even if it does not use it itself. Remember in the partially deregulated market MEDB is the only purchaser, but it has to sell electricity at subsidised rates and has to recover costs from the Manila Metropolitan Government, itself funding MEDB subsidies from shared tax revenue it receives from the central government. Without such an undertaking from MEDB, and possibly with a third party guaranteeing MEDB's performance (such as the city, state or central government), BPC might not attract project lenders, and may itself thus not be willing to proceed with development of the plant in the first instance. Further, it is logical to expect that there will be contractual consequences following off-take agreement defaults (for example, payment defaults), which we will also say more about in following sections. And perhaps more seriously, we have a currency mismatch as explained before: BPC will earn income in one currency and pay at least for its main input, coal, in US dollars, so there is the further risk that exchange rates will move against BPC. We summarise these points into Table 4.3.

Other commercial risks

Analysis of supply and demand risks and risk management options illustrates the nature of the project finance venture business model introduced in Chapter 1 particularly well – the aim of a project finance transaction is to isolate and manage supply and off-take risks, in order to stabilise project earnings from as early as possible after commencement of operations. In Chapter 3 we pointed out the importance of stabilised cash flow as a principal determinant of how attractive a project company would be to project lenders. Stable cash flow and effective risk management are crucial to the bankability of projects – how attractive it is to lend to the project. It would seem that risks to net income generated by BPC are neatly managed with supply and demand agreements, but of course we know all ends have not been secured. Apart

Table 4.3 Demand and off-take risks.

Demand and off-take risks	Risk management mechanisms
Demand risk: volume	Long-term purchasing agreement (off-take agreement) Put or pay arrangements Multilateral third party guarantees Diversify customers and buyers
Demand risk: price	Fixed prices, or agreed pricing formula Multilateral third party guarantees
Foreign exchange risk: output sold internationally	Hedge medium-term sales risks, forward sale contracts, currency swaps. Hedge short-term sales risks, currency futures, currency options Diversify sales in same currency jurisdiction
Interest rate risks	Hedge medium-term risks with interest rate swaps, forward swaps and swaptions Insure short-term interest rate risks with futures and options
Local political environment	Best practice corporate governance BOT arrangements Government and local interest equity participation Multilateral co-financing agreements
Regulatory bodies and competition	Highest government's express agreement Accepting local jurisdiction is a sign of commitment to the local community International dispute arbiters.
Legal system, and judicial review and enforcement	Accepting local jurisdiction is a sign of commitment to the local community International dispute arbiters.

from supply, demand and price risks, further operational risks are generated in BPC's relations with local communities, regulators, loan financiers, in its industrial relations, relationship with competitors (if any), local communities, and more. For our purposes, the principal interface of these operational interactions with BPC is through its board of directors. We will explain briefly how risks may be generated, but from here our analysis becomes cursory. As has been shown with the analysis of supply and off-take risks, management of risks in the operational environment depends in great measure on the nature of the project that is being analysed. Because many project finance ventures are large irreversible fixed capital investments, it becomes clear that many operational risks are related to the project finance venture's geographical location. A fixed location also greatly influences the next systems analysis level, namely the institutional environment. It would thus require substantial elaboration of the operating environment of any proposed project finance venture at its physical location to identify risks with industrial relations, local communities, and so on. We discuss selected further operational environment risks, illustrated in Figure 4.6.

INSTITUTIONAL ENVIRONMENT

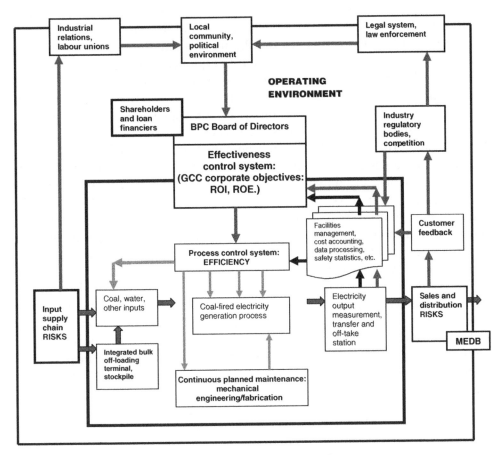

Figure 4.6   BPC's coal-fired power project: operating environment.

For convenience we may start on the left of Figure 4.6 with labour as an input. Of course, we have identified BPC as an extremely capital intensive venture, which employs highly skilled technical personnel primarily in facilities management roles, including maintenance and process support roles such as accounting and administration, data processing and so on. We stated also that the plant's core process management is likely to be a team drawn from GCC's pool of experienced plant operation managers. Further, BPC is an automated processing system from when coal is continuously loaded onto supply conveyors from the stockpile to electricity being drawn into the MEDB grid. Altogether, BPC may perhaps then employ directly as few as 100 personnel, likely to be highly skilled and paid, given how critical it is that process interruptions are absolutely minimised. In all, it is unlikely that BPC process operating personnel present risk of industrial disputes serious enough to damage BPC financially, but this of course does not consider the political dimension – BPC is a physically immovable asset in a foreign country, majority owned from abroad. It is thus an obvious target for political action, and it is clear that it would be foolish for GCC and BPC to ignore this. In the first instance it would seem that this risk can be managed if it is clearly demonstrable to the Philippines union movement that best international labour practices apply to any local employees – there is no room for error here, any other practices would invite industrial relations problems. Despite this, international labour movements may well have influence in the Philippine union movement, assuming the central government is not oppressive, so there always remains a residual industrial relations risk, here as everywhere else.

While we separate the local political environment from industrial relations, it is clear that the two are not entirely separable. Despite all else, it is quite likely that there will be political opposition to BPC, for reasons that may be real or opportunistic, and that local political interests could find BPC an attractive cause. GCC and BPC might have to communicate credible commitment expressly to the local community. Developing the plant under a BOT arrangement with a concessionary period of (say) 30 years signals temporary custodianship of an asset that is ultimately the property of the local community, while technical skills transfer arrangements also may be seen to benefit the skills pool of local communities. But be sure that there are also persuasive arguments that present such actions as self-serving and exploitative. It would seem that short of proving to the local community that BPC is a good corporate citizen, all political risk simply cannot be managed. By considering the local community's politics, we observe the first potential risk from outside the operating environment – the political process. We summarise these points into Table 4.2. We return to this in Section 4.5, where we consider the institutional environment.

Let us consider now the right side of BPC's operating environment, starting with sales and distribution. We have already briefly analysed price and off-take risks, and will not dwell on these. We do have to return to mechanisms under which BPC will produce and sell electricity to MEDB, however. We outlined a power supply and distribution sector that was starting to transform from a totally public sector controlled supply and distribution system to one with privatised competitive power suppliers, a regulated monopoly grid and consumer choice of power suppliers. The system we described is only conducting the first experiments towards developing a privatised system, by allowing BPC as the first MPP to sell to the otherwise state (MEDB) controlled system. As assumed, BPC has a long-term off-take agreement with MEDB, but while this may be sanctioned and supported by current regulatory development thinking, BPC is also concerned with regulatory stability. For example, while a long-term price off-take agreement deals with quantities sold and price risks, regulatory bodies are also responsible for periodic renewal of operating licences, occupational health and safety matters, emissions standards and inspections, and several such operational aspects. A 30 year concession agreement under a BOT arrangement will include arrangements concerning volume and price changes if it includes an off-take agreement, and will also cover performance over a range of regulatory, fiscal, labour and governance matters – all subject to risks. We introduced the nature of concession agreements in Chapters 1 and 3.

It would seem fair to expect that if agreement is reached under one set of regulatory requirements, that these be renegotiated with reasonable concern for disadvantages created should the regulatory frameworks change. But regulatory bodies are typically public agencies, also subject to political influence, and where regulatory frameworks are not stable, regulators may behave opportunistically to the detriment of BPC (recall our discussion of the fundamental transformation in Chapter 2). BPC must be fairly certain that MEDB will not use another MPP (possibly developed after BPC, deploying more advanced technology and with lower cost of production) to assert its monopsony power and force BPC to accept lower tariffs for supplying the MEDB grid. Similarly, other regulatory matters such as emissions and safety standards, or interim operating licence renewals, all may form the subject of regulatory powers. It will be extremely foolish, for example, if GCC expected no change in BPC's future status with environmental regulators, given that coal-fired power stations are not particularly efficient whilst also one of the worst greenhouse gas emitters. Radical international regulatory action is not beyond probability, and expensive mandatory retrofitting, or even early closure, are definite risks. While the risks associated with plant decommissioning were alluded to in Section 4.1, regulatory predictability remains an operational risk everywhere, but it remains subject to greater political manipulation when societies undergo major changes, as we are hypothesising with the case of BPC in its host location.

As with regulatory bodies and the risks they may introduce into the operational environment, we must also remember that all contracts entered into, whether on the supply or demand side, and despite the spirit of the agreements, rely on enforcement for their credibility with potential lenders to project companies. Thus we have to consider together with the predictability of specific regulatory mechanisms also the legal system and law enforcement in the project's jurisdiction. Legal risks may be mitigated by agreeing to alternative jurisdictions, such as London or Hong Kong, for dispute resolution mechanisms acceptable to all contracting parties; but while this may signal a credible commitment to agreements by a host government, these commitments remain compromised where local politics is unstable, and may simply be politically unacceptable in the first instance – so risk mitigation may be possible by suggesting dispute settlement conducted in offshore jurisdictions. Some countries, including the Philippines and Pakistan, have attempted to overcome regulatory and legal enforcement problems by creating an entirely separate legal and regulatory environment specifically to encourage foreign investors to develop project finance ventures similar to the fictitious BPC concept, with significant success. We summarise these points into Table 4.2 and see from Figure 4.6 that both regulatory and legal systems and legal enforcement include residual risks that originate in the institutional environment of the project location. The PPP's Board of Directors forms the interface for all these operational risks, but it also becomes clear that these are risks that have to be identified, analysed and addressed before construction begins.

From the analyses presented above, it would seem that while much attention had been given to the details of supply chain and off-take risks, these do seem less complex than operational risks emanating from considering industrial relations, regulatory agencies, legal systems and legal enforcement. All these also have important political dimensions, which determine that BPC's fixed location exposes it to instability in any of these areas. While these are operational risks, they certainly cannot be neatly separated from the institutional environment within which BPC has to operate. We return to institutional matters in Section 4.5.

Up to here we have largely considered BPC as an operating venture, a going business concern, so to speak, albeit a single asset project company. But of course, before it can operate, generate earnings, service its debt and pay dividends to shareholders, it has to be constructed. This raises a raft of serious risks that precede any of the internal and operating risks analysed so far.

4.4   Risks in the construction phase

We started this chapter by assuming BPC owned a completed power generating facility, and that it was fully operational. We outlined common risks in its internal environment and its operating environment. In effect we did not draw on the representation of any project life-cycle phases as introduced in Chapter 1, and it is here that we turn now. We mentioned repeatedly that there is no project before it is constructed and commissioned, and from a systems viewpoint the management of risks in design, planning, construction and commissioning precede any operational risks. It is hard to overemphasise the overall riskiness of the construction phase, as project feasibility as a business venture relies so closely on managing cost, time and quality in constructing the plant.

This section does not aim to outline and explain the management of risks that originate during construction of the project facility; instead the assumption is that companies engaged in construction projects are expert at managing logistics, supply chain management, site risks and other risks that arise from the construction process. To commence, it is possibly correct to state that there are only a handful of companies with the experience, expertise and record of success to construct a complex process plant costing billions of dollars, within stringent quality, time and cost constraints. In this group are companies like Bechtel, Fluor, ABB, and selected others. Also, given that so many project ventures are constructed in often remote and inaccessible international locations, often with serious site-specific physical complexities and risks, it is often the same set of companies that deliver these projects in many parts of the world. Typically these are integrated whole facility design, engineering and construct (EDC) companies that compete on delivery of a completed facility designed to operate to the agreed performance requirements of the client project company. These are giant companies in their own right, experts in systems design and integration, and specialise in delivering large and complex projects in remote, difficult and often very dangerous locations.

Thus it is no surprise that EDC companies, with their financial resources and design, engineering and project delivery expertise, have inevitably also become important project promoters, rather than mere contractors. Since the design and construct model is the typical mode for delivery of project company facilities, we will concentrate on this method to illustrate risks and risk management in the construction phase, because this model allows the responsibility for delivery of the facility together with all risks associated with design, engineering and constructing to be concentrated with one agent – the EDC company. For an indication of this form of project delivery, see Box 4.11.

Let us return to BPC in order to consider construction risks and risk management measures. In order to identify clearly how risks are identified and managed in this phase, let us assume GCC, on behalf of the BPC being formed, had invited three large EDC companies to present proposals and budgets for the design, engineering and construction of the BPC plant, the associated coal terminal and all related infrastructure, and that our fictitious EDC company the Bagatelle Corporation was the

Box 4.11   A typical design and construct (DC) transaction.

Design and construct (DC) transactions (often referred to as turnkey contracts) refer to a family of approaches to construction contracts where the intention is for a contractor to deliver a completed project at a specified date, for an agreed cost, and to a specified quality and performance standards. This can entail any scale of project from a small house to a large process engineering facility. Simplistically, one principal contractor undertakes to design, engineer and construct a facility in its entirety, typically (but not exclusively) for a pre-determined price, which may be a fixed sum in the case of simple projects, or a complex formula with major schemes. In this manner the customer purchases forward a whole facility for delivery at a future date, and technically the customer simply receives the facility at the agreed date, as with the purchase of a manufactured product as explained in Chapter 2. The principal contractor might be a full-service DC contractor (such as Fluor), which directly employs all the resources to complete the design and engineering phase of a facility, and typically employs directly project and construction managers to manage the construction phases of the facility. A main contractor will manage the physical construction of the facility, and will in turn employ specialist subcontractors to complete all specialist features, for example fabrication and installation of steam distribution pipework in a power generating plant. However, given the scale and capital intensity of many whole facilities in the infrastructure, process engineering and manufacturing industries, a single contract between two parties to deliver a facility cannot comfortably cope with the logistical complexity of most construction contracts, whether large or small. It has been traditionally common for transactions involving scores of subcontractors to be fraught with incentive conflicts in execution, between principal contractor and client, principal contractor and main contractor, between main contractor and subcontractors, and between different subcontractors. Many careers have been established and fortunes made from the necessity to resolve disputes between all these parties in transaction governance.

Modern project finance techniques have increasingly demanded that these incentive conflicts be guarded against institutionally by making all major contracts coterminous and irrevocably linked, so that the costs of inaction or poor performance are realised in the project, rather than by any single investor, sponsor or contractor. Thus instead of a sponsor seeking compensation (contractually or legally) for the performance deficiency of a contractor, it shares that contractor's incentive through linked contracts. While disputes between main contractors and subcontractors are often serious and can paralyse the progress of projects, we are more concerned with incentive conflicts between the DC (principal) contractor and its customer, following the transaction cost framework presented in Chapter 2. Also, following the usual transaction governance arrangements, this is where conflicts will affect the project company. First, we know that without the completed facility, we have no operational project company. This places the project company at risk of being unable to commence trading to schedule, thus losing business and facing default on concession agreements, input and off-take agreements, and other pre-operational agreements entered in order to obtain construction and project loans. These may each have direct or indirect financial consequences. Project companies typically manage the financial risks associated with delayed completion through four mechanisms.

- Principal contractors typically must prequalify as reputable and with excellent performance records, with sufficient resources and be sufficiently capitalised to complete the project.
- Principal contractors are required to provide third party guarantees of completion (performance guarantees or some other form of financial surety).
- Late delivery will almost always incur escalating financial penalties.
- Customer–principal contractor incentives are aligned by requiring the contractor to make a non-trivial equity investment in the project company.[5]

---

[5] Chapter 8 contains a case study involving two principal contractors taking substantial financial stakes in a project company, but which may have been insufficient to provide incentives to their involvement in subsequent negotiations over the project's future after it had become victim to deleterious external events.

The equity requirement must often be funded before any construction loan tranches are made available and made subject to lock-up periods which typically stipulate that the principal contractor may not dispose of its stake in the project for several years or until the fulfilment of certain project hurdles. Further, the principal contractor is often required to arrange back-up loans or equity infusions to ensure that the sponsor's project cost commitments remain fixed, ensuring that delays and accumulated interest costs remain the risk of the principal contractor. Quality and performance risks in the facility itself will usually be managed by a contractual provision whereby the principal contractor maintain the project for a period (a form of warranty known as the maintenance period), during which defects are to be corrected. This can be coupled with a maintenance bond, held by the sponsor to ensure compliance with the requirements of the maintenance period. Maintenance bonds are often financed by retention money covenants, where the principal contractor withholds part of periodic payments due during construction, which are released only at the end of the maintenance period. A simplified DC transaction is illustrated in Box 4.11 Figure 1.

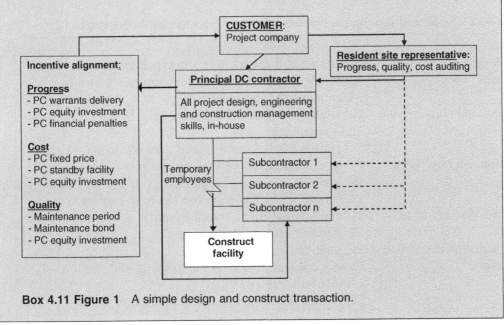

Box 4.11 Figure 1   A simple design and construct transaction.

Note: The incentive conflicts indicated above exist in all forms of construction contracts, of which DC is only one, albeit the most commonly used form in the procurement of large whole facilities. Variations on this form of contracting include guaranteed maximum cost contracts, target cost contracts with sharing of savings or extras, and so on.

successful bidder.[6] Bagatelle is a global EDC company with some 50 years of experience in design and construction of very large whole facilities, primarily in the power generation and oil and gas sectors, and has an exemplary record in delivering whole facilities on time, within budget and to the highest quality. Bagatelle itself has also initiated several large power plant projects in developing countries, including the Philippines. Despite this, and to labour the point even further, GCC remains aware of the fact that there is no BPC project company until there is a completed and commissioned power generating facility. All GCC has achieved so far is to obtain a contractual undertaking from Bagatelle to deliver the facility at a future date and to the quality required. In transactions terminology, while Bagatelle's reputation for delivering is obviously critical, BPC has also entered into both supply and off-take agreements that depend on Bagatelle's performance. Thus, it may be concluded that delivery of the facility remains *the* critical risk to BPC (and thus GCC). We summarise this risk into Box 4.12.

It is clear that the risk of delivery of the facility on time is not a risk that BPC can bear efficiently – least of all because it has no control over the project's construction despite the fact that it will have its own expert representatives monitoring construction. Generally speaking, there are several mechanisms which allow BPC to align

Box 4.12   Risks and risk management: project engineering and construction.

Design, engineering risks	Reputable and experienced principal contractor, with sufficient human and capital resources
	Meaningful PC equity participation
Physical construction risks	Reputable contractor, with sufficient human resources and experience
	Separation of construction and take-out loans
Non-delivery risks	PC to guarantee delivery or surety
	PC credible equity participation
	PC arranged standby construction loan facility
Late delivery risks	PC standby facilities, standby equity
	PC penalties for delivery delays
Quality/performance risks	PC quality guarantee
	Project company resident engineering and site inspection representatives
	PC maintenance period and maintenance bond

---

[6] It is not material to the narrative to discuss matters such as the documentation which forms the basis of the bidding process, although there are serious risks in this process – which we assumed are managed by dealing with reputable EDC companies only.

Bagatelle's incentives with its own, and we mention two direct financial mechanisms and a corporate guarantee from Bagatelle that function to insure delivery. Let us commence by analysing the last mechanism first: the efficient bearer of the delivery risk for the whole facility is Bagatelle – successful delivery of whole facilities is of course its expertise and the service in which it trades. It therefore makes sense that Bagatelle contracts to deliver the facility, and that it demonstrates its commitment to this transaction by bonding all of the Bagatelle Corporation to this commitment to deliver, in the form of a corporate guarantee from Bagatelle to BPC. In principle, this aims to put BPC in a position where it would have been had it not entered into this transaction with Bagatelle at all, and technically places Bagatelle in the position where it has a contingent liability to BPC, which converts into a real liability should it not deliver the project as agreed. It is likely that late delivery will incur financial penalties, but the guarantee is for delivery of the whole plant, not mere late delivery. In this manner BPC insures itself against non-delivery (see Box 4.13 for the outline of a typical guarantee of project delivery in these circumstances).[7]

This returns us to the first of two financial mechanisms to manage delivery risks. This is a requirement that Bagatelle is a non-trivial equity investor in the project venture. In the first instance this is a fundamental requirement and signal commitment, as explained in the agency treatment of transaction costs (Chapter 2). Recall also we explained that the prominent EDC companies often also will initiate projects, and also that they are very large companies with significant financial resources of their own. Typically, however, these companies do not aim to own a portfolio of project assets (as GCC does), although they might develop a project venture, stabilise its earnings and then sell it as a portfolio asset to a company such as GCC or an investment fund with the intention of making a capital profit. While they typically do earn significant corporate revenue from facilities maintenance and management contracts post delivery, their core business activity is design, engineering and project delivery, not portfolio investment. It does follow that if Bagatelle commits a significant equity contribution to the development of BPC, its incentives are aligned with BPC in that Bagatelle risks an equity stake as well as having the contingent liability of the delivery guarantee. A further developed version of this mechanism also would impose an equity lock-up period post delivery, sufficiently long for Bagatelle to demonstrate its commitment to delivering the facility that unequivocally meets quality and operational specifications.

---

[7] Of course, such a guarantee is only good if the EDC has the financial resources to honour it. Contractors in lower leagues than Bagatelle also guarantee delivery of projects, but often are also required to provide a performance guarantee from a third party, or a surety to provide funding for contract completion in the case of termination of contract for non-performance or insolvency. A surety bond is often provided from their bank or an insurance company for a fee and, if they have some track record. For insurers, this is actuarial, while the PF field is more akin to event insurance.

Box 4.13   Performance guarantees and sureties in construction contracts.

As explained earlier, insurance is a contract (policy) under which one party (the insurer) agrees to compensate the other party (the insured), upon an occurrence of an event to the subject matter (insurable interest) where the other party suffers loss as a result. The risk is thereby transferred from the insured to the insurer through the payment of a premium. A *guarantee* is a contract to make good an obligation, e.g. a debt, made by one party (guarantor) to another party (guarantee)[8] if the third party who has the obligation does not honour it. Therefore, in a specific event when there is a default by an obligated party to pay the debt, the contract of guarantee is very much similar to the contract of insurance. However, there are also many differences between insurance and guarantee. Insurance is not a collateral obligation to pay for the debt of a third party. The three main differences between insurance and guarantee are as follows.

- Insurance has a speculative factor. The insurer bears the risk in consideration of the premium.
- A contract of insurance is made between the insurer and the insured based on the information (proposal) submitted by the insured whilst a contract of guarantee may be a unilateral contract in form of a letter of guarantee to whoever deals with the third party who makes the promise.
- In a contract of insurance, the insured must have an insurable interest in the subject matter of the insurance. The court would look at the substance rather than the form in determining whether or not a contract is one of insurance or guarantee. (Denton's Estate, 1904)

In ensuring that a construction contractor will not default, the use of contract surety bonds is widespread in the construction industry, but surety bonds are widely used in all sectors of the business world. A party may assume liability for the debt or default of another party by surety at common law. A surety bond is simply a contract that incorporates the terms and conditions of such agreement. In construction contracts, a surety bond is not to protect a party against loss, e.g. as with an insurance policy, but rather it guarantees the *performance* of a construction contract. There are a few types of surety bonds.

- Bid bond, which guarantees the holder, i.e. the obligee, that the contractor will honour his bid and will abide by all the contractual terms upon being awarded the contract. If the contractor fails to honour his commitments, then the contractor and the surety are liable for the damage suffered by the holder. The damage is normally the difference between the contractor's bid and the next higher bid.

---

[8] Guarantee has two meanings. Guarantee also may mean that person to whom a guarantee is given.

- Performance bonds, which guarantee the holder that the contractor will complete all the terms of the contract. If the contractor defaults, the holder may require the surety to honour the contract.
- Payment bonds, which guarantee the holder that the beneficiaries, i.e. subcontractors and suppliers, will be paid in full by the contractor. The holder derives the benefit from such bond as the performance of the subcontractors and suppliers is assured. If the contractor defaults, the subcontractors and suppliers may hold both the contractor and surety liable.

The second direct financial mechanism functions to reinforce the effect of the delivery guarantee and Bagatelle's commitment of equity participation in the project venture. This mechanism relies on the overall intended capital structure of the completed project venture, which typically has a high debt component (Figure 4.2). It must be kept in mind that banks and other project company debt financiers compete keenly to lend to successful project companies, so they too are affected when there are project delivery failures. Consider a further scenario: assume Bagatelle commences construction, several tranches of the construction loan from the Fibonacci Commercial Bank (FCB) have been paid over to Bagatelle as progress payments, but in the end it fails to deliver the facility. Technically the performance guarantee will be called in, but the beneficiary of the guarantee is BPC. FCB, however, will manage its exposure in two typical ways. First, it will require the performance guarantee to be assigned to it, so that Bagatelle technically becomes liable to FCB in the event of failure to deliver. It further manages this risk through the structure of the loan, the conditions under which it provides capital to the project and in the sequence or timing with which it is made available, namely the loan tranches (we return to construction lending in Section 4.6 and tie it in with permanent project debt). These risks and risk management mechanisms are summarised in Box 4.12.

4.5   The institutional environment and risk

This section introduces the institutional context within which project companies, and indeed all commercial activity, are conducted. We identify important institutions in (normally) well-functioning economies, and point out their relevance to risks they create for project companies, as well as their relevance to the success of project ventures, particularly in an international context. We address the importance of government, legal, corporate, political and regulatory institutions that influence project ventures, and also consider the risks generated by unstable or poorly developed institutions.

4.5.1 Irreversibility, asset specificity, and project location: risks in the institutional environment

From the standpoint of risk and risk management, institutional aspects play signifi-cant roles. First, as discussed above and below, the institutional framework provides important tools for risk management, especially in terms of contracting tools, finan-cial instruments available (such as various forms of derivatives) and legal structures available (such as companies and collateral – both of which are rooted in the institu-tional framework). Second, in addition to providing the foundation for a variety of risk management tools and approaches, the institutional framework also raises new risks. Fundamentally, we can think of these as primary risks and consequential risks. Primary risks include political risks (including nationalisation, sequestration, non-payment and corruption) and legal risk, with the most critical factors being hostile changes in legislation, unwarranted potential legal liability and uncertainties in judi-cial enforcement. At the same time, the use of various risk management techniques dependent upon the legal and institutional framework also raises consequential risks, that is risks that the mechanism chosen may fail. Examples include contracts which are breached yet are not enforceable due to problems within the contract enforcement framework chosen, collateral devices which are unavailable or unreliable in different institutional environments, and limitations on financing techniques due to legal or institutional weaknesses (for example, underdeveloped capital markets, problematic financial information systems and the like).

4.5.2 The project company in its institutional environment

Many of these risks arise from the location and institutional environment in which the project itself is sited. In looking at these issues, at the outset, a fundamental dis-tinction can be made between four very different institutional contexts: developed, emerging market, developing and transition. At the same time, however, there is a high level of overlap between these categorisations, with actual circumstances varying from location to location.

Essentially, developed institutional environments are to be found in developed economies, such as those of North America, advanced Asia Pacific (Australia, New Zealand, Japan, Hong Kong and Singapore) and the European Union. In this sort of environment, political arrangements are stable, governments are relatively predict-able and bound by their decisions, legal systems support the major part risk manage-ment tools and do not introduce large levels of uncertain primary or consequential legal risks, and financial systems provide access to major financial instruments and techniques. In general, projects situated in such environments can be structured in a relatively straightforward manner, focusing on the financial fundamentals of the project itself.

Emerging market and developing economies in almost all cases will have less sophisticated and developed institutional environments. At the same time, these sorts of economies provide both the demand and the hosting environment for a substantial number of projects. Generally speaking, an emerging market economy may be classified as one in the process of integrating into the international financial and economic system and often with a functioning institutional framework. In this environment, political arrangements and governments have generally achieved a level of stability but not necessarily the permanence one would find in a developed economy. At the same time, the legal and financial systems are frequently incomplete and in the process of development. As such, political risk tends to be higher than in developed economies, as political arrangements and governments may change with greater frequency. In addition, because legal and financial systems are not complete, there may exist significant gaps and uncertainties in relation to both primary and consequential legal risk as well as available and reliable financial techniques.

Developing economies are typically poorer countries, often with underdeveloped political, economic, legal and financial arrangements. In these sorts of contexts, political risk can be very high, and both legal and financial systems prevent significant uncertainties and barriers to approaches adopted in more developed environments.

Transition economies are a special case and depending on the level of development may now have many of the characteristics of developed (for example, Poland), emerging market (for example, China) or developing economies (for example, Vietnam). At the same time, the environment in a transition economy is often sufficiently different from others to merit separate consideration. Essentially, transition economies are countries in the midst of a transition from economic, financial and political systems based on central planning, central ownership and central control. As such, and unlike developed, emerging market or developing countries, transition economies (at least at the start of the transition process, in most cases dating back to the late 1980s) did not have any sort of market oriented institutional framework. As a result, approaches developed in the context of developed environments may simply be unavailable or available to widely varying degrees within a transition environment. This is often especially true in relation to market augmenting legal and financial systems.

As a result, the environment in which a project is situated plays an extremely significant role in determining the risk management approach and structure.

Political risks: development, stability, and the rule of law

The form of risk most commonly considered in financial arrangements relates to political or sovereign risk. Essentially, these are risks that arise from political and governmental circumstances and behaviour in the jurisdiction at issue. At the extreme, these would include civil war and general economic and political collapse (in this time of failed states, a concern in many developing countries and one that may increase over time with global climate change, especially in Africa). At the same time, they also include less severe circumstances such as change in government (which may occur at fixed intervals in many contexts but may be less predicable in other contexts), legal changes (for example, as a result of new legislation or treaties) and the classic consideration of nationalisation (that is, government seizure or expropriation of assets).

At the outset, the most effective management technique is simply to locate projects within stable political environments. Unfortunately, however, this is not always possible nor always desirable, as additional levels of risk should bring with them the possibility of higher rates of return. At the same time, the approach frequently adopted is to work closely with the host government of the project and to secure various forms of assurance regarding the future of the project. In many cases, the host government may be directly involved in the project from the outset and this may be an important consideration. Nonetheless, political risk is not completely ameliorated by a close relationship with the government of the time, as governments and political arrangements can and do change. While it is a general principle that successor governments and political arrangements should carry on with the commitments of previous governments and regimes, this may not be politically acceptable in the actual circumstances. At the same time, it is very difficult to force governments to meet commitments made, especially in the institutional environment of many emerging market, developing and transition economies.

As a result, while direct involvement and commitment of the host government are generally essential, at the same time, depending upon actual circumstances, this may not be sufficient to manage and mitigate political risk concerns.

Other mechanisms frequently considered include involving other parties in the transaction which have greater bargaining power in the host political and economic environment, such as domestic firms and interests, international agencies and insurance.

In relation to international agencies, perhaps the best situated to mitigate political risk concerns are the multilateral development agencies such as the World Bank, Asian Development Bank, African Development Bank, Inter-American Development Bank and European Bank for Reconstruction and Development. Essentially, these institutions are established on the basis of international treaties and operate solely in the context of emerging market, developing and transition economies in their respective spheres of activity. The advantage of such involvement comes from the concept of preferred creditor status. Essentially, each of these agencies is likely to be paid first in the context of financial or political stress in a given country, as in each case, further lending and assistance are often predicated upon payment of existing obligations. As a result, involvement of one or more of the multilaterals can have an important risk mitigation function in projects involving problematic political environments. At the same time, such agencies may also bring other benefits, such as the possibility of addressing certain institutional impediments to specific transactions in ways unavailable to private sector entities. A prudent option, if available in any particular context, is political risk insurance (see Nevitt and Fabozzi, 2000).

Legal system: development, enforcement, rule of law, jurisdiction

As we mentioned, the level of development of the host legal system has important consequences for risk and risk management in any project. Generally speaking, project finance transactions are based upon complex contracting arrangements designed around matrices identifying and assigning risk in the transaction. At the simplest level, such complex contracting arrangements depend upon the possibility of enforcement: after all, if the constituent contracts are not enforceable, then any risk management structure based upon contracting may be completely irrelevant (a clear illustration of consequential legal risk). As a result, to the extent possible, the contractual framework underlying a project finance transaction will be based on a developed legal system providing predictable contract enforcement and dispute resolution procedures. Therefore, many projects situated outside developed countries will to the extent possible seek to have their contracts governed by English or New York law – two jurisdictions that are well-practised in matters of international commercial and financial dealings, and whose courts will commonly hear cases involving commercial contracts even if the underlying connection between the transaction and the jurisdiction is slim.

Unfortunately from the standpoint of contractual risk management frameworks, certain governments will be unwilling to subject themselves or certain transactions to a foreign legal system. In addition, depending upon the actual location of project assets, physical assets can only rarely be separated from the legal system of the place in which they are located (especially important in the context of property, collateral and insolvency). As a result, it will rarely be possible to structure a project involving an emerging market, developing or transition economy under legal arrangements which completely avoid the host legal system.

At the same time, it is frequently possible (and indeed advisable) to structure as much of a project finance transaction under one or more developed legal systems as possible, thereby minimising uncertainties and impediments arising in host or other less developed legal systems. In circumstances where this cannot be done, mechanisms to address political risk issues become of even greater importance. In addition, in some circumstances, it may be necessary to actually support legal development in a given jurisdiction in order for a particular transaction to succeed (for example, it may be necessary to enact a company law, a law dealing with collateral or an insolvency law). In these circumstances, the involvement of multilateral development agencies can be especially beneficial, as these sorts of entities specialise in such processes.

Financial sector: development and stability

As political and legal development play an important role in the structuring of risk management and financing techniques, so does the level of development of the financial sector. As with legal structure, financial development also can be separated into multiple systems in order to best address the specific context of the project.

At an initial level, even in projects situated in developed environments, it is frequently the case that financial arrangements may take place in the international financial system as well as (or instead of) the domestic financial system, frequently because the international financial system will provide a greater spectrum of investors and products than is available even in the most developed domestic financial system. Issues respecting the role of the international financial system are discussed in the following subsection.

In relation to domestic financial systems, at the outset, it is important to realise that there are a variety of structures around the world, each with implications for potential financial arrangements available therein.

Generally speaking, the major models of financial structure include:

- government-dominated financial system (the traditional model in most of the former centrally planned economies and the People's Republic of China, the limitations of which have been exposed as a result of the collapse of the Soviet bloc)
- bank-dominated financial system (the traditional model in Germany – with limitations being exposed through Germany's economic stagnation through most of the 1990s)
- government/bank-dominated financial system (the traditional model in Japan, South Korea and France, the limitations of which have been exposed as a result of the east Asian financial crises)
- family-dominated systems (prevalent in Asia and much of continental Europe)
- securities-based financial systems, with widely dispersed ownership (the Anglo-American or Berle and Means model, which has been receiving new scrutiny following several corporate governance scandals including the collapse of Enron).

In addition to basic structure, market based financial systems also are characterised by different levels of development (Arner, 2007). A financial system can be basic, functioning, developed or sophisticated. A basic financial system typically comprises simple currency, simple payment, simple banking, and simple insurance activities. A functioning financial system provides financing functions beyond the basic level – namely currency, payment, banking and interbank, insurance, simple securities, and simple derivatives transactions. Such a system will also provide a basic level of risk management. A developed financial system provides for effective allocation of resources via market pricing, as well as a variety of instruments and risk management functions. A sophisticated financial system will provide a full and ever-changing range of products and services. Many countries possess access to sophisticated markets, or act as host to a range of organised markets or large numbers of financial intermediaries, but none – not even London, New York, Tokyo or Hong Kong – is complete in this form of sophistication.

Unlike political stability and legal development which both largely parallel economic development, financial structure and financial development vary considerably across developmental levels. At the same time, however, while financial development and economic development do tend to be correlated, financial structure and economic development tend to be less correlated, with different economies adopting different models at different points in time.

The fundamental lesson is that financial techniques available in a given economy will vary depending upon financial structure and financial development and one cannot assume (outside the developed, Anglo-American securities based jurisdictions) that financial techniques (such as private equity or derivatives) will necessarily be available in less developed or otherwise differently structured financial systems.

In addition to financial development and structure, an issue which almost invariably arises in cross-border project finance transaction involves currency. Today, most (though importantly not all) currencies are tradable in international markets with varying degrees of freedom and availability. At the same time, outside developed country projects, finance parties will frequently prefer cash flows to be structured in major currencies (most typically US dollar or euro) and most internationally traded commodities are traded in US dollars. Nonetheless, construction expenses (outside major items of imported technology) will frequently be in local currency; revenues and expenses in the operation phase may also be in local currency to a greater or lesser extent. Many projects will therefore require mechanisms to transfer and manage currency risks (as discussed above). While the international financial system is rich in such tools and instruments, they are not always available (for example, when dealing with the Chinese currency) or available at reasonable cost or reliability.

Is culture relevant?
In addition to political, legal and financial structure and development, questions may arise as to the relevance of culture. While we will not dwell on this issue, it is important to realise that different cultures approach relationships in varying ways, and this applies as much in the context of project finance as any other.

4.5.3 *Project finance within the international financial system*

As noted above, project finance frequently makes extensive use of the international financial system, in addition to or instead of the host financial system. Generally speaking, the international financial system can be characterised as a sophisticated securities based system, operating generally on the basis of the highly developed legal systems of English or New York law. Although transactions do not necessarily involve either jurisdiction directly, in most cases, financial intermediaries based in either or both jurisdictions will be involved at least to some extent. Generally speaking, while the international financial system offers access to the wider possible range of instruments, techniques and institutional investors (retail investors are frequently only accessible through recourse to the appropriate domestic financial system and regulatory framework, for example in the United States and generally the European Union), use of the international financial system also brings with it a certain additional legal complexity and uncertainty, as intermediaries and investors may be domiciled in a wide range of different jurisdictions. While a wider discussion is outside the present context, it is important to realise that even in the international financial system, concerns regarding appropriate choice of governing law and especially jurisdiction of domicile or incorporation of various entities in the transaction become not only significant but complex, raising especially issues of consequential legal risk.

In addition, it bears mentioning that an increasing number of intermediaries (especially banks) involved in the international financial markets are regulated under a variety of international regulatory standards which can have important implications for structuring of financial arrangements, related incentive framework and regulatory concerns. Of note are the international standards regarding regulatory capital treatment of different classes of assets of the Basel Committee on Banking Supervision: the 1988 Basel Capital Accord and its recent revision, the Basel II Capital Accord. The latter especially provides detailed standards addressing special types of lending such as project finance which will impact directly in many cases on approaches taken in the international financial markets.

4.5.4   Development initiatives, multilateral participation and guarantees

We mentioned previously in the context of political risk the usefulness in certain circumstances of involving a variety of multilateral agencies. In addition to their use in managing political and other institutional risks, multilateral agencies also have a direct interest in many project finance transactions. Generally speaking, agencies such as the World Bank and Asian Development Bank were formed to support economic development in their countries of operations and today have a primary focus in funding transactions which will enhance welfare and reduce poverty. In this role, the multilateral agencies are very active in promoting, supporting, structuring and financing projects which will enhance welfare and reduce poverty in emerging market, developing and transitional countries around the world. As a result, these agencies have wide experience not only in operating in challenging institutional environments but also in pushing projects which fall within their remit and working with governments to make them work. Overall, therefore, these agencies may be involved in all aspects of project finance transactions.

In addition to the multilateral agencies, there are also a number of bilateral agencies which may be involved in project finance transactions in various roles and can play a useful role in allocating and managing risks. Of these, the most significant are the European Investment Bank (which operates largely in the same way as the multilateral development agencies, but focuses on the European Union and its former colonies) and a series of bilateral aid, investment promotion and trade guarantee agencies. Of these, those of the major developed countries (United States, Germany, Japan, United Kingdom, France, etc.) tend to be most frequently involved in project finance transactions. Generally speaking, the bilateral aid agencies (for example, USAID, GTZ) may be involved in the feasibility study stage or providing financing for various technical experts (typically tied to the country providing the financing through citizenship or residency). In addition, investment promotion agencies may provide a variety of political risk guarantees and insurance to transactions involving firms from the country concerned. Finally, a variety of trade guarantee agencies (typically called export-import banks or eximbanks) provide financing, guarantees and risk management tools (for example, insurance) for technology and capital goods being exported from the home jurisdiction to other jurisdictions.

4.6   Risk management and project lenders

The circumstances surrounding project venture risks and how these manifest themselves over the project phases were broadly outlined in Sections 4.2 to 4.5 above. It was stated in the introduction to Chapter 4 that the principal point of departure for Sections 4.2 to 4.5 is the risk position of the project owners – the equity owners (which includes promoters with equity invested). This is because under normal corporate finance assumptions and most legal frameworks, we know that the equity investors have residual rights to all assets after all company liabilities have been settled; thus they own the net assets of the company and they are the ultimate risk-takers. With project companies lenders are much more exposed, given the typically high proportion of project debt relative to overall project capital investment, and given that project finance loans are ideally intended to be non-recourse or limited recourse. As explained before, this means that the project lenders are generally more exposed than equity investors by scale of capital invested. This also means that project risk identification and management is of absolute paramount importance to project lenders, too important to leave to project promoters. Indeed, most, if not all, of the project company risk management measures put in place and described in Sections 4.2 to 4.5 are likely to be insisted upon by potential lenders as formal preconditions to entering into any project loans with project promoters and equity investors. This is aimed at creating acceptable credit risk for project lenders given the scale of capital provided. In project finance terminology, if project promoters cannot demonstrate to potential project lenders that acceptable project risk management measures are in place, project lenders will likely not consider the project proposal a bankable credit. Recall that we pointed out that banks specialise in analysing and taking credit risk, and not equity or venture capital risk (despite the fact that there remains residual business risk in every lending decision).

These observations necessitate that we make a number of summarising observations about loan characteristics from a project lending bank's point of view. These observations have all been dealt with in some form elsewhere, and where relevant we will refer to the relevant sections in chapters 1, 2, 3 and 5. This section aims to group in one place broadly how project lenders' risks are typically covered in loan contracts including how loan amounts are determined, so that project company risk management generally can also be placed in its wider context. We approach this section from the perspective of the economics of contracts as explained in Chapter 2.

4.6.1   Project phases and lending

In project lending, it is possibly a fair assumption to make that lenders prefer to act as lenders to both the construction phase as well as the operational phase of a project

– obviously earning fees and interest from two transactions is preferable to one. As has been pointed out, the risks associated with the construction phase have typically led to separation of debt financing of the construction phase from the operational phase. These are effectively treated as two independent lending transactions, with the further condition that the second transaction will not proceed before the first is successfully completed.

The separation of project lending into separate loans to the construction phase and the operational phase may be described as a fundamental risk management strategy practised by financial intermediaries in project finance lending. This fact may be evidenced by the fact that lenders often conditionally commit themselves to what has been referred to as permanent project lending, based on the condition that the construction and commissioning phase is completed successfully – hence the separation between construction lending and permanent (or take-out) lending. Although it greatly depends on who the parties are, the risk management logic is simply that the construction loan liability must be substantially settled before a permanent project loan contract is finalised. This means that if the construction and commissioning phase is not successful, the construction loan becomes due and no permanent loan will be entered into. While the construction loan and the permanent project loan formally are separate and sequential loan contracts, in practice the construction loan is often converted into a permanent project loan to the project company after the construction and commissioning phase has been successfully concluded. It would be extremely risky for project promoters to enter into a construction loan without a commitment of permanent loan finance conditional upon successful completion of construction and commissioning.[9] We comment below in more detail on how this specific mechanism translates into different details for construction lending and project lending.

4.6.2 Permanent project debt

Again, we will commence by assuming the completed project is the subject of analysis, and that the lender is considering permanent project lending – which of course refers to a long-term loan to the project company, influenced overall by the project's expected life. Typically loans to project companies classify as term loans, although project financing has also been provided through bonds and the bond markets and private placements,[10] mostly in the United States. In the United States, project bonds

---

[9] Of course, it does happen. The risk then becomes that there is no permanent loan forthcoming to relieve the construction loan repayment commitment due at project completion. This explains why permanent project finance is also often referred to as take-out finance.

[10] Private placement refers to securities that are not listed on an organised exchange. In the United States, the term has a more specific meaning under securities laws, and relates to exemption from certain disclosure rules of the Securities and Exchange Commission. Because less disclosure and formalities are required than for public issues, arranging a private placement is typically less costly to arrange than a public issue, but may involve higher yields due to their lack of liquidity.

have also been used to finance transportation infrastructure by some local authorities (these are sometimes referred to as revenue bonds). (See Chapter 2, Appendix 1: Long-term Debt Instruments for more details.)

The term debt obligations of project companies are more commonly encountered as non-recourse or partial recourse loans rather than bonds. Term loans are either made directly or arranged by banks, which usually syndicate large transactions in order to reduce their own exposure to any single obligor (Chapter 2, Appendix 1 includes a discussion of loan syndication). It has to be kept in mind that particular items of equipment (such as turbines in a power plant) and some other assets (for example land) are often financed through leasing from manufacturers and landowners, or financial lessors. There are essentially two types of leases, briefly explained in Box 2.2, Chapter 2. For our purposes the capitalised value of financial leases may be viewed as purpose specific alternative debt contracts, and thus form part of project company debt, while rental payments on operating lease contracts (such as for land or industrial assets with long economic lives) may be viewed as interest only payments. A major advantage of a term loan over bond finance is the relative speed with which a term loan can be made, also relative to the complexity of its terms. The finance made available by term loans to project companies is typically funded once the engineering, construction and commissioning are successfully completed, and amortised usually according to a pre-determined schedule over a period influenced by the project company's expected life. In this respect it resembles a corporate term loan, except of course that it is to a project company. While it is not uncommon to release permanent debt finance to the project company in tranches, it is likely to be advanced when the project's operational phase commences, and so refinance or 'take out' any remaining outstanding construction lenders. Banks are by no means the only investors in project finance loans; insurance companies loans view project finance, and other investors, also as attractive investments.

The most fundamental risk to lenders is credit risk, and credit risk appraisal is the principal risk management mechanism employed by project lenders. Specialist credit risk analysis is considered to be the most fundamental skill of commercial banks (Greenbaum and Thakor, 1995) but it also has to be appreciated that banks are active in various different types of lending (trade finance, consumer lending, term lending to business, etc.). Each loan category has its own particular risk characteristics and credit risk analysis frameworks, and so it is with project lending. In essence project loans are special term loans to special types of businesses – project businesses. Term lending to business of course involves several concerns, including the industry and its economics, financial resources available to the company, quality of its management, the composition and stability of earnings, assets available for collateral, constraints on financial resources relating to restrictive covenants from other loan obligations, and so on, generally similar variables as those that bond ratings agencies analyse. We are not going to deal with the total phenomenon of credit risk analysis, instead we proceed from the assumption that a bank has satisfied itself with the credibility of the promoters and equity investors, with the project's feasibility and its bankability, and has conducted analyses of all the matters mentioned above, including due diligence.

To round off this section, we mention two important further matters relevant to credit risk analysis of project lending, namely interest rates and credit ratings. Two observations are important in considering the level of interest rates for permanent project loans, be these project term loans or bonds. The rate of interest at which the loan is contracted is important to both borrowers and lenders. Recall when considering a project company's debt capacity (following Section 3.2), the present value of a project company loan is inversely related (though not linearly) to the rate of interest. To borrowers it is of immediate importance, because the lower the loan interest rate the higher the loan amount the project company may secure for the same set of project cash flows, but also the lower its interim interest charges, which directly affect its free cash flow and operational profits. As with all lending, lower project credit risk is expected to reflect in lower cost of borrowing through lower transaction fees and lower rates of interest on syndicated project company loans, typically through lower margins over the loan benchmark (Libor, Hibor or another benchmark). To project lenders, there is an obvious moral hazard risk in this situation. Because well-conceived and executed projects benefit from lower rates of interest, there is a clear incentive to promoters, and risk to lenders, that project promoters may misrepresent project details in order to suggest less project risk. Project lenders' due diligence procedures therefore include extensive in-house simulation and stress-testing of project assumptions and feasibility to overcome credit risk from such actions. Clearly this is an asymmetrical information problem, and demonstrates again the importance of dealing with reputable counterparties that have demonstrated credible commitment, for example through non-trivial equity participation. The second observation is that the main historic difference between corporate bond financing and large term loans (whether for project financing purposes or not), is that historically the interest cost to the borrower in a bond transaction was typically fixed, while modern project loans are typically at variable or 'floating' interest rates. While this distinction is rendered largely irrelevant by modern derivatives markets, especially interest and currency swaps, as explained in Box 4.3 and Chapter 2 Appendix 1, it still has to be observed that if unmanaged, interest rate risk with fixed rate bonds lies with lenders, while with floating rate debt it lies with borrowers.

Before considering typical project loan conditions, it is also appropriate to discuss briefly credit ratings agencies and their activities in rating projects. Traditionally credit ratings agencies independently analysed and rated the risk of corporate bond issues in capital markets according to their ratings frameworks, focusing originally on bonds that were to be traded publicly. It is fair to say that independent rating of a bond issue is critical to a bond issue's marketability and tradability in bond markets everywhere, and particularly so because many financial institutions are by regulation not allowed to invest (or are limited in how much they may invest) in unrated bond issues for portfolio purposes. The most influential international ratings agencies are

Standard and Poors, Fitch, and Moody's. Pioneered by Standard and Poors, since around 1990 ratings agencies have expanded activities into rating project risk, first in the United States but now also internationally. Project risk ratings have become an important development for project promoters in developing countries, since it has facilitated access to international lenders and capital markets for project debt in developing countries, but it is also worth mentioning that project ratings activities received an important setback during systemic financial instability following the Asian financial crisis in 1997–1998.

We conclude discussion of typical term loan contracts to project companies by repeating the behavioural assumptions that underlie all transactions (and the written evidence of transactions, namely the written contract), all as discussed in Chapter 2. The assumptions are of course bounded rationality, imperfect contracts and counter-party opportunism. We will use these concepts to consider typical conditions of long-term loans to project companies, because it will allow us to concentrate on the logic of standard terms and conditions of international loan contracts and so significantly simplify presentation. To recap, we concentrate on the economic logic that influences the form of a particular complex transaction, and use the search, screen, contract, bonding and monitoring provisions, and transaction governance framework introduced in Chapter 2 to describe a typical term loan, rather than typical legal structure which normally concentrates on parties, covenants (financial, positive, negative), and defaults and remedies. Obviously the search for and screening of the lender's counterparty had been completed by the time we arrive at this, the 'contracting' stage, and details of the contracting parties, their status and capacity to contract, and so on will all be part of the formal contract, so it will not be raised again. We are thus concentrating on the contract as the formal representation of the loan transaction, of typical bonding, and of the overall monitoring provisions and transaction governance arrangements. Because we have introduced most of the relevant considerations in earlier sections of this chapter, in Table 4.4 we present in summarised form only the nature of typical term loan conditions. Remember that we are concerned with the spirit of the transaction, and while detail is important and good drafting of legal documents is critical, we consider counterparty opportunism to be the enemy of the spirit of any contract, given its inevitable incompleteness. The spirit of the transaction is what concerns us under the column 'Transaction cost logic'.

4.6.3   Construction lending

We have explained on several occasions that the most risky period of any project is the construction phase, simply because so much can go wrong and because there is no project before successful completion of construction. Risks in the planning, design and engineering phase of the project may be managed adequately through choice of experienced promoters and facility designers, choice of proven construction technologies and adequate investment in uncovering particular access and physical risks of

Table 4.4 Summary terms of model project term loan transaction (following Nevitt and Fabozzi, 2000).

Transaction activity	Typical contract conditions	Transaction cost logic
Search and screen	Description of parties, definitions, and jurisdiction. Key results of search and screen process explicitly stated, as part of due diligence. This includes the disclosure of all relevant borrower information, such as incorporation, subsidiaries, associates, domicile, legal standing and capacity to contract, status of legal title to assets, compliance with laws and regulatory requirements such as approvals, taxes and licences, the financial standing of the borrower, and the status of pending litigation. The borrower will make warranties and representations as to the accuracy of disclosed information.	The logic of summarising all these representations and warranties is simple. If misrepresentations are made over any relevant aspect of the transaction (facts which may influence materially the debtor's ability to pay) the contractual result will be to release the lenders from certain obligations, for example to make further loans. The credit agreement will include a test of materiality in these respects.
Contract	Credit transaction described in detail: nature, type, and purpose. Commercial terms of the transaction: commitment amounts, availability and conditions precedent to the drawing of the loan or any single part or tranches. Repayment provisions, grace periods, provisions for accelerated performance related repayments, interest rate basis, credit spread and basis of calculations, permissible variations in payments, grace periods in the event of delinquent payments, prepayments and prepayment penalties, if any. Seniority or subordination of loan, accounting compliance, disclosure and reporting information, compliance with regulatory and legal requirements such as corporate governance, permissions, licences, and taxes, agreement to maintain project company property and assets. Limits on management actions over sale of assets and businesses, changes in business purpose, sale of project company, takeovers, mergers, acquisitions and significant shareholder transactions. Restrictions on further borrowing and granting of collateral. Restriction on alienation or assignment of property, plant, equipment and the lease of assets. Assignment of earnings and insurances, performance bonds or sureties (related to project output quality or construction performance).	Describes the transaction, and the agreement the counterparties have made as to its detail. In receiving agreement to the project loan, the project company management relinquishes managerial discretion, for example over use of free cash flow, and agrees to manage the company according to prescribed contractual rules.

Table 4.4   *Continued*

Transaction activity	Typical contract conditions	Transaction cost logic
Bonding	Execute formal charges, and if necessary create informal liens over property, plant, equipment and leases as security for the loan. Prescription and assignment of loan performance guarantees, if the lenders have limited recourse to project sponsors. Prescription and assignment of input and off-take risk management transactions. May include supply agreements (e.g. put or pay, price agreements, volume agreements, associated third party supply or demand guarantees, demand agreements (take or pay, volume agreements, third party guarantees, access to infrastructure (rights of way), and leases. Prescription and assignment of facility maintenance management contracts, other long-term contracts to supply services or associated with operating the project company, such as operational management contracts.	Intended to bond counterparties to their contractual undertakings and promises and monitor subsequent performance while the transaction is extant. Except in respect of payment defaults and cross-defaults on other funded indebtedness, this is largely a device to permit and safeguard lender control.
Monitoring and general transaction governance	Transaction monitoring arrangements, general availability of and rights of access and inspection of accounting and other management reporting information, information relating to upkeep of regulatory requirements, including approvals, licences, taxes, and service fees. Periodic financial reporting, including management compliance statements as to financial covenants such as leverage, interest coverage, working capital requirements, and liquidity requirements. Other financial reporting, such as financial ratios; arrangements to inspect plant, equipment, maintenance plans, schedules and contracts; administrative arrangements for managing default and remedies. Circumstances under which the transaction may be accelerated, terminated or otherwise restricted in the event of defaults or outside occurrences such as force majeure events.	Monitoring and reporting on borrower actions to be taken to ensure contractual performance. In principal the necessity for observation of, or monitoring, performance is to ensure transaction execution meets intentions, and to act as a control mechanism to permit early corrective action. Serial breaches may signal a deterioration in the parties' interest to bring the transaction to its intended conclusion. Contractual remedies are then specified, including access to full institutional enforcement machinery. Transaction governance may include rules to allow changes to the agreement, and rules for making decisions in unforeseen circumstances.

the proposed facility, all coupled with appropriate guarantees. Most large commercial banks have project finance departments that employ technical experts (including engineering and construction consultants, cost consultants, legal consultants, and so on) that are able to audit independently what is proposed in the promoter's project memorandum, including identifying further difficulties foreseen in executing project construction.

The construction phase nevertheless remains an extremely risky period of the project life cycle. The circumstances under which project loans may be made without separation of the construction and operational phases in lending are rare indeed, irrespective of how good the reputation of the promoters and construction companies is and how good their track record is. Similarly, it will be rare if construction lending is not conditional upon a performance guarantee from the contractor (its own, or a third party) or formal surety, plus some form of retention account or maintenance bond. This is still not necessarily of great comfort to project promoters and lenders – a half completed technologically complex facility in a remote location that calls for the replacement of a principal contractor for non-performance is a disaster of immense magnitude in terms of delays, cost overruns, managerial time and for political relationships that might suffer as a consequence. While problems during the construction phase happen and can be managed by reputable contractors, events leading to dismissal of the principal contractor during the construction phase are simply so serious that absolute prevention is practically the only risk management option. This is primarily the reason why it is crucial that the engineering and construction phase be the responsibility of highly reputable companies. But while lenders are aware of the risks in the construction phase, returns to construction phase lending are then also expected to be higher.

From Chapter 3 we know that in essence a construction company places a highly leveraged bet when it contracts to deliver a project. It invests corporate financial resources in the start-up of the project, which is subsequently leveraged by receipt of interim progress payments as the project progresses. In financial terms project success for the construction company invariably depends on how well it manages its outflows (project costs) relative to inflows (interim progress payments) – the better its operational financial management during the project, the better will be the returns it earns on the corporate financial resources it invests in the project start-up (see Chapter 3). But operational cash flow management cannot dominate the necessity to construct the facility, and often is the source of incentive conflict between project companies concerned with progress and project managers attempting to manage construction cash flows to optimise returns.

Given this simplification, we are able to describe a number of critical concerns that are typical in construction lending to project companies. Typically the loan will be made to the project company, which will in turn employ a design and construct company to complete the design, engineering and construction phase (commissioning may be part of this arrangement, but is often separated for confirmation of facility quality and performance). As explained, often construction companies are very active

as project promoters, and are thus frequently also equity investors in project companies – and as has also been stated, in many cases project promotion is part of their
business model, in order to secure construction contracts but also in many cases to
secure post-construction business in facility management activities. Interest rates for
construction loans reflect the higher risk of the construction phase. While repayment
arrangements are not standard, the usual case with construction lending is that it is
in principle due upon completion, with interest often accumulating on the outstanding balance, as it increases progressively with project progress.[11]

For our purposes a simplified view of construction loans is thus as a term loan of
a corporate finance variety, made to a project company to finance the physical construction of the project facility. The loan is to be repaid to the lender upon satisfactory
completion of construction of the project facility, before the subsequent operational
phase and irrespective of the operational success of the project. Typically construction
loans will be advanced to the project company in tranches, in turn to be used to pay
the construction company based on construction progress. Thus the project company
technically pays the construction company as work progresses on the project, but in
this respect the project company effectively acts as a flow-through mechanism – it
pays the construction company with proceeds of the construction loan tranches it
receives from the bank which has provided the term loan (see Chapter 3). This provides the lender with some ability to monitor interim progress, and withhold funds
to the project company to ensure quality.

Disbursement of construction term loans in tranches also functions to align construction company and project company incentives. Assuming the construction
company is a shareholder in the project company, it is typical to make the release of
the first tranches of the construction loan conditional upon the construction contributed company (in this case including the construction company) having at least part
of its investment of its equity commitment in order to signal commitment to the
construction of the project as well as its operational success. This further explains
why the equity commitment of the promoters must demonstrate credible commitment to the project. As a further risk management mechanism by the lender, the
project company may be required to guarantee repayment of the loan (in turn typically
guaranteed by its parent, or a third party), while the construction company may be
required to provide a performance bond or surety. Both these contingent liabilities
will only be discharged upon satisfactory completion and commissioning of the facility (often it partly remains in place for a maintenance period after completion). Of
course, the operational phase is intended to be financed with a separate permanent
term loan. Once the conditions to accept the project as operational are fulfilled following successful construction and commissioning, the construction loan will be

---

[11] In most jurisdictions, for balance sheet purposes accumulated construction loan interest is treated
as part of the capital cost of a facility.

repaid by the project company, usually by using the proceeds of a new term loan, the permanent project loan. Thus the construction loan is converted into a term loan to the project company for the operational phase of the project.

We will approach summarising conditions encountered in typical construction loan contracts again by repeating the behavioural assumptions that underlie all transactions, namely bounded rationality, imperfect contracts and counterparty opportunism. As before, we will use these concepts to consider typical conditions of construction loans to project companies and their promoters, rather than take a legalistic contractual approach, and as with the approach taken with term loans, we use the search, screen, contract, bonding and monitoring provisions, and transaction governance framework introduced in Chapter 2 to describe a construction loan. We thus concentrate on the contract as the formal representation of the construction loan transaction, of typical bonding, and of the overall monitoring provisions and transaction governance arrangements. Because we have introduced most of the relevant considerations before, in Table 4.5 we present in summarised form only the spirit of typical construction loan conditions. (Chapter 2, Appendix 1 contains a more detailed explanation of project related long-term debt.) Remember again that we are concerned with the spirit of the transaction, and while detail is important and good drafting of legal documents is critical, we consider opportunism to be the enemy of the spirit of any contract, given its inevitable incompleteness.

Table 4.5 Summary terms of typical construction loans (following Nevitt and Fabozzi, 2000).

Transaction activity	Typical contract conditions	Transaction cost logic
Search and screen	The first requirement is a detailed description of the parties, definitions, jurisdiction, expenses, etc. In the contract the key results of the search and screen process will be explicitly stated, essentially as part of evidence of due diligence. Assuming the borrower is the project company, this will include all the warranties and representations of relevant borrower information such as incorporation and controlling shareholders, domicile, legal status and capacity to contract, legal status of titles to assets, compliance with laws and regulatory requirements such as taxes, licences, etc., financial standing of the borrower, no pending adverse legal actions, and so on. Similarly, lenders will also make representation as to its status and intentions with the transaction, its good standing, and so on.	The logic of summarising all these representations and warranties for both parties is simple. If any misrepresentations are made by either party concerning any relevant aspect of the transaction (such as facts which may influence construction company credibility), the guilty party may be easily identified and action taken in law OR according to transaction governance rules.

Table 4.5   *Continued*

Transaction activity	Typical contract conditions	Transaction cost logic
Contract	The actual transaction itself is described in detail: what the transaction is (a loan by the lender to the borrower), what the nature of the project company business is (project company), and what the purpose of the loan is (construction finance for the project facility). This will further include details of the financial transaction: the overall amounts, tranches, and associated amounts, conditions precedent to payment of any tranches. Further it will include details of interest rates, repayment principal and interest, which benchmarks are applicable, prepayment and penalties (if any), and grace periods (if any). Agreement on seniority/subordination of loan and rights of other lenders; maintenance of accounting and other project cost auditing/ reporting information; compliance with all regulatory and legal requirements such as corporate governance, licences, taxes. Limitations on managerial actions with respect to sale of the project company, take-overs, mergers, acquisitions, shareholder transactions prior to completion of construction.	This describes the transaction, and what the counterparties have agreed to with respect to its detail. It is concerned with what the parties are going to exchange. In exchange for the construction loan to the project company, the project company management relinquishes managerial discretion over the application of construction loan tranches, and agrees to manage the construction project according to prescribed reporting and auditing arrangements. This is to ensure that loan proceeds are strictly applied to constructing the project facility.
Bonding	Assignment of the construction project and all titles to fixed assets (land titles, leases, etc.) and the construction contract. Prescription and assignment of loan guarantee to lender. Assignment of construction company performance guarantees and/or completion. Bond/surety/maintenance bond to lender. Restrictions on further borrowing (or particular types of borrowing, e.g. other short-term construction lending). Prescription and assignment of fire, property, casualty, regulatory insurances; performance bonds or sureties (related to project completion and/or construction performance).	These conditions are all intended to bond counterparties to their promises. The intention of course is that the party that breaks their bond has to suffer a financial consequence. If the project company fails during construction, the lenders may take over project construction; similarly if the construction company does not perform, the lender may force the project company to replace it.

Table 4.5   *Continued*

Transaction activity	Typical contract conditions	Transaction cost logic
Monitoring and general transaction governance	Transaction monitoring arrangements: general availability of and rights of access to the construction site. Requirements for reporting on construction progress: cost, quality and time. Arrangements for disbursement, accounting and auditing of construction loan advances. Inspection of accounting and other auditing information regarding disbursement of project loan tranches. Information relating to upkeep of regulatory requirements, licences, taxes, service fees, etc. Circumstances under which the transaction may be terminated or otherwise impaired, such as lack of progress, financial distress, force majeure.	There are two purposes with this category. One is reporting on those borrower actions that have to be taken under the contract to ensure performance during the construction of the project. In principle the necessity for observation, or monitoring, of construction progress is both to ensure that the project is delivered on budget, on time and according to quality requirements, but also to act as an auditing function of the use of construction funds, and as control mechanism to ensure early corrective action. In the event of serial breaches, of course it may signal a deterioration in the parties' interest to complete construction of the facility, and then contractual remedies are specified, including access to full institutional enforcement machinery. Transaction governance may further include rules to follow to modify the agreement, and rules for making decisions in unforeseen circumstances affecting project construction.

Key concepts

The following concepts are considered sufficiently important to memorise as key vocabulary for use in subsequent chapters.

Construction loans	Put-or-pay agreements
Over-the-counter transactions	Take-or-pay agreements
Permanent project company debt	Commissioning risks

Construction phase risks
Credit ratings
Demand risks
Design and construct transaction
Design and engineering risks
Distribution risks
Diversification
Economic development initiatives
   and project finance
Exchange-traded contracts
Financial system risks
Forward contracts
Guarantees
Hedging
Insurance

Irreversibility risks
Legal system risks
Options
Performance guarantees and sureties
Political risks
Risk and uncertainty
Risk identification, risk analysis, risk
   assessment
Risk transfer
Risk management process
Risk phases
Supply chain risks
Sureties
Swaps
Technology and technological risks

5

Continuing Evolution: from PF to PFI, PPP and beyond

5.1   Introduction

The discussion in chapters 1–4 illustrated the development of project finance over time. The central role that the private sector has played in project financing provided the inroad that enabled both privately financed projects, whether as part of the private finance initiative (PFI) in the United Kingdom or public private partnerships (PPPs) both there and elsewhere, to develop as derivative models. While PPP is a generic term that encompasses a number of partnership options for service delivery, PFI is singularly associated with the United Kingdom. Other countries with PPP programmes include Australia, Canada, Chile, Czech Republic, Finland, Germany, Greece, Hungary, Ireland, Italy, Japan, Korea, the Netherlands, Portugal, Spain, Singapore, and the United States. International Financial Services, London, a private sector organisation promoting United Kingdom based financial services internationally, reports that as many as 70 countries worldwide are currently taking steps to develop their own PPP programmes. Further expansion of PPPs is expected in Europe under the European Union (EU) Growth Initiative and both the World Bank and the Asian Development Bank support PPPs through various initiatives as well. Because it is well documented, we draw in large measure on British PPP/PFI experience in this chapter.

In PFI transactions a private sector service provider is typically given responsibility for designing, building, financing and operating assets from which public services are delivered. The central objective of a PFI scheme from the public sector point of view is to create a structure that produces and optimises value for money (VfM); in short the solution that consumes the fewest resources to achieve the intended service outcomes. VfM is premised upon private sector innovation and management skills as well as the efficiencies realised from optimum risk allocation between the parties over the anticipated whole life of the asset in service delivery. While VfM is expressed in financial terms it may represent other forms of costs or benefit, e.g. social, environmental, health, safety, etc. Relative value is determined by comparing service delivery options and the business case analysis against these forms. When these benefits are coupled with the lower demands on government operating budgets and possible reduction of reduced fiscal deficits the case for PFI is strengthened. Research by the United Kingdom Treasury in 2003 indicates that better VfM arose on PFI projects versus those traditionally procured by between 5% and 40%. It should be pointed out that much of the argument for PPPs depends upon accepting these efficiencies of the private sector though in actual fact there is not a great deal of empirical evidence to support it. Looked objectively at what can be said is that efficiencies do derive from competition and thus where it is increased the outcomes involved are often improved. This element of contestability is an important part of the promise and success of PPPs and when coupled with new providers and the element of diversity that they bring, the benefits are further reinforced. Hence one may look at PPPs both in terms of how competition is introduced within the private sector as well as between the private and public sectors.

From a conceptual standpoint PFI marked a change in procurement from the need to build or acquire an asset to the purchase of a service. Thus assets typically procured around the fulfilment of a construction, or design and construction, contract over a finite and fairly short of period of time became transformed through the addition of concomitant financing arrangements and extended periods of operation or facilities management of those assets as well. The addition of the finance component in particular changed forever the approach to the project and the degree of sophistication needed to assess the risks where private funds were at stake. Over time what has become most evident is that this transformation has underscored the shift in concept from where assets were used to house service providers, to where those service providers themselves have become the focus of attention. The full range of asset procurement options is set out in Figure 5.1.

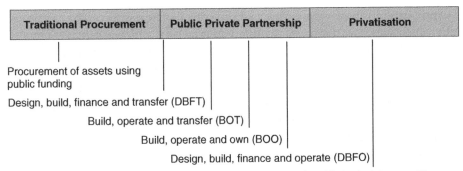

Source: KPMG, KLegal

Figure 5.1 Full range of asset procurement options.

While PFI may be framed in terms of procurement it is also central to the challenge facing countries today to fund or invest in new infrastructure development or refurbishment to meet demographic trends, and close in effect a funding gap which has arisen in response to this need. The funding gap can be seen in particular with the under-investment that the London Underground experienced over a period of decades. The London Underground PPP is looked at in Chapter 9.

Governments today are better placed to understand not only what best promotes successful procurement but also how the public interest should be served. Thus development has to reinforce such public goals as efficiency, access and high-quality services. As infrastructure is funded from tax revenues and should reflect that taxpayers are entitled to have their projects delivered on time and on budget with appropriate safeguards, VfM and PFI have become new cornerstones for procurement in many of the countries referred to above which are now pursuing PPP programmes.

Closely related to VfM is the appropriate role for the public sector in either controlling or owning assets which have been delivered through PFI. PFI contracts are thus on the cutting edge of delineating and redelineating the lines between the public and private sectors. At that intersection, and given that public funds are at risk, tendering processes must be scrupulously fair, transparent and efficient. Clear guidelines and full public disclosure have thus become central principles in relation to the process. In practice this has meant that governments should reveal as much as they can about the real costs and likely benefits from their PFI and PPP projects to ensure long-term public support for their programmes. Finally, and as these projects are of such long duration, it has been shown that there must be real accountability for them to succeed. The lack of accountability has been a significant weakness in traditional procurement when assets have often deteriorated faster than expected with either cost increases or impairments in the service delivery over time. Strong accountability entailing accurate and complete forecasting is thus a very important factor in judging the efficacy of PFI project frameworks and is touched upon below when lessons are considered.

The public private split in privately financed projects also bears upon what at times are divergent objectives for the participants when acting independently. Thus a government will place more emphasis on political or public opinion than may a private sector partner. Similarly the achievement of social goals may not feature prominently on private sector projects. In contrast, the private sector partner will typically be preoccupied by financial returns, retaining or increasing its competitive position in the market and realising corporate goals. Thus, in privately financed projects, when the two participants come together a true partnership must be struck in relation to this split. Compromises on these and similar types of issues by the participants then lay the foundation for success, particularly where their interests strongly overlap.

A brief review of PFI development is now outlined.

5.2 PFI origins

The 1980s in the United Kingdom were a time of increasing involvement of the private sector in the provision of what had typically been public sector projects. This coincided with a renewed interest in both privatisation and project financing and the early success of build-operate-transfer (BOT) projects in a growing number of countries. It was coupled with other trends, such as outsourcing, which accustomed the public to seeing the private sector in unfamiliar roles. Seeing these developments in context the United Kingdom Treasury established a committee chaired by Sir William Ryrie to consider and report on the role of private finance in public sector works. Releasing its report in 1981, the so-called Ryrie Rules became the basis on which private sector funding could be used by the government. In short, the rules stated that to be approved, privately funded projects had to be tested against a publicly funded alternative and shown be more cost-effective and that, save in exceptional circumstances, privately funded projects would result in equivalent reductions in public expenditure.

In practice, rather than enhancing the role of private finance in public sector projects, the Ryrie Rules largely had the opposite effect. The dual requirement of having to show not only that private finance was more cost effective than public funding but was also displacing it as well, meant that few projects could qualify. Those qualifying included: the Channel Tunnel in 1985 which involved no public sector financing; the Queen Elizabeth II Bridge or Dartford Crossing in 1987; and the Second Severn Crossing in 1990. As the projects were concentrated in the transport sector it was becoming clear that more would be required to expand the role of private finance to other sectors.

The change came in 1992. It was at this time that the Conservative Government formally launched the Private Finance Initiative (PFI). PFI was designed to involve the private sector in both the financing and management of infrastructure and other projects which to that point would have typically been delivered by the public sector. While primarily associated with roads, bridges, railways and utilities, or what may be termed *hard infrastructure*, the last decade or so has also come to associate the term with schools, prisons and hospitals, or what is sometimes called *soft infrastructure*. According to the then Chancellor of the Exchequer in 1992, Norman Lamont, PFI would '*allow private financing of capital projects*' in the public sector. Advocates of PFI would argue that the private sector would also bring a greater degree of innovation and business acumen than would have been the case under solely public sector delivery. Momentum gathered and in November 1994 the Chancellor stated that the '*Treasury will not approve any capital projects in future unless the private finance options have been explored*'. Between 1992 and the Labour Government coming to power in 1997 the United Kingdom government policy on PFI was shaped by a Private Finance Panel within the HM Treasury. The Private Finance Panel comprised public and private sector representatives who could mobilise support for the new direction. The new direction, in contrast to the previous policy under the earlier Ryrie Rules, no longer required private sector projects to be compared against a publicly funded alternative. Apart from this key change other directions and forms of private sector participation emerged. Significantly, the private sector was encouraged to lead joint ventures with the public sector as well as take advantage of greater scope for the use of leasing. In exchange for a larger private sector role the government sought real risk transfer on the projects and the showing of VfM as that term came to be understood.

The creation of the Private Finance Panel coincided with the release of a Treasury report, entitled *Breaking New Ground – towards a new partnership between the public and private sectors*, that outlined the move to public private partnerships. While the change was not fundamental it underscored the fact that PFI, going forward, was but one form of a more generic type of public private partnership. As a means of encouraging government departments to go forward with PPPs, the Treasury announced in 1994 that it would not approve capital projects unless the department concerned had considered PFI as one option. This requirement, which came to be expressed as universal testing, meant that bidding based upon the PFI option increased dramatically. The result was a flood of projects pouring forth, and it was only on those projects where private finance was considered to not produce VfM or was otherwise inappropriate that public sector financing was to be adopted. However, what might have been viewed as a success by the private sector was instead viewed as overwhelming. The sheer number and novelty of projects being proposed so challenged the private sector that it found it difficult to assess which projects would be likely to prove the most attractive over time. This challenge, coupled with increasing bid costs, eventually led to the government ending the need for any form of universal testing. In its stead was to come a move to a form of testing against a hypothetically costed project – the public sector comparator (PSC) – but the mandatory function was gone. The timing in this change was opportune as it coincided with the election of a Labour Government and its stated intention to revisit the policy.

The new government tasked Sir Malcolm Bates to revisit PFI and its underlying assumptions. Sir Malcolm, a member in good standing with the Private Finance Panel, produced the Bates Review, as his report came to be called. Importantly, the Bates Review endorsed PFI and then went on to set out 29 recommendations to revitalise the whole process. The Review's key recommendations fell effectively into four categories: institutional structure; improved processes; lessons learned; and bid costs.

Taking the recommendations forward was a newly formed Treasury Taskforce that replaced the Private Finance Panel and centralised the overall policy role within government. The Taskforce's remit also included helping to identify significant projects for support; in effect providing an early stamp of approval. By identifying these projects and then going on to outline their early critical success factors the Taskforce was able to serve as both a source of best practice as well as a catalyst for their launch. Certain of these key factors included affordability, VfM, risk allocation, resources requirements and bankability. To the project finance professional the factors were familiar. To government departments not used to such specificity in their procurement role, they were challenging. Working deliberately to standardise some of the processes for the benefit of departments, the Taskforce began to publish extensively on the subject. Its publications ranged from generic guidance and policy statements to technical notes and case studies. The worth of the publications was proved time and again as the number of projects, or deal flow, as project financiers refer to them, increased. The effect of the publications served other ends. Initially they served to promote a better informed and more common understanding of the risks and their optimum allocation in projects. They also enabled a more consistent approach to pricing of similar projects and over time served to reduce some of the transaction costs involved in bidding and negotiating their terms. Lastly, the publications coincidentally laid down standards that would eventually be picked up by other countries pursuing their own PPP programmes. It is perhaps this last function that has actually had the greatest impact given the number of countries taking forward their own PPP programmes.

In overview PFI and PPP as it came to develop in the United Kingdom reflected a diverse cross-section of views on the political spectrum. While it may have begun as Conservative Party philosophy, some of the benefits which the policy came to show enabled the Labour Party to adopt and expand it. Central to the success of the policy was the unquestioned need to further invest in the physical infrastructure of the United Kingdom. That investment was manifested in new or refurbished hospitals, roads, and schools to begin with and eventually touched a much wider range of sectors. The progress that was shown once again to attract interest across Europe and elsewhere.

5.3   Types of PFI

There is no agreed classification mechanism to categorise what constitutes a PPP or how to classify PFI projects, though some consensus exists around a three-fold classification comprising financially freestanding projects, joint venture or public private partnerships and services sold projects. While this classification may be considered useful it should be borne in mind that the actual project agreement structure may entail a wider range of forms or commercial mechanisms inclusive of: concessions, franchises, leases, ownership, licences, rents or rights. The ultimate choice of commercial mechanism is dependent upon the objectives for the project, the principal's organisational structure, promoter's preferences and commercial objectives, the service to be provided, and the degree of recourse available.

5.3.1   *Financially freestanding projects*

Financially freestanding projects are those projects with future revenue streams that wholly offset the project costs, thus overcoming the need for any VfM test. The revenue streams typically come from the end users with public involvement limited to covering only those risks beyond the sponsor's or promoter's specific capabilities. It follows that the public sector will retain very few risks on this type of project. Generally, the sponsor or promoter will design, privately finance, build, operate and maintain the asset typically through one or more special purpose or project vehicles (SPVs), or stand behind it. By specially incorporating vehicles to serve in these roles a wider range of expertise may be drawn upon through consortia. Mature markets insuring many of the obligations undertaken by the SPVs have now been developed.

5.3.2 *Joint ventures or public private partnerships*

This type of PFI involves contributions from the public and private sectors. By combining the respective skills of the participants a more desirable solution may be achieved which might not have been as attractive from either a solely public or private point of view. In the case of a joint venture majority control will usually be held by the private sector. The selection of the private sector partner typically takes place competitively. Joint ventures are usually set up to exploit the commercial potential of public sector assets. Regarding PPPs, the European Commission has lately defined them as the transfer to the private sector of investment projects that have traditionally been executed or financed by the public sector. The agreement between the parties in joint ventures and PPPs is closely defined in terms of both contributions and limitations, and there is clear apportionment of risks and rewards between the parties. Equally there is significant and often new emphasis placed upon service provision. The public sector contribution may vary considerably, from consents to equity or loan participation and asset transfer. Financing is matched with future revenue streams, and any shortfall is offset, for example, through grants, subsidies, transfer of assets, loans or otherwise. Contract award usually follows a competitive process with formal VfM testing. Today a wide range of facilities may be commissioned using these types of PFI including those for healthcare, accommodation, telecommunications, schools, lighting, fire, defence, police, leisure, and waste management.

5.3.3 *Services sold*

The services sold form of PFI can again take various forms which may entail the provision of a capital asset as an integral part or not. Whether a capital asset is provided or not is secondary to the provision of a service. Those services may be discrete, as with the provision of an individual laboratory service, or not. The revenue stream to the PFI provider may be generated directly or indirectly from the public, e.g. tax and spend. As in the case of joint ventures or PPPs a SPV may be formed to deliver the services. Similarly once again to joint ventures or PPPs and the range of facilities that may be commissioned in respect of them, the services sold model may involve many of the same examples albeit often without the provision of a capital asset which can be a distinguishing feature. For completeness franchising and outsourcing could also be mentioned, as well as leasing, as all of these modes are occasionally referred to in the context of PPPs even though they may not strictly qualify for the term. Leasing, for example, has been a relatively more popular approach in the United States as a result of their tax structure.

5.4   PFI features

While there are a range of contractual structures that may be employed to deliver a PPP the most common form of partnership for new or improved capital assets typically involves design-build-finance-operate (DBFO) schemes. Supporters of DBFO schemes point out that, by combining responsibility for all of these features, increased efficiencies result thus justifying their use in PPPs. Under such schemes the public sector specifies the services that it wishes to be delivered and the private sector then carries out its role in delivering them. In contrast to traditional procurement where the public sector owns the assets, in PPP the public sector now pays for their use with ownership or title being vested in the private sector. In general there will be a fairly long contract or concession period for the arrangement – usually around 25 years. Central to the use of the asset are four inter-related features: real risk transfer; the use of an output specification; whole life performance of the asset; and performance related payments. The features may be briefly looked at in turn.

5.4.1   Risk transfer

Risk management is reasonably well understood today as are many of the principles on which risks are allocated. The subject has been introduced in earlier chapters, although its importance to PFI transactions necessitates a further summary here. Typically, risks are now allocated to the parties best able to manage them. In PFI projects this means the transfer of real risks from the public sector to the private sector involving the aforementioned design, building, financing and operation of the asset. On the operational or facilities management side, the real risk transfer may extend between the same parties and in the same direction as level of occupancy, volume or usage. To a large degree the risk transfer here reflects the extent to which the public sector controls use. From the private sector's perspective, and conversely in exchange for accepting the agreed risks, they are priced and managed in accordance with sound risk management practices. It may be noted that the risk transfer which is sought is intended to be optimal rather than total given that VfM itself is expected to generally decline once an optimal point for risk transfer in each case is passed. From the public sector's perspective the VfM which it seeks is both objective and as a corollary of the real risk transfer.

5.4.2 Output specification

In PFI contracts, clearly definable and measurable output specifications will be used to set out the service outputs which are expected to be delivered by the private sector operator. The specifications may be framed in terms of opening hours, fitness, cleanliness and otherwise, without unduly restraining the private sector operator's discretion to innovate in service delivery. It is the use of an output specification that creates real opportunities for the private sector to be innovative in the design, construction and service delivery or in use of the asset in question.

5.4.3 Whole life performance of the asset

PFI contracts typically will require the operator to assume responsibility for performance of the asset over the entire contract or concession period. In this way it is assumed that the operator will act in its best interest and thus construct the asset in the first instance in such a way that the costs of whole life performance will be minimised and thereby creating some efficiencies and potentially larger financial returns. The centrality of the whole life performance aspect follows from the long-term nature of the contracts which often involves time frames over 20, 25, 30 years or even longer.

5.4.4 Performance related payments

Payments to an operator under a PFI contract will commonly comprise a regular or unitary component or charge as well as a performance component as measured against a series of benchmark criteria. These criteria will be derived from the output specification and the standards outlined. It is common for payments to be made for both availability of the asset to deliver the output contracted for and the performance of the operator in terms of what it has achieved or the volume of usage. The final payment mechanisms ideally match the contract and performance monitoring regime as well as the requirements of the output specification. It should designate the overarching framework for realistic, demanding but nevertheless achievable performance standards. To achieve the maximum effort of the bidder and ultimate service provider a portion of the unitary component or charge is generally placed at risk and subject to increasing penalties when not met. Striking precisely the right balance between incentive and disincentive in this regard is how the payment mechanism should ideally be structured.

A brief comparison between traditional public sector procurement and PFI procurement is outlined in Table 5.1.

Focusing on the payment profile brings out the contrast in these two forms of procurement in another way.

Table 5.1 Brief comparison between traditional public sector procurement and PFI procurement.

Project stage	Traditional procurement	PFI procurement
Design	procures detailed design	specifies service exclusive of design
Specification	invites priced tenders to build	private sector pays for and manages construction
Construction	party to and constructs	pays for services post completion
Operation	operates	monitors service performance

5.5 Procurement process principles

Underlying both PFI and PPP projects is a growing body of practice and experience related to procurement of privately financed projects. The collective experience of hundreds of PFI projects in the United Kingdom and thousands of PPP projects across a wide range of countries confirms the importance of a rational stepped process and the principles which underlie it. Typically the process is broken down into a series of steps that can be modified to reflect individual project characteristics over a time-frame that often takes between 18 and 24 months. The steps can be seen across a range of precedents that are available from developed economies, e.g. the Office of Government Commerce in the United Kingdom; Partnerships Victoria in Australia; and even the Public Private Infrastructure Advisory Facility of the World Bank. In an ideal world the available precedents would be used to develop a unique procurement process that fully reflected the aims and objectives of the key participants. Drawing upon these and other precedents the typical PFI or PPP process or sequence will involve these or very similar steps.

5.5.1 Establish the need

The first step in any project is to clearly establish the need for a particular service, asset or improvement at a policy or strategic level. This may follow from a project review or otherwise. Typically it will be an extension of the public sector's overall plans (corporate or asset management) and policies in the sector or area and its performance assessment of them. There may be local or national standards or benchmarks that come into play with inventories or surveys, studies or reports all being drawn upon. A range of factors will drive the needs assessment which would include demographics and anticipated changes across programme delivery, technological, environmental, economic, and social characteristics. The needs assessed should be both medium and longer term inclusive of maintenance costs and funding options. If an infrastructure need has been identified and prioritised the needs assessment and planning process would then move toward appraisal of the options and development of a business case. It may be noted that while this process is typically driven by the public sector, occasionally unsolicited proposals that anticipate a particular need are developed and submitted to the public sector. Various jurisdictions deal with these at times expressly, as in the case of Taiwan, or in an ad hoc manner.

5.5.2 Appraise the options

Before proceeding it will have to be clear that taking the project forward is the best available option and certainly preferred to doing nothing at all. The technical inputs of each option should be compared and contrasted and costs involved and resources used for each option or programme set out. Other options may include doing the minimum, life cycle investment leading to full replacement, fast-tracking, full replacement or refurbishment of stock. Each appraisal will be underpinned by a clear understanding of what the project is intended to achieve comparable to systems design objectives mentioned in Chapter 1. Particular emphasis will be placed upon the outputs and how each of them enables the service objectives to be achieved. To the extent there is an existing service being provided the appraisal will consider it systematically and critically as well.

5.5.3 Develop the outline business case

The outline business case (OBC) is central to the process and is intended to prove or disprove that project objectives and investment criteria are supported financially, technically, legally and commercially. The OBC may build upon any informal expression of interest (EOI), which might have been sought, or even some soft market testing which could underpin it. Some countries encourage unsolicited proposals for projects as well. Ultimately the process is only practicable where there are potential service providers who have both the capacity and interest to participate in the project. The privately financed project will have to be justifiable and affordable for the principal given any relevant budget constraints. Justifiability and affordability are routinely demonstrated through comparisons to reference projects or scheme profiles, the best known type of which is again described as the public sector comparator or PSC.

If a PSC is employed it stands as the benchmark solution against which bids may be compared for VfM. In practice whether a fully costed PSC is released to the bidders varies from country to country. Though the PSC has been endorsed and criticised at various times many still support the notion that hypothetically costing a project can serve a useful function. The reasons why the use of PSCs has been controversial may be attributed to various factors including that at times they seem to have been used more as if they were a pass/fail test on projects when their accuracy does not really permit it. Further, the need for complex financial modelling and the reliance upon uncertain forecasts in preparing PSCs dictates that they be used prudently as well and at times they have not. The PSC's construction accounting tools including net present cost, net present value, risk adjustment, tax adjustment, optimism bias and even affordability may all be employed and factored into their development. Other outputs in the OBC may include a final VfM calculation, the preliminary output specification, payment mechanism, performance standards, model project agreement and risk register for the reference project. Overall evaluation will extend to both financial and non-financial (often social) indicators. The financial models in particular have to be fully developed, tested and robust to withstand a variety of different scenarios. The OBC will highlight whether the projects are affordable and bankable, or able to raise finance for the project on reasonable terms. By closely matching affordability and bankability the foundation for an effective partnership should be laid. Lastly, the OBC may suggest a preferred procurement route and project management strategy as well as the overall impact of the project.

5.5.4 Set up the project team

Once the OBC has been confirmed the formal procurement process may begin. The project will proceed under the auspices of a project team. Though terminology varies somewhat the team will be led by a principal on the public sector's part, who is ultimately responsible for choosing the commercial mechanism to be applied to the project and may be the final owner of the asset upon either transfer or termination. The principal's opposite or counterpart will be the sponsor or promoter. The principal may also serve as the key representative of the users of the service to be provided. Formal consultation mechanisms with the users or stakeholders in general may be put in place. The principal may appoint a project director or manager who could be responsible for the remaining selection of members, any process auditor, development of the project team and any other external advisors involved in evaluating the bids. Aside from the important technical role the principal and project director may play, they also often serve as local public sector champions for the project. The importance of these champions both in the public and private sectors cannot be overestimated. In part this reflects competition for resources across markets. Other things being equal participants will tend toward those countries where clear political support exists and fair, transparent bidding procedures are in place. Ideally the more senior the public sector champion, the more likely the success of the programme. Returning to the project team, other members may include project managers, financial, technical and legal specialists, insurers, managers and other specialists as called for. One of their first actions may be to refine the privately financed project procurement process and timetable.

5.5.5 Advertisements

Most countries will have established local or international obligations to promote competition in public sector contracting. The best examples are those imposed upon members of the European Union (EU) through the European Community Treaty to publish contract notices in the Official Journal of the European Community. The EU and other countries may similarly be bound to advertise pursuant to World Trade Organisation commitments. Details as to the nature of the project and whether involving private finance or otherwise may be called for. The objective of the notice requirement is to ensure that suppliers are not discriminated against on the grounds of nationality and that competition in the provision of supplies, works or services is enhanced. Advertisements may set out the overall criteria for award including financial, technical and commercial with any respective weightings. Participants expressing interest in the project may be provided with a more detailed information memorandum that will produce much of the detail developed for the original OBC. Advertisements for PPP projects on a global basis may routinely be seen in newspapers such as *The Economist*.

5.5.6 Determine selection strategy

Once the advertisements have run the project team will develop a final view on selection strategy. An important aspect of this is whether to choose a preferred bidder. In the process as a whole, a preferred bidder will often be selected after narrowing down from those prequalified through to the shortlist. Bidders, it may be noted, may come forward individually, as consortia or as joint ventures.

5.5.7 Prequalification

If the project is attractive it can be expected that there will be a long list of potential bidders. The prequalification process is designed to shorten that list by highlighting those who are best qualified and most competent to participate. Prequalification questionnaires (PQQs) are fairly standard in this respect. They will also give the project team some indication of the level of interest or lack thereof in the market, the general, financial and technical capabilities of bidders, their potential weaknesses and even possible problems. The prequalification process will lead to short listing but may also serve as reason for changes to the original method of procurement. Latterly, it may assist in contract development and profiling of the best-qualified participants. A report setting out the formal results and conclusions derived from the PQQ evaluation is often produced to support the reasoning for the decisions taken and stands as an audit trail.

5.5.8 Short listing

The prequalification exercise is intended to develop a long list of bidders who are generally competent. In contrast, short listing is directed at identifying from the long list perhaps three or four bidders who would have specific competence for the project at hand. Thus the short listing may move beyond the general capabilities outlined to how the bidders would use them to carry out the project itself. This would extend to their willingness to assume risks, finance or price the project. The shortlisting exercise may overlap with what is referred to at times as an invitation to submit outline proposals (ISOP). The ISOP is employed if additional information needs to be sought. Once again, the intention is to elicit from the bidders how they would address specific project issues as outlined in any information memorandum, for instance in relation to quality, cost estimates or otherwise while maintaining a level playing field throughout. Formal interviews and presentations may form part of the process.

5.5.9   *Project revisions*

Prior to any formal invitation to negotiate (ITN) with shortlisted bidders it is not uncommon for the project team to revisit their original project plan and assumptions. This allows the project team the opportunity to benefit from the feedback which the bidding process thus far has provided to them. Changes may be called for in relation to any number of aspects of the project ranging from refinement of the OBC to the output specification and reference project. The output specification is central to the process as it forms the basis on which the principal states what is expected from the bidders in clear and comprehensive output terms coupled with the relevant standards to be achieved. A firm foundation in the output specification will also serve as the basis upon which innovation may evolve.

5.5.10   *Invitation to negotiate*

The invitation to negotiate (ITN) phase is designed to refine the contract and control structures that will be put in place and upon which bidders will submit their detailed designs, technical, financial, legal and general responses. It formalises the further close evaluation of bidders along narrower lines and often builds upon the information memorandum. If any parts of the formal documentation have not been released they will be produced at this stage. From a technical point of view this will include the output specification, performance standards and input specification. The financial documentation will include the payment mechanism and pro-forma response. The legal documentation will be based upon the project agreement. Additional information released may include how the project team will deal with the ITN, timelines, format, standard and variant bids. To the extent that clarifications are called for, workshops or presentations may too be included. From the project team's perspective very detailed provisions will pertain to their formal evaluations across all categories.

5.5.11   *Formal negotiations*

During this stage the principal and project team carry out formal parallel negotiations with each shortlisted bidder both to clarify the details in each proposal as well as ensure that the output requirements are being met. A further objective of the negotiations is to narrow the commercial terms upon which the project would be awarded and operated thus seeking to ensure the most efficacious terms. Negotiations will be broad based and inclusive of the proposed final draft terms of the project agreement which would be enhanced by the competitive process. Toward the end of the formal negotiations each shortlisted bidder may be asked to submit a best and final offer (BaFO). This will typically follow the original ITN quite closely but will be narrower in scope as the process of negotiation will have concluded certain terms and conditions between the parties and which can allow them to focus exclusively on the remaining areas of disagreement or non-compliance. The BaFO may operate almost as another stage in the process as the principal works to refine its contract structure and controls, sometimes adjusting its scoring of the remaining bidders as it proceeds while the individual bidders look to areas of compromise and opportunity.

5.5.12   Final negotiations

Following the formal negotiations with the shortlisted bidders a selection will be made of a preferred bidder or sometimes two preferred bidders, one of whom waits in reserve. The second preferred or reserve bidder will understand that their position is secondary but may nevertheless be asked to keep their BaFO on the table in case the first preferred bidder and the principal are unable to come to final agreement. At this stage both parties work toward not only the final award of the contract but also the improvement of the final terms. In this regard the OBC, which the parties have proceeded upon, may be finalised as a full business case. The project team may work with other specialists or groups who stand behind the project or with whom the power of final decision-making rests. As with the earlier negotiations pertaining to the detailed designs, technical, financial, legal and general responses closure in each respect is required. It is at this stage that outstanding items will also have to be addressed. These items may range from insurance to related property sales, statutory approvals or even employment concerns for any affected staff. Both parties will continue the process with extensive due diligence as they each prepare to contract. Due diligence is itself a process which tests, verifies or validates the underlying assumptions that have been made with respect to elements and evaluation criteria used to date in relation to the project as well as the material mitigation measures and resulting in either their confirmation or rejection.

5.5.13   Financial closure and contract award

The final negotiations ideally result in agreement between the parties on all the relevant terms, the financial closure and contract award. Financial closure occurs when the principal and the promoter/preferred bidder sign the project agreement. The principal thus commits itself irrevocably to the project and thus the purchase of the service, asset or improvement from the promoter/preferred bidder who agrees to provide them. In many cases that purchaser is the government. The project agreement then forms the basis upon which finance can be secured based on the predicted revenue streams over the life of the project. By reaching financial closure the parties agree that the project agreement becomes unconditional and that all outstanding conditions precedent have been met or waived.

5.5.14   Contract operation

Although not technically a step in the procurement process, it is worth noting that the objective is the contract operation during the life of the project agreement or concession. To this end the contract and control structures will have to be both precise and robust enough that the parties' intentions are ideally fulfilled. It will be expected that the day-to-day operation proceeds along the lines envisaged and the overall fair and genuine commercial intentions of the parties. To the extent that the intentions are not being fulfilled, contractual mechanisms normally enable swift and effective sanctions or penalties to be imposed as well as more formal dispute avoidance and resolution mechanisms between the parties to be activated.

5.6 Contract and control structure

The contract and control structure of a PFI establishes the legal and operational relationships between the principal and promoter or sponsor. While each type of project will exhibit individual characteristics a generic list of the typical key documents or components may be set out for completeness:

- project agreement and riders
- payment mechanism
- commercial mechanism
- output specification
- performance standards
- lenders' direct agreement
- service provision direct agreement
- escrow accounts.

To these documents may be added a further series of legal and operational relationships between the sponsor and others involved in the delivery of or support for the project:

- construction and contractor agreements
- architectural and engineering design agreements
- off-take agreements
- supply agreements
- operation and maintenance or facilities management agreements
- guarantees, warranties and bonds
- inter-creditor agreements
- debt agreements
- shareholder agreements
- standby credit facilities
- corporate formation documents.

From the range of contracts and relationships referred to it can be seen that complexity is a central feature of PPP and PFI.

5.7 The special purpose or project vehicle and financing

The special purpose or project vehicle is a company separately, independently and especially established for the sole purpose of delivering the project at hand. The SPV may be formed by individual groups of bidders, sponsors or promoters to carry out the terms of the project agreement and all supporting collateral agreements. The SPV enables the creation of a separate legal entity that can operate solely according to the confines of the relevant project documentation. The SPV plays an especially important role as a vehicle for financing and may in fact be central to how lenders co-ordinate their capital contributions on a project. The SPV often contemplates non-recourse finance in the ideal albeit very often in practice limited recourse finance but also with the possibility of full recourse as well. The recourse referred to here is to the funds or assets of those who stand behind the SPV, typically the sponsor or promoter. Hence in the case of non-recourse finance the lenders can look only to the SPV and its future revenue stream for repayment. A limited recourse financing would presuppose that the lenders would have some degree of access to the funds or assets of the promoters beyond that which they hold in the SPV. Very often the limited recourse finance is indirect or given as some form of credit support as with bonds, guarantees or other types of collateral. Full recourse, as the name suggests, would contemplate unrestricted access to the funds and assets of the promoter. As such when full recourse is given some of the underlying rationale for the formation of the SPV is withdrawn and, on a comparative note, it is then tantamount to traditional corporate versus project finance. However, while aspects of PPP transactions are indeed similar to project finance-type arrangements they may be conceived and arranged without due consideration of one critical aspect of project finance arrangements, namely high asset-specificity.

As with the majority of project finance transactions the use of the SPV is intended to be off-balance sheet to those who stand behind the vehicle. Those standing behind it may be either private sector or public sector entities or both. If the public sector is involved it may be by way of an equity stake if there is profit sharing or by way of a guarantee, loan or subsidy. Once again, in this way the off-balance sheet aspect may be either in relation to the public or private sectors. By being able to fund the project off-balance sheet significant risks are shifted away from the key participants. To the extent that the public sector has contributed to the project and thus earned a right to some part of future project revenues, it may even seek to securitise the claim or essentially sell its future right to revenue or participation. With regard to the formal accounting treatment and whether off-balance sheet treatment will be accorded or not, this will turn upon the specific percentages of risk involved and the relevant accounting rules. From a broad and historical point of view this off-balance sheet treatment has served as an important rationale for not only private sector participants but the public sector as well. For the public sector the consequences are that it can treat payments made under PFIs as revenue expenditures rather than capital expenditures and amortised over the life of the PFI contract. Private sector standards for accounting treatment often dictate that the accounts of the SPV are consolidated with any controlling entity. Lastly, it may be assumed that the SPV will work throughout to minimise bid costs on projects, provide a return on the investment made and if possible retain a measure of liquidity in relation to that investment.

5.8 PFI/PPP and rational privatisation

PFI and PPP need to be differentiated from contracting out and privatisation. Privatisation refers to those situations where the total responsibility for providing a particular service is given to or assumed by the private sector. Traditional examples could be found in the transport, telecommunications, energy and water sectors though private inroads have been made in all of these areas with telecommunications having gone the farthest. Case studies following look at transport in relation to the Sydney (Chapter 6) and Hong Kong tunnel (Chapter 7) projects and Dabhol (Chapter 8) in terms of power. For the most part these sectors have traditionally been subject to detailed regulatory regimes which often employ either rate of return regulation or price regulation as means to control prices. Contracting out or outsourcing refers to the situation where the private sector is given or assumes responsibility for the day-to-day carrying out of a service. PPPs are not privatisations although they do offer what might be held out as some of the benefits of privatisations. While the 1980s is a decade associated with increased privatisations the trend began to wane during the 1990s. The reasons for this are not agreed upon. At the same time, during the 1990s private investment in infrastructure projects increased as an alternative to privatisations. Evidence of this can be seen in both developed and developing countries. The best evidence in respect of the latter comes from the World Bank's Private Participation in Infrastructure Database. That evidence reveals that during the 1990s private activity in infrastructure increased each year save for 1998 and 1999 and that most developing countries introduced private investment in infrastructure. This increase in private investment in infrastructure was coupled with the emergence of PPP in some developed countries as well. Since 2000 and in more recent years, Central Asian and European countries have experienced a significant decline in interest from international private operators and investors in domestic infrastructure projects, and it can be argued that the PPPs currently emerging are becoming one of the most important ways to offset this decline and rekindle interest in domestic infrastructure. The PPP model contemplated involves the full range of participants which have typically been involved in past project financings during the 1980s and 1990s.

The key differences between the various modes of procuring assets or services are illustrated in Table 5.2.

PPP differs from privatisations in two key ways. First, PPP often operates outside a formal regulatory regime and rather governs the parties' relationship contractually.

Table 5.2   Key differences between the various modes of procuring assets or services.

Type	Provision of services	Provision of assets	Service definition
Traditional	Public	Public	Public
Outsourcing	Private	Public	Public
PFI/PPP	Private	Private	Public
Privatisation	Private	Private	Private

The output specification provides the primary means of control over the service provider rather than a regulatory body. Second, PPP does not involve a permanent change in ownership of the assets involved. Operational control and ownership revert to the public sector at the close or termination of the PPP concession or contract. Thus the nature of the control, its temporal quality and its continuation distinguish PPP from privatisations.

5.9   Risk management

In recent years greater attention has been paid to risk management. The need for more structured approaches to projects of all sizes has been highlighted by occasional expensive and highly publicised failures. The somewhat chequered record of the construction sector across many jurisdictions has further brought to the fore the imperative for better project planning and risk management. While risk and uncertainty are an inherent part of all projects, it has become clear that success in managing both correlates highly with better outcomes across the traditional linked parameters of time, cost and quality. Research by the United Kingdom Treasury in 2003 indicates that PFI leads to better project delivery than those projects traditionally procured, and in this regard PFI projects show that 88% were being delivered on time or earlier and with a more consistent quality of services. Commercial pressures too and the increasingly competitive environment today dictate that companies need to develop their ability to take risks and innovate, while delivering projects on time and on budget as well as meeting ongoing daily operational requirements. The public sector, it would seem, is aware of these pressures and has sought to harness them through the PFI process. Responding to change in a resilient and reliable way that ensures services are provided and that improvements are sustained is challenging in any environment, let alone one that runs over 20 and 25 year timeframes as with PFI and PPP projects. As a result it is in this setting that some of the most innovative applications and approaches to risk management are being developed.

5.9.1   Key stages of risk management

There are numerous models of risk management. The models bring out that risk management involves a series of actions taken by individuals or companies endeavouring to alter the risks which confront them, as introduced in Chapter 4 and Box 4.2. It is a formal orderly process that systematically identifies, analyses and responds to (or identifies, evaluates and mitigates) risks or risk events arising throughout the project life cycle, all with a view to realising the optimum or most acceptable degree of risk elimination or control. Although key stages may be outlined the most effective risk management is carried out as a seamless iterative process. The stages may be described as follows.

Risk identification

In risk identification both short- and long-term risks are identified by available reliable and comprehensive information. While all parties to PFI and PPP projects will share in risk identification, to some degree there is little doubt that the greater burden rests upon the private sector and is reflected in its completion of risk matrices for projects. As experience has been gained with privately financed projects, better understanding of, and more uniform approaches to, the identification and allocation of risks have taken place. Thus, in many jurisdictions, it is now reasonably well agreed which parties will carry which risks exclusively or on a shared basis. Some of this familiarity is attributable to the more explicit risk management which PFI and PPP have brought to the fore. To the extent that this greater understanding has been developing it has left the parties to devote relatively more time to those risks which may prove more problematic in the negotiations. It would also appear that the more systematic identification of risks by the parties enables their closer pricing and hence raising of the competitive floor that is set. The occasional failure of a private sector service provider is in itself some evidence that real risk transfer is taking place. Whilst it is not the policy of the public sector per se to anticipate or expect failure on the projects, it is nevertheless to be expected that it can and will undoubtedly occur at times given the public sector's expectation that risk transfer is indeed genuine.

Risk identification techniques on PFI and PPP projects fall broadly into three categories although many techniques cross the boundaries:

- inductive: e.g. failure modes and effects, criticality analysis
- deductive: e.g. fault trees
- intuitive: e.g. brainstorming.

Other means include use of the Delphi technique, checklists, thorough site visits, drawing upon past experience, research, interviews, surveys, experts and, perhaps most commonly, risk registers. The risk register may be a document or database that records each risk on a project or asset and may resemble or be used in the same way as a checklist which is software ready. Specific data in relation to the risk, its identity, and the activities associated with it, its precedence and likely value may all be recorded. The likely values are often developed based upon qualitative and quantitative analysis that predicts their varying impacts as outlined below.

In relation to projects various attempts to classify the different categories of risk may be referred to and which might include the following:

- construction
- financial
- performance
- demand
- residual value risk.

These categories of risk would be present whether it is a PFI or PPP project.

Risk assessment

Risks are assessed and recorded in terms of their current status and potential to have an adverse impact. As the risks are identified and the risk management process unfolds through the use of certain of the techniques described, their assessment becomes more important. It has to be taken for granted that almost any risk may or may not materialise on any given project. If one accepts this then it can be seen that, in the first instance, dealing with the materialisation of those risks becomes largely a matter of probability. It will be seen that these probabilities can be estimated. However, the risk assessment exercise is then only half-complete, because to fully appreciate the effect of the risk materialising one needs to understand its impact as well. It will be seen that impact too may be estimated. It is this combination of probability and impact that is central to the risk assessment stage of risk management. Accurately forecasting these two traits for the risks enables the participants to decide upon their most appropriate course of action.

Risk assessment may be either qualitative or quantitative. In qualitative risk assessment both probability and impact are assessed subjectively while in quantitative risk assessment both the probability and impact are assessed objectively. Midway between these two stands what may be referred to as semi-quantitative risk assessment where probability is assessed subjectively but impact is assessed objectively. Two of the most commonly used tools for quantitative risk assessment are sensitivity analysis and probability analysis carried out in computer simulation programs. The choice between these turns largely on the availability of data. Thus a qualitative risk assessment will typically be used when there is uncertainty, as in the absence of data, and so subjectivity prevails. In those cases where some but incomplete data exist then the semi-quantitative risk assessment may be used. Lastly, in those cases where full data are available a fully quantitative risk assessment will be preferred. Once again, as experience on PFI and PPP projects grows so too does the weight of quantitative or empirical data on which analyses may be carried out and probabilities determined.

Without turning to how individual organisations will assess the impact of individual risks or even whether they seek to address every one or only the main risks, or some spectrum in between, it may be noted that very often those impacts are translated into financial terms. In this way the organisations may ultimately bring their risk management to bear on their commercial terms.

As a footnote to risk assessment it may be observed that it is an exceptionally difficult enterprise. The vagaries of language and the uncertainties associated with legal interpretations complicate the process notwithstanding trends toward simpler and more standardised templates.

Risk response

Project staff have or are given the capability and supporting tools inclusive of contingency plans to manage risks. The idea of responding, addressing or mitigating risks is essentially a process of finding solutions to counter them. Thus, given a choice between pricing for the potential probability and impact of a risk event occurring, or simply avoiding or indeed even transferring it to another, staff are able to select the best options. Also, following Chapter 4, risk response essentially involves four strategies:

- elimination/avoidance
- reduction/mitigation
- transfer
- retention.

Risk elimination or risk avoidance may range from not bidding for a project to spending heavily to abort it. Risk reduction or mitigation too may involve taking extra steps or precautions which would serve either to reduce the probability of the risk occurring or, if it were to occur, reducing its impact or magnitude. The transfer of risk is perhaps most common as contractual structures are used to pass on risks to other parties on the project or even to insurers. The optimum timing of risk transfers from the public sector to the private sector on PFI projects though has been subject to differing views and practice. Risk retention may be what is conceived as simply the risks that a party is left with, post negotiation. In PFI projects the public sector may operate upon the assumption of retaining only a narrow band of risk and that all other risks are to be transferred to the private sector (Figure 5.2). Risks should be allocated to the party best able to manage them.

The pricing of risks is central to their transfer. This stems from the fact that governments have to be able to compare the cost of delivering services themselves versus purchasing the services from the private sector. It is here once again that governments seek to relieve themselves of those risks that the private sector is better placed to handle. In working out this complex task it is important for both parties to distinguish between project-specific risks and market risks. The former reflect different outcomes on the projects themselves while the latter reflect the underlying economic developments that affect all like projects. Market risks are not diversifiable and thus pricing must be carried out properly. The public and private sectors will adopt different approaches to pricing market risks. Thus governments tend to use risk-free rates such as the social time preference rate to discount future cash flows when appraising

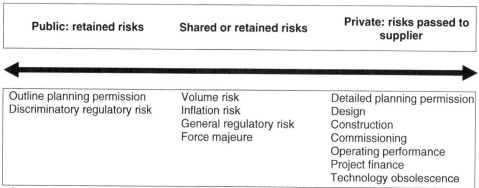

Public: retained risks	Shared or retained risks	Private: risks passed to supplier
Outline planning permission Discriminatory regulatory risk	Volume risk Inflation risk General regulatory risk Force majeure	Detailed planning permission Design Construction Commissioning Operating performance Project finance Technology obsolescence

Source: PricewaterhouseCoopers.

Figure 5.2 Efficient risk management.

projects while the private sector in contrast will follow weighted average cost of capital principles to select a discount rate that it applies to appraise future revenue streams (see Box 3.2).

From a project point of view various contractual and other means are employed to mitigate the risks including the use of: guarantees; bid, performance, adjudication and surety bonds; insurance and letters of credit. With respect to the instruments and tools available to mitigate certain of the financial market risks regarding exchange rate, interest rate and price fluctuations, financial derivatives are heavily drawn upon. To recap briefly the main types of derivatives used include forwards and futures, options and swaps (Box 4.3, 4.4 and 4.8). Forwards entail obligations to purchase or sell assets at a specified forward price on an agreed date. Swaps in turn enable a party to obtain a fixed price for an asset even though changes in its cash price have occurred. Their inherent flexibility seemingly allows holders to simultaneously purchase and sell cash flows in multiple currencies yielding differing interest rates with varying underlying assets. Futures are standardised and exchange traded forward contracts forms of forward contracts. Options provide the freedom of action for their holders to either act or not in relation to the assets, thereby avoiding adverse price changes in relation to them. This type of freedom of action is lending credibility to financial risk management also on PPP and PFI projects through hedging strategies.

Review and reporting of risks

Regular checks are carried out to ensure that risk assessment remains up to date and reliable and that risk management remains essentially fit for purpose. This process follows naturally from the first three stages of risk management. The careful review, documentation and reporting of the risks over the project life cycle lead to better information for risk management purposes in subsequent projects. It parallels very closely the ongoing monitoring of risks undertaken by financiers who in turn use it to inform subsequent credit committee decision-making.

5.10   Financial risk in PFI and PPPs

The privately financed project is exceptional in that it combines two risk-laden endeavours: construction and project finance. While the two areas have always had a degree of overlap it is the advent of PFI in particular that has seen them become something altogether greater. In the majority of cases PFI and PPP projects will be structured on a limited or non-recourse basis. While typical project finance transactions in the power sector, for instance, have been limited to project revenue streams or cash flows it is PFI that has seen those templates applied to the range of assets underlying service provision in new sectors from hospitals to schools, prisons and otherwise. Given that the credit risk associated with these types of projects is low where the public sector participant provides the revenue stream or is the ultimate purchaser of the service provided, financiers devote more time to where the other principal risks reside, e.g. performance, construction and completion risks among others. Financiers will draw upon the full range of risk identification approaches and tools outlined above in carrying out their analysis. However, financial models are developed drawing upon sensitivity analyses working off base cases as the financiers' key risk assessment tool. In effect the identified risks are weighed and expressed mathematically in relation to the project's cash flows and respective financial impacts on the repayment streams. Working off various scenarios the financiers are able to reach decisions in relation to their margins and the gearing they will permit on the projects. It is a carefully orchestrated process that once again brings risk and privately financed projects full circle.

Historically the main form of capital raised for project finance has been senior bank debt, although over time other forms of capital both from the banking sector as well as institutional investors and the capital markets have diversified the sources. Today privately funded projects reflect a broad mix of sources of finance of both debt capital and equity, and even though the form and documentation for the various forms of finance will vary all participants expect to earn a return. Lenders primarily will seek to pass down as much risk as they can onto those in the contractual chain who are the most bankable. At the same time financiers will seek to limit the risk which is retained in the SPV whether through caps or other mechanisms. The pricing of the various forms of capital will closely reflect their relationships not only to their profitability and liquidity but also the risk each will bear. As a result it can be said that the differing risk profiles of individual projects and the participants themselves dictate that a wide range of sources of finance may yield the best terms and outcomes.

5.11   Challenges for PFI and PPP and the responses

The growing interest in PFI and PPP has focused attention on various challenges. While each country's experience will vary the nature and structure of privately financed projects, they share sufficient similarities that these challenges should be taken seriously where programmes are being taken forward.

5.11.1   Costs of finance

The cost of finance is a somewhat contentious topic. From one perspective private sector participants in privately financed projects raise finance on commercial terms. Thus to the extent that the public sector competes with the private sector it holds a comparative advantage given that its relative cost of funds is less. That cost advantage varies between 1% and 3% in many jurisdictions. However, while this is seen as a challenge by some, and though the private sector will often concede it, its strategy is to seek to compete and offset these higher costs by demonstrating VfM on other bases. These bases may include fully valuing risk transfer which in traditional procurement may linger as an uncosted residual risk on the project. A more radical view would hold that other things being equal, the public sector should have no financial advantage in relation to the costs of finance. This view holds that the cost of capital is dependent only upon the project characteristics and not mode of financing even though the source of financing could influence project risk. As governments can spread their risks across the whole tax base it would seem that they have an advantage over the private sector. However, given that the private sector can spread its risks across financial markets as a whole it would seem that it should have the advantage. As this does not mirror present practice some other explanation must be sought. One possible explanation for the additional cost of finance for the private sector may be from the increased risk of borrower default. Given that the public sector may offset this increased risk through its power to tax, it benefits from a relatively overall lower rate. That lower rate may be reflected in the 1–3% cost advantage referred to and is likely one that will continue to be used when discussing this issue.

5.11.2   Off-balance sheet financing hides longer-term, open-ended public sector liabilities

The rules with respect to when the public sector may treat privately financed project liabilities as off balance sheet continue to be refined to the point today where this feature no longer serves as the main rationale for projects. As fewer projects are accounted for off balance sheet the scope for this challenge diminishes. Nevertheless it merits a response which today is that many jurisdictions have already moved to account for their long-term liabilities for unitary payments on projects over the life of concession agreements. By so responding it can be argued that public sector liabilities are more clearly revealed than on an annual budgeted basis and that the true whole life costs of the asset inclusive of financing and operation become more transparent. That said, there is as yet no single or defined comprehensive fiscal accounting and reporting standard for PPPs. This means that there will continue to be variations until one is agreed on when or whether the public sector may treat liabilities off balance sheet. The development of such a standard would promote transparency in the process and should increase efficiency. In the meantime, as practice and experience grow, and as scrutiny has increased, particularly through rating agencies, the risk of abuse by the public sector in the use of PPPs in this respect is diminishing.

5.11.3   Public sector staff suffer while terms and benefits decrease

A common challenge both in practice and in public debate centres around terms and conditions for public sector staff who may be transferred to a private operator or have to work alongside their staff. At times this has occasionally led to two-tier benefit models. However, while this can occur it is an issue that the public sector can address if it wishes through either the terms of the concession agreement or background legislation, as has been done in the United Kingdom with The Transfer of Undertakings Protection of Employment Regulations. One appropriate response of the public sector is to challenge the private sector to better its terms and conditions through enhanced productivity, innovation and the like. In fact the same position can be taken with respect to any disbenefits that may accrue as a result of the privately financed project and, for that matter, for the public sector to insist that they are all offset by the benefits held out. Current experience confirms that this can be done.

5.11.4   Loss of public sector flexibility

It is argued that by committing itself to fund service delivery over 30 year timeframes into the future, the public sector is giving up the flexibility that it needs for programme planning and delivery. While it is true that these are the timeframes involved, it is a false premise in that the public sector commits to whole life maintenance etc. of the asset notwithstanding the form of procurement because the work must still be carried out whether costs are accounted for or not. Where PFI and PPP mark an improvement then is that these costs are for the first time revealed and accounted for upon approval. Privately financed projects also take out of the equation both the bargaining and competition for funds that occur during annual budget processes. This may mean that the investments required to maintain assets to their expected standards are available and being made as opposed to being traded off, for example in the run-up to public sector budget exercises.

5.11.5   Bidding costs

The bidding costs on projects are higher than has been the case using traditional procurement. In part this reflects the larger role that private sector advisors are playing in the process but in part it also reflects that these costs are now being amply illustrated. This can be seen in the lengthier and more rigorous private financed procurement process inclusive of fully costed business cases outlined above, which stands in contrast to traditional procurement and compulsory competitive tendering. Depending upon the stage that individual jurisdictions are in, many are showing falling bidding costs as experience increases, templates and standard forms develop, risk allocation is better understood and projects are let in bundled packages as well.

5.12   The lessons

The lessons from PFI and PPP programmes are slowly beginning to emerge. The slow pace in part stems from how widely individual countries' experience varies. In addition, the discussion is complicated in part due to variations in vocabulary. As noted, some groups wrongly draw no distinction between PPPs and privatisations. Others see PPPs as no more than traditional procurement with greater margins. This confusion, in addition to the widely varying experience of countries and sectors, detracts from and is curtailing full understanding of the lessons. However, with those caveats, some observations can still be made. First of all it would appear that privately financed projects are not a panacea for either the public or private sectors. In short, while there is no *magic bullet* in terms of how projects should be structured some best practice is emerging and much of it has been outlined above, in particular with the attention devoted to the procurement process principles. It can also be noted that the use of privately financed projects is increasing at a time when private investment in infrastructure as a whole is declining in developing nations. The projects that do proceed reveal considerable resiliency in terms of how they are being adapted to suit local conditions. Regarding the appropriateness of the model to various sectors PFI and PPP seem to yield better results when the private sector is one step removed from the ultimate end-users. Thus strength comes more from the delivery and maintenance of the physical infrastructure that supports service delivery rather than what might be called core government service delivery itself. Two possible explanations for this may be suggested. First, however well positioned the private sector provider is it is unlikely to support traditional public sector (e.g. social or non-profit making) goals as fully or unequivocally as the public sector does. Second, the private sector is often better placed to deliver certain services one step removed from the public because the services themselves may be inherently non-contractible, e.g. national defence and the foreign service. Thus it is not core public sector service delivery roles which might be best suited to the PPP or PFI model but non-core public sector roles, for instance building or maintaining the physical facilities or assets out of which the roles are fulfilled.

The role of governments is also becoming clearer based upon some partial successes. Thus successful initiation of PPP programmes typically requires strong political and occasional legislative support. Political commitment is important at the policy level to ensure that market participants will come forward and invest in their own capabilities and the projects themselves. The absence of political commitment creates political risk that detracts from market interest in the potential projects. Legislative support varies with the countries involved, but at a minimum it should clarify that jurisdiction exists to take the programme and projects forward. Numerous countries, e.g. the United Kingdom, Italy and Spain, have all had to revise their legal frameworks to ensure they had the requisite jurisdiction. Brazil has taken this issue furthest by introducing legislation that would govern all forms of PPP at all levels of government. It would appear that the stronger the legislative and institutional framework, the more likely the PPP programme is to be successful. Tied to these pillars governments should delineate clear lines of communication, responsibility and accountability to limit the possibility of corruption. Central to this are consistent approaches to tendering procedures and contract administration. Perhaps less important has been the ability of governments to prioritise projects and ensure a predictable number of projects into the future. Again this reinforces market interest and stimulates investment. Perhaps underestimated in the success of privately financed projects, at times, is the overall macroeconomic climate within countries. To the extent that greater over lesser stability pertains in the economy, the prospects for PPP are better.

Where contracts or standard templates can be developed it is repaid with greater confidence in the markets and bidding. A moderate number of projects going forward also build confidence and experience among advisors and participants that too can contribute to success. Supporting these projects is also best achieved with ongoing public communications, community and stakeholder engagement beyond the project contract, control and monitoring mechanisms. These mechanisms ideally assume an underlying legal framework which recognises and protects concession agreements and the resolution of disputes under and in relation to them.

One very important component of PPPs involves expertise or the full cross-section of skills needed to develop, appraise and prioritise projects as well as administer such a programme in the public sector. Governments involved in developing PPP programmes successfully have often invested heavily in upgrading public sector skill sets to respond to the challenges presented. Formal training, staff exchanges, insourcing and other means all contribute to an upskilling in the public sector and contribute once again to programme successes. In the United Kingdom in particular various organisations have specialised in fulfilling such a role. Two that may be mentioned are Partnerships UK and the Public Private Partnerships Programme (4Ps). Partnerships UK serves not only to promote PPP projects but also provides specialist legal, financial and technical advice on projects. The 4Ps plays a like, albeit more limited, role in relation to projects and on behalf of local (or municipal) authorities.

For privately financed projects to be successful there is another aspect less tangible than some of the above observations that is also required. Just as the differing objectives of the public and private sector participants were outlined and the compromises that would be required from them to excel noted, participants are also essentially being called upon to act almost against nature. Thus the private sector participant may have to take on greater responsibility for public sector goals than it normally otherwise would while the public sector has to manage the private sector service provider along lines which it would be familiar with, e.g. paralleling private sector management criteria. The parties thus seemingly change roles.

In the end and notwithstanding the fairly limited timeframe in which this model has developed and that few projects in turn have run their full course, the accumulated experience shows not only that appropriateness of the model varies across jurisdictions but also within individual jurisdictions depending upon the sector involved. Further, within countries and sectors, views toward privately financed projects polarise at times depending upon the degree of success or failure that is being demonstrated. This likely has a self-reinforcing effect which then either tends to increase or lessen the likelihood of more projects going forward in the future in those countries or sectors.

Key concepts

The following concepts are considered sufficiently important to memorise as key vocabulary for use in subsequent chapters.

BaFO
Bates Review
BOT
Design build finance operate
Due diligence
Treasury Taskforce
ISOP
ITN
Outline business case
Output specification
PQQ
Principal
Private finance initiative
Private Finance Panel
Privately financed project
Public private partnership
Public sector comparator
Ryrie Rules
Sponsor or promoter
Value for money (VfM)

6

The Relevance of Sound Demand in Infrastructure Project Finance: the Sydney CrossCity Tunnel

The Sydney CrossCity Tunnel case is a good example of an infrastructure project finance development where all indications were for a successful public-private sector partnership with efficient risk distribution, sound financial partners, and project promoters such as Cheung Kong Infrastructure and Bilfinger Berger BOT GmbH – with support of a number of first-division international banks, lead by Deutsche Bank. Furthermore, the Australian state of New South Wales (NSW) provides an institutional and legal environment as stable as anywhere in the world, although state politics in Australia can on occasion be spectacularly populist. What is not stated in the case narrative is that the Project Company went into liquidation at the time the case narrative ends. What went wrong?

6.1   Infrastructure finance: the Sydney CrossCity Tunnel[1]

In September 2000, the directors of Cheung Kong Infrastructure Holdings Limited (CKI) were excited about an unprecedented opportunity to invest in a transportation project in Australia. CKI had a vision to become an international infrastructure

---

[1] Mary Ho prepared this case under the supervision of the authors for class discussion. This case is not intended to show effective or ineffective handling of decision or business processes. Although the utmost care has been taken in preparing the history and fundamental background in this case, the investment decisions here are fictitious but typical. The opportunity discussed in this case does not represent an actual specific company opportunity. Ref. 07/324C, Copyright © 2007 The University of Hong Kong. This material is used by permission of The Asia Case Research Centre at The University of Hong Kong (http://www.acrc.org.hk).

enterprise and was keen to extend its footprint beyond mainland China and Hong Kong. The group found that Australia, with its stable regulatory environment and sound economic prospects, provided an excellent environment for infrastructure investment. In 1999, CKI made its first foray into energy assets in Australia with the acquisition of a 19.97% stake in Envestra Limited, the largest natural gas company in the country. Envestra proved to be a prime asset and generated robust financial returns to CKI.

In 2000, CKI was vying for further opportunities to invest in transportation projects in Australia when the Roads and Traffic Authority (RTA) in New South Wales invited tenders for the Sydney Cross City Tunnel (CCT) Project (see Table 6.1 for the chronology of events). CKI was particularly interested in tolled transportation investments because they were usually regulated under a finite life concession period and carried relatively low political and regulatory risks. The nature of these investments usually determined that competition for the services generated by the assets was also limited, which meant investing in such projects could possibly provide stable and predictable returns to the group. Although transportation investments could entail significant construction and patronage risks, CKI believed that these risks were manageable and could be mitigated through careful planning and negotiation with the relevant government authorities. If CKI won the tender for the CCT project, it would be no small accomplishment, as the group's existing transportation assets were all located in greater China. Given CKI's strategy of globalisation, was the project too good to miss?

6.2   History of the CCT project

6.2.1   Background

As in most big cities, road traffic congestion had long been a significant problem in central Sydney. In 1998, the Minister for Transport of NSW produced a manifesto entitled *Action for Transport 2010*, which aimed to expand the road infrastructure and thus improve traffic flows in central Sydney. The manifesto was followed by a public consultation report prepared by the RTA in October 1998. The report proposed the construction of an east–west road tunnel that would cross under the heart of the Sydney Central Business District (CBD). The primary objectives of the CCT project were to relieve traffic congestion in central Sydney, improve the reliability of public transport and provide a safer environment and improved amenities to vehicles, cyclists and pedestrians. By removing east–west through-traffic on surface streets and reallocating road space for public transport and pedestrian use, the CCT was expected to provide a number of benefits to the Sydney community, including (RTA, 2003):

- reduced congestion, as drivers could bypass 18 sets of eastbound traffic lights or 16 sets of westbound traffic lights on the old routes

Table 6.1 The Cross City Tunnel chronology.

Date	Event
22 Oct 1998	The Premier (Mr Carr) and the Minister for Roads (Mr Scully) released an exhibition for comment on the initial concept (the 'short tunnel') in a 16-page report entitled: The Cross City Tunnel: Improving the Heart of the City. AU$2 toll is flagged.
April 1999	The City of Sydney Council released the Cross City Tunnel Alternative Scheme. This was a longer tunnel than the one proposed in the 1998 *Improving the Heart of the City*, running to the eastern end of the Kings Cross Tunnel and included narrowing William Street.
15 Sept 2000	The RTA invited registrations of interest from private sector parties for the financing, design, construction, operation and maintenance of the Cross City Tunnel project.
23 Oct 2000	Closing date for registrations of interest to construct and operate the tunnel.
Feb 2001	The Minister for Roads (Mr Scully) announced that three consortia had been short-listed to prepare detailed proposals: Cross City Motorways (CCM), E-TUBE and Sydney City Tunnel Company.
Oct 2001	Detailed proposals for implementation of the project were lodged by the three consortia and reviewed by assessment panel.
Feb 2002	The Budget Committee of Cabinet approved CCM to be selected as the preferred proponent and for the CCM 'long 80 tunnel' option to be selected as the preferred proposal.
27 Feb 2002	The Minister for Roads (Mr Scully) announced that CCM was the preferred proponent. The tender submission from CCM incorporated changes to the approved activity that the Minister for Roads believed would provide more benefits and reduce construction related impacts on the community. As a result of the proposed changes, a number of additional environmental impacts would occur. A supplementary EIS was prepared.
14 Mar 2002	A letter was sent from the Treasurer (Mr Egan) to the Minister for Roads (Mr Scully) stating 'A key objective of the project has been its development at no net cost to government' and 'it is not certain at this time that the project can achieve a 'no net cost to government' outcome. If the project cannot proceed without a government contribution, any such contribution would need to be funded out of the RTA's existing forward capital program'.
16 Dec 2002	Approval was given by the Treasurer (Mr Egan) to sign the project deed under the Public Authorities (Financial Arrangements) Act 1987.
18 Dec 2002	A contract between the CCM consortium and RTA was signed to finance, construct, operate and maintain the CCT. Differential tolling was set, AU$2.5 per car and AU$5 for heavy vehicles.
28 Jan 2003	Major works started on the AU$680 million Cross City Tunnel.
21 Dec 2004	The Treasurer (Mr Egan) approved the RTA to enter into the Cross City Tunnel Project First Amendment Deed with the CCM under section 20 of the Public Authorities (Financial Arrangements) Act 1987. This deed included provision that '*in consideration for the CCM's agreement to fund and carry out certain changes if required by the RTA, CCM might increase the Base Toll to be collected from motorists on the terms set out in the First Amendment Deed*'.
23 Dec 2004	The First Amendment Deed was entered into by RTA and the CCM, enabling AU$35 million of additional works to be paid for through a higher base toll (increased by AU$0.15).
28 Aug 2005	The Cross City Tunnel opened.
Nov 2005	The Summary of Cross City Tunnel Project Deed was made public.
19 Dec 2035	The Cross City Tunnel was due to be returned to public ownership.

Source: Parliament of New South Wales (February 2006) *Joint Select Committee on the Cross City Tunnel, First Report*.

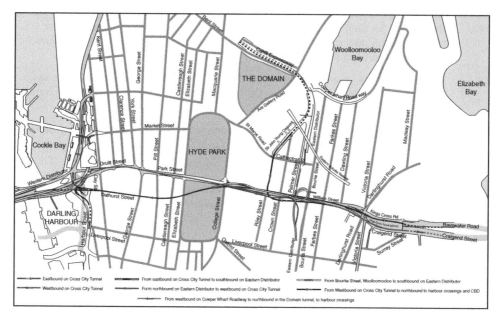

Figure 6.1   The CCT horizontal alignments.

- improved travel times, with estimated savings of up to 20 minutes during peak hours
- higher service reliability for buses through the introduction of bus priority measures
- improved access to and movements for pedestrians and vehicles
- safer and more pleasant environments with better urban designs, wider footpaths and the removal of intrusive through-traffic
- better air quality and reduced traffic noise levels.

In August 2000, following extensive environmental investigation and consultation, the RTA issued an Environmental Impact Statement (EIS) detailing the potential environmental impacts. After considering responses from the community, the RTA made some amendments to the original design concepts. These modifications were approved by the Minister for Planning in October 2001.

According to the construction plan, the CCT project comprised two stages. Stage One encompassed two east–west tunnels that spanned 2.1 km under the Sydney Central Business District and Darlinghurst/Woolloomooloo area. The tunnels would run between the eastern side of the Darling Harbour and Kings Cross. Stage One works also involved the construction of associated tunnelled links to Sir John Young Crescent, the Cahill Expressway and the Eastern Distributor (Figures 6.1–6.4). It was envisaged that Stage One works would be completed in 2004 or 2005 (Parliament of New South Wales, 2006).

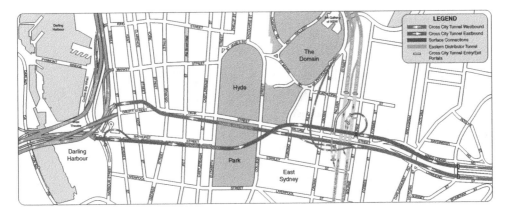

Figure 6.2 The CCT route map.

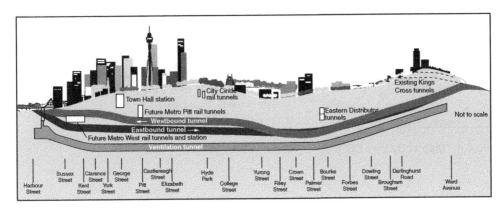

Figure 6.3 Indicative vertical alignments of the main tunnels.

Most of the tunnel would be excavated using a driven tunnel method, a proven and widely used construction technology. This involved the use of tunnelling machines that cut their way through the underground rock mass. The remainder of the tunnel would be constructed using the cut-and-cover method, whereby rock was excavated and then covered with concrete beams or planks to form a tunnel (CrossCity Motorway Pty Ltd, 2006). No extraordinary construction risks were envisaged, though it was fully understood that this was a complex and risky project from site access, logistics and project planning perspectives.

Subsequent to the opening of the tunnel, Stage Two surface works would begin to take advantage of the opportunities afforded by reduced traffic congestion. There would be improvements to surface roads, including new bus and bicycle lanes and other improvements to pedestrian facilities. Table 6.2 presents the different proposals for the CCT.

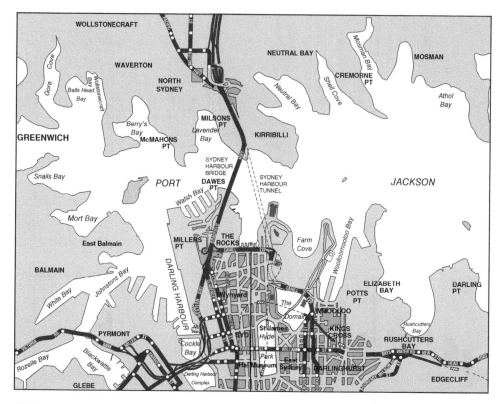

Figure 6.4   Map of Sydney CBD.

Table 6.2   The CCT tunnel models.

Model	Details
The short tunnel	1.2 km tunnel, exiting William Street near Museum of Sydney. Taking approximately 40 000 vehicles. Two-way toll of AU$2. Cost estimated at AU$273 million. Published in: Transforming the Heart of the City (1998).
The long tunnel	Approximately 2 km tunnels exiting in the Kings Cross Tunnel to the east and connecting to the Western Distributor in the west. Two-way toll of AU$2.5. First described in the initial Cross City Tunnel Environmental Impact Statement (2000).
The long 80 tunnel	Approximately 2.1 km tunnels exiting east of the Kings Cross Tunnel to the east and connecting to the Western Distributor in the west. Two-way differential tolling of AU$2.5 and AU$5 (later increased to AU$2.65 and AU$5.3). First described in the Supplementary Environmental Impact Statement (2002).

Source: Parliament of New South Wales (February 2006) *Joint Select Committee on the Cross City Tunnel, First Report.*

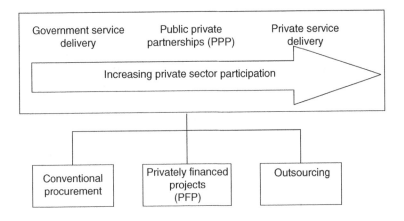

Source: New South Wales Government (November 2001) *Working with Government Guidelines for Privately Financed Projects.*

Figure 6.5 Relationship between PFPs and PPPs.

6.2.2 *Public or private financing?*

The NSW government considered a number of options for financing the CCT project. These included the utilisation of public funds or government borrowings. Alternatively, the NSW government could limit the commitment of public capital and finance the project through a public private partnership (PPP) arrangement (Figure 6.5). A privately financed project (PFP) was a specified form of PPP that involved not only private sector financing but also controlling ownership. Hence, PFPs differed from the outsourcing of infrastructure services or the procurement of design and construction services by the government. Under a PFP arrangement, the infrastructure would be created with a concession agreement whereby the private sector provided the capital to finance, construct, operate and maintain the assets for a specified contract period (the concession period). Such concession agreements typically formed part of the overall regulatory framework under which such investments were operated. When the contract period expired, the operation and maintenance of the asset would revert to public (government) ownership. While construction and operating risks could be transferred to the private sector through the use of a PFP arrangement, the market risk could still be shared by the government through an appropriate compensation arrangement. This essentially followed a project finance model where the assets were returned to the NSW government after the concession expired.

Although a number of infrastructure projects in NSW had been financed through public funds or other forms of PPP, the NSW government considered that PFP was the most appropriate method for financing the CCT project. This was largely because the NSW government was pursuing a debt reduction strategy at the time and thus, at least for the purposes of this project, imposed a budgetary constraint on itself (Parliament of New South Wales, 2006). If the project was financed through the use of public funds or government borrowings, the construction of the tunnel could be delayed due to competing demands for public funds. Alternatively, financing the project through PFP could avoid public debt altogether and result in early delivery. The RTA's total costs on the CCT project were expected to be AU$98 million (Lee, 2006). The sum represented project preparation costs, work on utility networks and other ancillary works associated with the project and could thus be viewed as normal spending on infrastructure improvement. Furthermore, by allowing non-conforming proposals during the tender process, the government could possibly bring in innovative approaches from the private sector and find ways to recoup its project expenses. The ultimate objective of the government was to minimise its financial exposure, delivering the project at no cost to itself.

6.3   Cheung Kong Infrastructure Holdings Limited

Listed on the main board of the Stock Exchange of Hong Kong in 1996, CKI was one of the leading companies in the infrastructure sector in mainland China and Hong Kong. The group has a diversified portfolio of investments in energy infrastructure, transportation infrastructure and infrastructure-related business (see Table 6.3 for a list of project assets of CKI). In 1997, CKI acquired a controlling interest in Hongkong Electric Holdings Limited, the sole electricity company serving residents on Hong Kong Island and Lamma Island in Hong Kong. The acquisition substantially boosted CKI's recurring revenue and strengthened its capital base. CKI and Hongkong Electric's investments in ETSA Utilities and Powercor made the group the largest electricity distributor in Australia.

With a turnover of HK$3.345 billion and net profits of HK$3.228 billion[2] in 2000[3] including shares associates (Figure 6.6), CKI had a strong financial position and a good reputation for quality infrastructure projects. During the year, the energy division was the largest profit contributor, accounting for 78% of CKI's profit contribution. Investments in China and Hong Kong accounted for 86% of the group's overall profits, while investments in Australia, Canada and other countries accounted for the remaining 14%.

In 2000, CKI was actively seeking investment opportunities in three major infrastructure areas (Morgan Stanley, 2004):

- hedged electricity generation positions, wherein generation volume was contracted under a secure power purchase agreement structure

---

[2] AU$1 = HK$7.78 on 31 December 2000.
[3] Cheung Kong Infrastructure Holdings Limited consolidated income statement for the year ending 31 December 2000.

Table 6.3 Major projects of CKI in 2000.

Project	Business	CKI shareholding
Energy		
Hongkong Electric, Hong Kong	Exclusive right to generate and distribute electricity to Hong Kong Island, Ap Lei Chau Island and Lamma Island	38.87%
Envestra Limited, NSW, Australia	Distribution of natural gas in the states of South Australia, Queensland, the Northern Territory, Victoria and New South Wales	19.97%
ETSA Utilities, South Australia, Australia	Right to operate the electricity distribution network in the state of South Australia for 200 years	50% (another 50% held by Hongkong Electric)
Powercor Australia Ltd, Victoria, Australia	Right to operate the electricity distribution network covering an area of over 150,000 sq. km in the state of Victoria and retail operation in certain areas of Australia	50% (another 50% held by Hongkong Electric)
Fushun Cogen Power Plants, Liaoning, China	Operational	60% interest in the JV
Nanhai Power Plant I, Guangdong, China	Operational	30% interest in the JV
Qinyang Power Plants, Henan, China	Operational	49% interest in the JV
Shantou Chaoyang Power Plant, Guangdong, China	Operational	60% interest in the JV
Shantou Chenghai Power Plant, Guangdong, China	Operational	60% interest in the JV
Shantou Tuopu Power Plant, Guangdong, China	Operational	60% interest in the JV
Siping Cogen Power Plant, Jilin, China	Operational	45% interest in the JV
Zhuhai Power Plant, Guangdong, China	Operational	45% interest in the JV
Transportation		
Eastern Harbour Crossing Rail Tunnel, Hong Kong	Rail franchise period 1986–2008	50%
Changsha Wujialing and Wuyilu Bridges, Wunan, China	Operational	44.2% in the JV
Guangzhou East-South-West Ringroad, Guangdong, China	Operational	44.4% in the JV
Jiangmen Chaolian Bridge, Guangdong, China	Operational	50% in the JV
Jiangmen Jianghe Highway, Guangdong, China	Operational	50% in the JV
Jiangmen Jiangsha Highway, Guangdong, China	Operational	50% in the JV
Nanhai Road Network, Guangdong, China	Operational	49%–64.4% in the JV
National Highway 107 (Zhumadian section), Henan, China	Operational	66% in the JV
Panyu Beidou Bridge, Guangdong, China	Operational	40% in the JV

Table 6.3 *Continued*

Project	Business	CKI shareholding
Shantou Bay Bridge, Guangdong, China	Operational	30% in the JV
Shen-shan Highway (Eastern Section), Guangdong, China	Operational	33.5% in the JV
Shenyang Changqing Bridge, Liaoning, China	Operational	30% in the JV
Shenyang Da Ba Road and South-West Elevated Sections, Liaoning, China	Operational	30% in the JV
Shenyang Gongnong Bridge, Liaoning, China	Operational	30% in the JV
Shenyang Shensu Expressway, Liaoning, China	Operational	30% in the JV
Tangshan Tangle Road, Hebei, China	Operational	51% in the JV
Zengcheng Lixin Road, Guangdong, China	Operational	51% in the JV
Infrastructure materials and infrastructure-related businesses		
Anderson Asphalt, Anderson Asia, Hong Kong	One of Hong Kong's largest asphalt producers	
Asia Stone, Anderson Asia, Hong Kong	One of Hong Kong's four contract quarries	
Bonntile, Anderson Asia, Hong Kong	Exterior wall spray-coating system specialist	
Ready Mixed Concrete, Anderson Asia, Hong Kong	One of Hong Kong's largest concrete producers	
Green Island Cement, Hong Kong	Only integrated cement producer in Hong Kong	
Shantou Cement Grinding Plant, Guangdong, China	Operational	100%
Yunfu Cement Plant, Guangdong, China	Operational	67% interest in the JV
Siquijor Limestone Quarry, Philippines	Operational	40% interest in the JV
Polyphalt Inc, Canada	Developed and commercialised polymer modified asphalt technology, products and services. The company blended several of its asphalt technologies with plastics and rubbers, including recycled materials	63.7%
Stuart Energy Systems Corp., Canada	Developed and supplied hydrogen generation and supply systems through its proprietary water and electrolysis technology	12.9%
Shenyang LPG Business, Liaoning, China	LPG filling stations and vehicle conversion facilities	51% interest in the JV
Yueyang Water Plants, Hunan, China	Operational	49% interest in the JV
e-Smart System Inc, Hong Kong	Applications of the patented 'Eyecon' microprocessor-based contactless smart card technology in the Asia Pacific region	50%

Source: Cheung Kong Infrastructure Holdings Limited (2000) Annual Report.

Turnover (HK$ millions)

Group turnover	2,567
Share of turnover of jointly controlled entities	778
	3,345

Group turnover	2,567
Other revenue	1,373
Operating costs	(2,819)
Finance costs	(621)
Operating profit	**500**
Share of results of associates	2,413
Share of results of jointly controlled entities	588
Profit before taxation	**3,501**
Taxation	(288)
Profit after taxation	**3,213**
Minority interests	15
Profit attributable to shareholders	**3,228**
Dividends	(1,353)
Profit for the year retained	**1,875**
Earnings per share	HK$1.43

Source: Cheung Kong Infrastructure Holdings Limited (2000) Annual Report.

Figure 6.6   CKI consolidated income statement for year ending 31 December 2000.

- near-monopoly transportation business, including toll roads, bridges, tunnels, airports and possibly sea transport facilities
- regulated monopoly network businesses, including electricity wires, gas and water pipelines.

In mid-September 2000, the RTA invited tenders for the construction, financing and 30 year operation of the Sydney CCT. The project presented an unprecedented opportunity for CKI towards its global ambitions in transportation infrastructure project investment. In response to the invitation of Registrations of Interest from the RTA, CKI together with its major business partner, Bilfinger Berger Aktiengesellschaft (AG), decided to bid for the project in October 2000. Bilfinger Berger AG was one of the world's leading construction companies. Based in Mannheim, Germany, it had global operations in civil engineering and real estate, and had actively participated in privately financed build-operate-transfer (BOT) projects. With equity support from the minority superannuation trust investors, they formed the CrossCity Motorway Consortium (CCM) and obtained a commitment to provide financing from the Deutsche Bank AG, a major German bank. Bilfinger Berger AG and its wholly owned Australian subsidiary, Baulderstone Hornibrook Pty Limited, also played a sponsorship role for the tender. Baulderstone Hornibrook had a number of infrastructure projects in Australia, including the M5 East Freeway and Anzac Bridge in Sydney, the Western Link section of the Melbourne City Link and the Graham Farmer Freeway in Perth (RTA, 2003).

6.4 The bidding process

In February 2001, the RTA short-listed three consortia out of eight competing for the CCT project. The CCM consortium was one of the three short-listed consortia for the final bid. The other two short-listed consortia were the E-Tube consortium (sponsored by Leighton Contractors Pty Limited and Macquarie Bank) and Sydney City Tunnel Company (sponsored by Transfield Holdings Pty Limited and Multiplex Constructions Pty Limited). All three selected consortia had to submit their detailed proposals for implementation on or before 24 October 2001. The proposals would be reviewed by an assessment panel from the RTA, which would select the preferred proponent based on various pre-determined criteria. The panel would conduct a comparative value assessment against a public sector comparator (PSC). A PSC was a model of the hypothetical, risk-adjusted costs of delivering the project under a government-financed method. The panel would also evaluate the tender submissions based on various non-price determined criteria, which included (RTA, 2003):

- project structure, participants and organisation
- design and construction
- initial traffic management and safety plan
- initial project plans for quality assurance, project management, environmental management, design, construction, operation and maintenance, community involvement, incident responses, occupational health, safety and rehabilitation management and project training
- operation and maintenance.

6.4.1 Preparing the proposal

The CCM consortium had to submit detailed proposals showing how it planned to complete the design, construction and operation of the CCT. According to the tender documents from the RTA, business consideration fees were potential areas that were available for further exploration if there was potential excess revenue over cost during the term of the concession. All short-listed consortia could nominate a business consideration fee to the RTA, which was intended to contribute towards RTA costs associated with the project.

The CCM consortium was aware that the RTA was likely to prefer proposals that were consistent with its objective of delivering the project at 'no cost to government' (Parliament of New South Wales, 2006). Given RTA's policy, the CCM consortium felt that the payment of a reasonable business consideration fee would be effective in meeting the objectives that the government had set. By putting forward a proposal that would focus its bid on the business consideration fee, the CCM consortium was confident that its proposal could offer superior value for money over traditional methods of government delivery.

Since there was no mandatory requirement to submit a conforming proposal to meet all conditions set by the RTA, there was an opportunity for the CCM consortium to bring in critical changes to the original project plan to reap maximum returns and recoup the costs incurred by the business consideration fee. For environmental and traffic reasons, the CCM consortium considered that the length of the tunnel could be increased by 300 metres while the depth at the eastern end could be increased by 30 metres. Such changes would increase the tunnel's daily capacity by an extra 17 000 vehicles (Tunnels and Tunneling International, 2003). However, the extra work required for this 'long 80 tunnel' proposed by the CCM consortium could result in an increase in construction cost of US$135.7 million.[4] The total projected construction cost would therefore become AU$680 million.

6.5   Valuing the project

The value of the CCT project would depend principally on the revenues generated by traffic volume and tolls charged during the term of the concession. The CCT was composed of two main tunnels. Both of the main tunnels would have two lanes, while the other entry and exit tunnels would have single lanes. Both tunnels would be electronically tolled. Users of the CCT had to purchase an electronic pass or hold a toll account and have a valid electronic tag. The CCM consortium felt it was necessary to apply a differential pricing scheme. Under this scheme, the toll charged would depend on the size of the vehicle and the route that it took. Passenger vehicles such as motorbikes, sedans, station wagons, taxis and vehicles towing trailers were classified as Class 2.[5] Heavy vehicles were classified as Class 4.[6]

The CCM consortium considered that the initial toll charges for vehicles other than buses (Table 6.4) could be set as follows (RTA, 2003):

- for vehicles using the main tunnels to and from Darling Harbour, including vehicles entering from or exiting to the Eastern Distributor, AU$2.5 for Class 2 passenger vehicles and AU$5 for Class 4 heavy vehicles (March 1999 prices, including GST)
- for vehicles entering the westbound tunnel at Rushcutters Bay and then using the Riley Street tunnel to exit onto Sir John Young Crescent, AU$1.1 for Class 2 passenger vehicles and AU$2.2 for Class 4 heavy vehicles (March 1999 prices, including GST).

---

[4] AU$1 = US$0.5115 on 31 December 2001.
[5] Class 2: height less than or equal to 2.8 m, length less than or equal to 12.5 m.
[6] Class 4: height greater than 2.8 m, length greater than 12.5 m.

Table 6.4   Proposed toll charges.

	Eastbound Tunnel	Westbound Tunnel	Sir John Young Crescent Exit
	Darling Harbour to Eastern Distributor exit or Rushcutters Bay	Rushcutters Bay to Darling Harbour	From the east
	(AU$)	(AU$)	(AU$)
Class 2	2.5	2.5	1.1
Class 4	5	5	2.2

Note: all tolls are inclusive of GST.
Source: RTA (June 2003) Cross City Tunnel Summary of Contracts.

Buses providing public transport services were not required to pay tolls, but higher charges could apply to vehicles without electronic tolling transponders. The CCM suggested an initial administrative charge between AU$5 and AU$8 for each casual use of any of the tunnels on top of the standard tolls. In addition, higher tolls might be charged for traffic exiting from the westbound tunnel onto Harbour and Bathurst Streets, to help reduce congestion in the western CBD. The extra revenue collected could be used for public transport, pedestrian, cyclist, air quality and other amenity improvements.

The CCM consortium believed that it was imperative to secure the right to increase future tolls at its discretion. It therefore proposed a toll escalation scheme, under which the tolls could be increased in line with inflation or in line with minimum quarterly rates of increase equivalent to 4% per annum until the June quarter of 2012 and then 3% per annum until the June quarter of 2018 (RTA, 2003). From mid-2018, the maximum increases would be in line with inflation.

With advice from Hyder Consulting, the CCM consortium prepared traffic estimates for the CCT. According to the projections, the CCT would be used by over 90 000 vehicles per day by 2006 and over 100 000 vehicles per day by 2016 (RTA, 2002). By lengthening the tunnel, the consortium would be able to earn additional revenue of AU$10.98 million per year based on the traffic projections (Tunnels and Tunneling International, 2003). Figure 6.7 shows the traffic forecast of the CCT. Table 6.5 summarises the results of the likely economic performance of the CCT project.

Under the Land Lease Agreement with NSW, the winning consortium would have to make rent payments for the first 12 months of the lease, each successive six-month period during the lease and then the final period of the lease, as follows (RTA, 2003):

- AU$1, plus
- 35% of actual gross revenue – less any amount collected for GST or other taxes or government charges, other than income tax – from any non-toll business uses of

	Eastbound	Westbound	Sir John Young Crescent
Capacity	60,000	60,000	50,000

	Annual Average Daily Traffic (AADT)			
Year ending	Eastbound	Westbound	Sir John Young Crescent	Total
31 Dec 04	30,041	36,626	22,433	89,100
31 Dec 06	30,713	37,597	22,796	91,106
31 Dec 11	32,433	40,137	23,695	96,265
31 Dec 16	34,250	42,848	24,629	101,727
31 Dec 21	35,507	44,409	25,315	105,231
31 Dec 26	36,811	46,027	26,021	108,859

	Semi Annual Growth Rate (%)			
Year ending	Eastbound	Westbound	Sir John Young Crescent	Average
31 Dec 04	0.56	0.66	0.40	0.56
31 Dec 06	0.56	0.68	0.39	0.55
31 Dec 11	0.56	0.67	0.39	0.41
31 Dec 16	0.37	0.36	0.28	0.34
31 Dec 21	0.37	0.36	0.28	0.34
31 Dec 26	0.39	0.39	0.39	N.A.

	Annual Growth Rate (%)			
Year ending	Eastbound	Westbound	Sir John Young Crescent	Average
31 Dec 04	1.11	1.32	0.81	1.12
31 Dec 06	1.10	1.32	0.78	1.11
31 Dec 11	1.10	0.92	0.63	0.82
31 Dec 16	0.72	0.72	0.55	0.68
31 Dec 21	0.72	0.72	0.55	0.68
31 Dec 26	0.72	0.72	0.55	0.68

Source: Cross City Tunnel Parliamentary Notice, http://www.lee.greens.org.au/campaigns/crosscity.htm (accessed 7 October 2006).

Figure 6.7 Traffic forecast.

Table 6.5 RTA estimates of the likely economic performance of the cross city tunnel.

Discount rate (%)	Present value of initial and recurring capital costs and mainte- nance costs	Present value of road user and pedes- trian benefit	Net present value	Benefits : cost ratio		Net present value/capital cost
				Benefits O&M D&C	Benefits D&C +O&M	
	AU$ m	AU$ m	AU$ m			
4	693	2754	2061	5.0	4.2	4.0
7	576	1689	1114	3.4	3.0	2.4
10	495	1102	607	2.4	2.3	1.4

Taking account of initial and recurring capital costs, operation and maintenance costs, road user benefits (savings in vehicle operating costs, travel time savings and savings in accident costs) and pedestrian benefits, but not counting environmental externalities.

Source: RTA (June 2003) Cross City Tunnel Summary of Contracts.

Table 6.6 The RTA's share of any unexpectedly high revenues.

Actual revenue, as a percentage of forecast revenue	RTA's share of this portion of the actual revenue (to be paid by the trustee as part of its rent under the Land Lease)
(%)	(%)
Up to 110	0
110–120	10
120–130	20
130–140	30
140–150	40
More than 150	50

Generally over six-month periods, from tolls and administrative charges, including additional charges for casual tunnel users without electronic tolling vehicle transponders.

Source: RTA (June 2003) Cross City Tunnel Summary of Contracts

the tunnels or the land leased, such as the use of the tunnels or land for telecommunications infrastructure

- if the actual toll and administrative charge revenue for the relevant period – less any amount collected for GST or other taxes or government charges – was more than 10% higher than that forecasted by the private sector participants' base case financial model for the project, a progressively increasing share of this extra revenue (Table 6.6).

The CCM consortium considered that the toll escalation scheme could help to recover the business consideration fee. Therefore, in return for the RTA's granting it the right to undertake the project, the CCM consortium proposed payment of a business consideration fee in the sum of AU$100.1 million plus GST to the RTA.[7] According to the tender documents, the fee had to be over and above the costs associated with the project. The fee could be used by the RTA to cover its ancillary costs associated with delivering the project, including work on utility networks affected by the tunnel and cost recovery for project preparation costs.

6.6 Assessing project risks

The CCM consortium was aware that substantial risks would be transferred from the government to the consortium if it won the tender. The consortium had identified various risks associated with the project. These included (RTA, 2003):

- risks associated with the financing, design, construction, operation, maintenance and repair costs of the project
- the risks that the tunnel might fail to deliver the anticipated traffic volumes or projected revenues
- the risks of over-estimation of the value motorists would place on the tunnel's benefits
- the risks that the capacity of the tunnel was not sufficient to allow for the projected traffic estimates
- the risks of over-forecasting asset use – a study conducted by Standard and Poor's about traffic modelling on 104 international toll roads, bridges and tunnels showed that on average, across all toll roads, bridges and tunnels, forecasts overestimated traffic in the first year by 20–30% (Lee, 2006)
- income tax risks
- the risk that their works or operational and maintenance activities might be disrupted by the lawful actions of other government and local government authorities or a court or tribunal.

To minimise patronage risks, the CCM consortium proposed certain changes to the road network or the introduction of traffic calming measures. One of the major changes involved removing access to the harbour crossings from Sir John Young

---

[7] The quantum of the business consideration fee changed over the course of the contract negotiations between CCM and the RTA. The original figure changed following the acceptance by the RTA of the consortium's non-conforming long 80 tunnel. As a consequence of the differing minister's planning conditions of approval, and later requirements imposed by more stringent air quality standards (the construction of a third tunnel for ventilation purposes, with estimated cost of $37 million) and through community consultation, CCM reduced the amount of the BCF they proposed to pay the RTA to AU$96.8 million.

Crescent and Cowper Wharf Road. Nevertheless, such changes would restrict road users' choice and could affect residents of the affected communities. The CCM consortium would thus also require the government to pay compensation if any changes to the Sydney public transport system had a material effect on the amount of traffic traversing the toll road tunnel. A material effect would include blocking a lane or otherwise directly hampering traffic flow into the tunnel (Scott and Allen, 2005). The compensation could be computed based on the consortium's expected profits every year until 2035. This could amount to as much as AU$100 million annually.

To mitigate the risks of additional costs that could arise due to a change in the scope of works directed by the RTA, the CCM consortium would require the RTA to compensate it for the costs that reasonably arose from the change, including those associated with the contractors' overheads and profits and any delay costs or equity holding costs. If the RTA decreased the scope of works and subsequent operation, maintenance and repair obligations, the CCM consortium would pay to the RTA 75% of the direct cost savings.

6.7   Capital structure

6.7.1   The equity investors

The CCM consortium had a complicated special purpose vehicle structure that involved the use of trusts and corporate vehicles that were set up by the equity investors. A new company, known as CrossCity Motorway Nominees No. 2 Pty Limited (the trustee), was incorporated to serve as a trustee of the CrossCity Motorway Property Trust, which was established in October 2001. The Property Trust was held by CrossCity Motorway Nominees No. 1 Pty Limited, in its capacity as trustee of the CrossCity Motorway Holdings Trust, another unit trust set up in October 2001. In turn, all the shares in the trustee and the Holdings Trust were owned and controlled by CrossCity Motorway Holdings Pty Limited (CCM Holdings), which was owned by the following equity investors (see Figure 6.8 for an overview of the structure of the CCT contracts) (RTA, 2003):

- CKI City Tunnel Investment (Malaysian) Limited, a wholly owned subsidiary of CKI (50%)
- Bilfinger Berger BOT GmbH, a wholly owned subsidiary of Bilfinger Berger AG (20%)
- DB Capital Partners, a private equity arm of Deutsche Asset Management (30%). It invested in the project on behalf of a number of Australian superannuation funds, including Development Australia Fund (CrossCity Motorway Media Release, 2002)
- all the units in the CrossCity Motorway Holdings Trust were owned in the same proportions by the same equity investors.

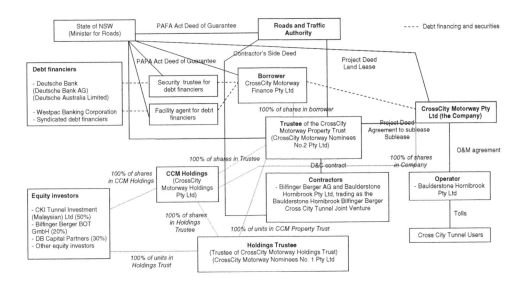

Source: RTA (June 2003) Cross City Tunnel summary of contracts.

Figure 6.8 Overview of the structure of the CCT contracts.

CCM Holdings set up a wholly owned subsidiary known as CrossCity Motorway Pty Limited (the Company). If the consortium won the bid, the trustee and the Company would be used as the vehicle to enter into the Cross City Tunnel Project Deed with the RTA.

6.7.2 The contractor

The contractor of the project would be the Baulderstone Hornibrook Bilfinger Berger Cross City Tunnel Joint Venture, a partnership set up by Bilfinger Berger AG and Baulderstone Hornibrook Pty Limited (RTA, 2003).

6.7.3 The operator

The operation, repair and maintenance of the tunnel and its surface works would be undertaken by Baulderstone Hornibrook Pty Limited until reversion to public ownership.

6.7.4 The borrower

Borrowings would be made by the CrossCity Motorway Finance Pty Limited, which was wholly owned by the trustee in its capacity as trustee of the property trust. The borrower would be the vehicle used to contract with the project's debt financiers.

Table 6.7 Projected consolidated balance sheet (2003–2008).

Six months ending	30/6/03	31/12/03	30/6/04	31/12/04	30/6/05	31/12/05	30/6/06	31/12/06	30/7/07	31/12/07	30/6/08	31/12/08
Period number	1	2	3	4	5	6	7	8	9	10	11	12
Assets (AU$ millions)												
Current assets												
Cash on deposit	—	—	—	—	—	—	12,528	—	—	—	—	—
Cash balance in equity account	—	—	—	—	—	68	319	245	196	163	160	166
Accounts receivable	—	—	—	—	—	—	25,169	25,169	—	—	—	—
Cash balances in ramp up reserve account	—	—	—	—	—	—	—	—	—	—	—	—
Cash balances in major maintenance reserve account	—	—	—	—	—	—	—	—	—	—	—	—
Cash balances in lockup account	—	—	—	—	—	—	—	—	—	—	—	—
Cash balance in equity holding account	—	—	—	—	—	—	—	—	—	—	—	—
Cash in debt drawdown account	—	—	—	—	—	—	—	—	—	—	—	—
Prepayment of tax	—	—	—	—	—	—	—	—	—	—	—	—
Total current assets	—	—	—	—	—	68	38,016	25,413	196	163	160	166
Non-current assets												
Future income tax benefit	1,338	2,869	7,602	12,623	18,895	33,902	36,260	37,829	39,373	40,965	41,903	42,842
Carrying value of fixed assets	243,166	362,627	493,665	653,250	738,500	758,583	785,022	763,212	740,370	717,527	696,805	676,050
RTA payment/tender process allowance	95,392	93,925	92,457	90,989	89,522	88,054	85,587	85,119	83,652	82,184	80,716	79,249
Total non-current assets	339,896	459,421	593,724	756,862	846,917	880,539	906,869	886,160	863,395	840,676	819,424	798,141
Total assets	339,896	459,421	593,724	756,862	846,917	880,607	944,885	911,574	863,591	840,839	819,584	798,308
Liabilities												
Current liabilities												
Accounts payable	—	—	—	—	—	380	1,269	1,105	926	939	907	904
Provision for income tax	—	—	—	—	—	—	—	—	—	—	—	—

						380	1,269	1,105	926	939	907	904
Total current liabilities	—	—	—	—	—	—	—	—	—	—	—	—
Non-current liabilities												
Deferred income tax liability	(110)	(282)	(519)	(849)	(1,261)	4,983	6,163	5,887	5,611	5,334	5,058	4,782
Senior debt	176,235	291,608	439,782	608,229	704,139	756,191	579,992	580,000	565,016	565,016	565,016	565,016
Working capital	—	—	—	—	—	—	—	—	—	—	—	—
Shareholder loans	—	—	—	—	—	—	—	—	—	—	—	—
Capitalised RTA payment	—	—	—	—	—	—	—	—	—	—	—	—
Total non-current liabilities	176,125	291,326	439,263	607,380	702,878	761,174	586,155	585,887	570,628	570,350	570,074	569,797
Total liabilities	176,125	291,326	439,263	607,380	702,878	761,554	587,424	586,992	571,554	571,289	570,981	570,701
Net assets	163,771	168,095	154,461	149,482	144,039	119,053	357,461	324,582	292,037	269,550	248,603	227,607
Shareholders equity												
Ordinary equity – original equity	163,790	158,756	153,721	148,686	143,651	138,617	211,207	210,860	210,472	210,068	210,068	201,895
Ordinary equity – retail equity	—	—	—	—	—	—	—	—	—	—	—	—
RPU	—	—	—	—	—	—	72,613	72,613	72,613	72,613	72,613	72,613
Bonds	—	—	—	—	—	—	90,000	68,015	44,017	28,474	9,989	—
Total ordinary equity	163,790	158,756	153,721	148,686	143,651	138,617	373,820	351,488	327,102	311,154	292,670	274,508
Retained earnings												
Retained earnings (opening balance)	—	(20)	339	740	796	388	(19,564)	(14,358)	(26,906)	(35,064)	(41,604)	(44,066)
Net profit after abnormals and tax	(20)	359	401	56	(408)	(19,952)	5,205	(12,548)	(8,158)	(6,540)	(2,042)	(104)
Dividends/distributions to equity	—	339	—	—	388	—	—	—	—	—	420	2,732
Retained earnings (closing balance)	(20)	339	740	796	796	(19,564)	(14,358)	(26,906)	(35,064)	(41,604)	(44,066)	(46,901)
Total shareholders equity	163,771	159,095	154,461	149,482	144,040	119,053	359,461	324,582	292,037	269,550	248,604	227,606
Total liabilities and shareholders equity	339,896	450,421	593,724	756,862	846,917	880,607	946,885	911,574	863,591	840,839	819,585	798,308

Source: Cross City Tunnel Parliamentary Notice, http://www.lee.greens.org.au/campaigns/crosscity.htm (accessed 7 October 2006).

6.7.5 Financing strategy

The total capital required for the CCT project was estimated to be AU$680 million. The consortium estimated that the project would be financed by AU$580 million in debt, and the balance by equity. Under the proposed debt financing plan, one-half of the initial debt finance would be provided by Deutsche Bank and Westpac Banking Corporation, with further syndicate participation expected to reduce their commitments if the project proceeded. CKI obtained further financing for its equity commitment by means of a 42 month equity bridging loan arranged by China Construction Bank.

6.8 Completion of the deal

In late October 2001, the three short-listed consortia submitted their proposals to the RTA. The CCM consortium was the only proponent that bid on the business consideration fee. Two other bidders did not bid an upfront payment of a business consideration fee. During the evaluation process, the RTA relied on estimates of traffic flow prepared by traffic consultant Masson Wilson Twiney Pty Ltd, to compare the value of the proposals against the PSC. The forecast produced by the consultant showed that the 'long tunnel' proposed by the CCM consortium would be used by 86 300 vehicles per day in 2006 and 101 700 vehicles per day in 2016. Such forecast figures were lower than the estimates prepared by the CCM consortium. Tables 6.7–6.9 present the projected financial statements and the assumptions used by the CCT for the period from 2003 to 2008.

In February 2002, the RTA's assessment and review panel concluded that the proposals submitted by the CCM consortium represented better value for money than the PSC and the proposals submitted by the other two proponents. On 27 February 2002, the Minister for Roads formally announced that the CCM consortium was the preferred proponent. The submission from the CCM consortium was a non-conforming proposal that had incorporated various changes to initial concepts of the tunnel. As a result of the proposed changes, a number of additional environmental impacts would occur and a supplementary EIS had to be prepared. The execution of the project's principal contracts was completed on 18 December 2002.

The CCT project heralded CKI's first transportation project outside mainland China and Hong Kong. Under the plan, the tunnel components of the project would be operated, maintained and repaired by the CCM consortium until they were returned to public ownership after 18 December 2035 or for 30 years and two months from the completion of the tunnels, if their completion was delayed (RTA, 2003). Analysts expected that the equity internal rate of return (IRR) on CKI's investment in this green field project would be around 15%. With sufficient cash on hand and reasonably low gearing, CKI itself had the necessary resources to finance this project. It was projected that in 2006, the CCT's annual contribution to CKI's bottom line would be HK$28 million.[8]

---

[8] AU$1 = HK$6.1405 on 31 December 2006.

Table 6.8 Projected consolidated profit and loss statement (2003—2008).

Six months ending	30/6/03	31/12/03	31/6/04	31/12/04	30/6/05	31/12/05	30/6/06	31/12/06	30/6/07	31/12/07	30/6/08	31/12/08
Period number	1	2	3	4	5	6	7	8	9	10	11	12
Revenue (AU$ millions)												
Toll revenue												
Eastbound revenue	—	—	—	—	—	2,956	19,106	20,603	21,327	21,924	22,239	22,835
Westbound revenue	—	—	—	—	—	3,007	19,246	20,656	21,700	22,638	23,301	24,278
Sir John Crescent: revenue	—	—	—	—	—	1,048	6,656	6,788	7,009	7,207	7,314	7,607
Total revenue	—	—	—	—	—	**7,011**	**45,008**	**48,046**	**50,036**	**51,770**	**52,854**	**54,720**
Sensitivity revenue	—	—	—	—	—	—	—	—	—	—	—	—
Other operating revenue	—	—	—	—	—	—	—	—	—	—	—	—
Total operating revenue	—	—	—	—	—	**7,011**	**45,008**	**48,046**	**50,036**	**51,770**	**52,854**	**54,720**
Operating and maintenance expenses												
Variable operating expenses												
Eastbound operating expenses	—	—	—	—	—	94	689	815	842	846	815	795
Westbound operating expenses	—	—	—	—	—	95	692	813	853	870	851	844
Sir John Young Crescent operating expenses	—	—	—	—	—	66	481	563	588	597	582	574
Casual user variable administration fee	—	—	—	—	—	895	3,290	1,778	930	571	556	570
Total variable operating expenses	—	—	—	—	—	**1,150**	**5,152**	**3,970**	**3,213**	**2,884**	**2,804**	**2,784**

Table 6.8 *Continued*

Six months ending Period number	30/6/03 1	31/12/03 2	31/6/04 3	31/12/04 4	30/6/05 5	31/12/05 6	30/6/06 7	31/12/06 8	30/6/07 9	31/12/07 10	30/6/08 11	31/12/08 12
Fixed operating expenses												
CCM costs	—	—	—	—	—	1,498	4,576	3,547	2,811	2,927	2,721	2,695
Finance and administration	—	—	—	—	—	762	2,059	2,123	1,576	1,847	1,642	1,653
Revenue collection	—	—	—	—	—	39	120	123	124	126	128	130
Operations and routine maintenance	—	—	—	—	—	835	2,550	2,596	2,588	2,632	2,677	2,720
Other	—	—	—	—	—	—	—	—	—	—	—	—
Contingency	—	—	—	—	—	82	236	242	214	230	222	225
Operator margin	—	—	—	—	—	258	745	762	675	725	700	709
Agency fee	—	—	—	—	—	—	—	29	29	29	125	66
Security trustee fees	—	—	—	—	—	—	—	17	17	18	18	18
Tranche C – bond facility LC	—	—	—	—	—	—	—	35	38	6	-	
Total operating and maintenance expenses	—	—	—	—	—	4,624	15,437	13,444	11,285	11,424	11,037	11,001
EBITDA	1,468	1,468	1,468	1,468	1,468	2,387	29,571	34,602	38,750	40,346	41,817	43,719
Depreciation/amortisation	(1,468)	(1,468)	(1,468)	(1,468)	(1,468)	23,722	23,727	24,314	24,314	24,314	22,227	22,227
EBIT	(1,468)	(1,468)	(1,468)	(1,468)	(1,468)	(21,335)	5,844	10,288	14,436	16,032	19,590	21,492
Repairs and maintenance – expense	—	—	—	—	—	—	—	—	—	—	—	—
Income tax expense/benefit on operation	—	—	—	—	—	(1,258)	6,895	8,430	9,674	10,144	11,238	11,792
Bank base case adjustment	—	—	—	—	—	—	—	—	—	—	—	—
Net operating profit less adjusted taxes (NOPLAT)	(1,468)	(1,468)	(1,468)	(1,468)	(1,468)	(20,077)	(1,052)	1,858	4,761	5,888	8,351	9,700
Loss on write-off of assets	—	—	—	—	—	—	—	—	—	—	—	—
Distribution to RTA	—	—	—	—	—	—	—	—	—	—	—	—

Net interest expense

Interest income

Other interest income	—	—	—	—	—	—	—	613	884	297	194	230
Total interest income	—	—	—	—	—	—	—	613	884	297	194	230
Interest expenses												
Interest expense – senior debt	—	877	2,101	3,828	5,624	7,380	1,817	21,825	21,831	21,267	19,572	19,572
Interest expense – RPU	—	—	—	—	—	—	—	3,451	3,451	3,453	3,451	3,451
Interest expense – shareholder loans	—	—	—	—	—	—	—	—	—	—	—	—
Interest expense – bonds	—	—	—	—	—	—	—	—	—	—	—	—
Line fee expense	—	—	—	—	—	—	—	17	17	18	18	18
Refinance fee expense	—	—	—	—	—	—	—	—	—	—	—	—
Total interest expenses	—	877	2,101	3,828	5,624	7,380	1,817	25,293	25,299	24,738	23,041	23,041
Total net interest expenses	—	877	2,101	3,828	5,624	7,380	1,817	24,680	24,415	24,441	22,847	22,811
Tax shield on net interest expenses	1,448	2,704	3,969	5,351	6,684	7,504	8,074	10,275	11,495	12,012	12,453	13,007
Net profit after tax (NPAT)	(20)	359	401	55	(408)	(19,952)	5,205	(12,548)	(8,158)	(6,540)	(2,042)	(104)
Abnormal items after tax	—	—	—	—	—	—	—	—	—	—	—	—
Net profit after abnormals and tax	(20)	359	401	55	(408)	(19,952)	5,205	(12,548)	(8,158)	(6,540)	(2,042)	(104)
Retained earnings (opening balance)	—	(20)	339	740	795	388	(19,564)	(14,358)	(26,906)	(35,064)	(41,604)	(44,066)
Net profit after abnormals and tax	(20)	359	401	55	(408)	(19,952)	5,205	(12,548)	(8,158)	(6,540)	(2,042)	(104)
Dividends/distributions	—	—	—	—	—	—	—	—	—	—	420	2,732
Retained earnings (closing balance)	(20)	339	740	795	388	(19,564)	(14,358)	(26,906)	(35,064)	(41,604)	(44,066)	(46,901)

Source: Cross City Tunnel Parliamentary Notice, http://www.lee.greens.org.au/campaigns/crosscity.htm (accessed 7 October 2006).

Table 6.9 Assumptions used for the financial projections.

	Base Case Financial Model – Assumptions book		
	Value		**Description**
Assumptions			
1. Transaction Parameters			
1.1 Analysis date	18 December 2002		Current model run date
1.2 Financial close	19 December 2002		Fulfillment of all CPs in loan documentation
1.3 Concession life	33		Years post financial close
1.4 Financial year end	30 June		Financial year end for Cross City Motorway SPV
1.5 Modelled entity	Op.co and trust		Entities which are modelled
1.6 Borrowing entity	trust		Entity that raises debt
2. Economic assumptions			
General CPI assumptions			
Last ABS published CPI	138.5		CPI index in last published ABS data
Date of last published CPI	30 September 2002		Date of last published ABS date
CPI in project deed base period	121.3		CPI index in the project deed base period
Project deed base period	30 September 1998		Base period specified in project deed
Long-term national CPI forecast	2.50%		Long-term national CPI forecast – assumed to be an effective rate

	CPI (%)	**AWE (%)**	
O&M Escalation			
Construction	16	84	As per the O&M contract
31 December 2006	27	73	As per the O&M contract
31 December 2007	41	59	As per the O&M contract
31 December 2008	47	53	As per the O&M contract
31 December 2009	50	50	As per the O&M contract
Thereafter	50	50	As per the O&M contract
Replacement expenditure escalation			
Average weekly earnings, trend, persons, total earnings, NSW		50	Steady state proportion
All mechanical services – Sydney		50	Steady state proportion
Blended rate		2.45	Calculated rate based on the above proportions and rates
Repairs and refurbishment expenditure escalation		33	Steady state proportion
Average weekly earnings, trend, persons, total earnings, NSW		67	Steady state proportion
All mechanical services – Sydney		1.92	Calculated rate based on the above proportions and rates
Blended rate			
Back office admin fee escalation		79	Steady state proportion
Average weekly earnings, trend, persons, total earnings, NSW		7.50	Steady state proportion
All mechanical services – Sydney		13.50	Steady state proportion
Blended rate		3.57	Calculated rate based on the above proportions and rates

Table 6.9 *Continued*

	Base Case Financial Model – Assumptions book	
Value		**Description**
3. Interest rate assumptions		
Swap rate during construction	5.32	Monthly rate used for swapped portion of base rate used during construction
Swap rate for first band of hedging post completion	5.99	Semi annual rate used for swapped portion of base rate used during first hedging period
First hedge band – months	42	Month for first swap post completion
Ongoing swap rate – five year swap rate	5.29	Semi annual on going hedge rate
Seven year swap rate – RPU	5.51	Case assumed for RPU
Floating rate – during construction	5.03	Three year monthly rate used for non-swapped interest rate during construction
Floating rate – post construction	5.08	Three year semi annual rate used for non-swapped interest rate post construction
Hedging		
% hedged during construction	95	Hedging profile
% hedged in first band	75	Hedging profile
% hedged in ongoing basis	50	Hedging profile
Cost of swaps – pre-refinance		
Liquidity	0.05	Premium on swap cost due to liquidity risk, before refinance
Basis	0.05	Premium on swap cost due to basis risk, before refinance
Credit	0.12	Premium on swap cost due to credit risk before refinance
Cost of swaps – post refinance		
Liquidity	0.05	Premium on swap cost due to liquidity risk, post refinance
Basis	0.04	Premium on swap cost due to basis risk, post refinance
Credit	0.09	Premium on swap cost due to credit risk, post refinance
30 day BBSW	4.87	30 day deposit rate

Table 6.9 *Continued*

	Base Case Financial Model – Assumptions book	
	Value	Description
4. Construction parameters		
Stage One construction		
Stage One construction start	January 2003	First month of construction
Stage One construction period	34	Stage One construction period
Stage One completion	31 October 2005	Stage One completion date
Stage One D&C contract – trust – including contingency	453,830	Allocation of D&C price for Stage One construction – construction
Stage One D&C contract – op co – including contingency	119,771	Allocation of D&C price for Stage One construction – plant and equipment
Stage Two construction		
Stage Two construction start	1 November 2005	First month of Stage Two completion
Stage Two construction period	8	Months of Stage Two completion
Stage Two completion	30 November 2005	Stage Two completion date
Stage Two D&C contract – trust – including contingency	34,511	Allocation of D&C price for Stage Two construction – construction
Stage Two D&C contract – op co – including contingency	0	Allocation of D&C price for Stage Two construction – plant and equipment
Total construction period	42	From financial close to completion
Stage One construction cost – including contingency	573,601	Summary of above costs
Stage Two construction cost – including contingency	34,511	Summary of above costs
Total construction cost – including contingency	608,112	Summary of above costs
Construction contingency allowed for	17,000	Contingency allowed by equity contributors
Construction contingency drawn	4,000	Amount of construction contingency assumed drawn
		Draw down on a pro rata basis according to the D&C contract
GST on D&C contract	10%	GST is payable on D&C contract Payments are lagged 90 days and are recovered in subsequent periods
Acquisition cost of tags	0	Commissioning tags provided by D&C contractor. Post this time provided through back office
SETA Equity contribution	1,200	One-off equity contribution to SETA. Contributed 30 June 2005

Table 6.9 *Continued*

	Base Case Financial Model – Assumptions book	
	Value	Description
5. Tolling assumptions		
Theoretical toll factor in modelled first period	138.89	Theoretical first operating period toll escalation amount as per project deed formula
Toll free months from Stage One completion	1	Number of months of legal toll free use
Differential tolling	yes	
Tolls		
Passenger		
Eastbound	2.5	Base tolls as per project deed, inclusive of GST
Westbound	2.5	Base tolls as per project deed, inclusive of GST
Sir John Young Crescent	1.1	Base tolls as per project deed, inclusive of GST
Commercial		
Eastbound	5	Base tolls as per project deed, inclusive of GST
Westbound	5	Base tolls as per project deed, inclusive of GST
Sir John Young Crescent	2.2	Base tolls as per project deed, inclusive of GST
Penalty fee charged to violators		
First round	10	Penalty fee charged to violators (in addition to toll) following round 1 notice
Second round	20	Penalty fee charged to violators (in addition to toll) following round 2 notice
Third round – police infringement notice	115	Penalty fee charged to violators (in addition to toll) following police infringement notice
Penalty fee CCM receives from violators		
First round	5	Dollar receipt to CCM (in addition to toll) following round 1 notice
Second round	5	Dollar receipt to CCM (in addition to toll) following round 2 notice
Third round	5	Dollar receipt to CCM (in addition to toll) following police infringement notice

Table 6.9   *Continued*

	Base Case Financial Model – Assumptions book	
	Value	**Description**
6. Operating and maintenance cost		
Variable opex charge	0.12	
31 December 2005	0.12	Cost per tag variable charge
30 June 2006	0.12	
31 December 2006	0.12	
30 June 2007	0.11	
31 Dec 2007	0.11	
30 June 2008	0.1	
31 December 2008	0.1	
30 June 2009	0.1	
31 December 2009	0.1	
30 June 2010	0.1	
31 December 2010	0.1	
30 June 2011	25,500	
Upfront insurance cost		Upfront insurance premia
7. Working capital days		
Accounts receivable		
First round penalty	60	Days assumed for collection of 1st round penalty
Second round penalty	90	Days assumed for collection of 2nd round penalty
Third round penalty	120	Days assumed for collection of 3rd round penalty
Accounts payable		
Operating and maintenance expenses	30	Days assumed for payment of accounts payables
8. Key financial ratios		
Minimum CLCR for a 33 year period	2.1	Concession life cover ratio
Period One minimum senior ICR	1.4	Debt sized off the minimum ICR in Period One
Period Two minimum senior ICR	1.5	Debt sized off the minimum ICR in Period Two
Period Three minimum senior ICR	1.6	Debt sized off the minimum ICR in Period Three
Minimum total ICR sizing test for RPU	1.45	Min total ICR for which RPUs are sized off
Date of minimum target for RPU sizing	31 Dec 07	Year of min total ICR for RPU sizing (3rd ratio date)
First ratio date	31 Dec 05	The first June or December 12 months post Stage One completion

Table 6.9 *Continued*

	Base Case Financial Model – Assumptions book	
	Value	Description
9. Senior bank debt assumptions		
Tranche A – equity bridge	253,238	Peak facility size
Facility size		Senior bank debt supported by guarantees or LCs provided by equity, refinanced by equity
Type of facility		Injection on completion, drawdown prior to tranche B – project facility
Term (months)	42	Construction period
Facility available	1 January 2003	Amount of principal outstanding at completion
Bullet	100%	
Credit margin		
Construction		Margin charged prior to completion
Operations	n.a.	Margin charged post completion (not applicable for this tranche)
Funding of interest during construction		Drawdown on the tranche A facility until tranche A facility limit is reached and then
		Drawdown on tranche B – project facility
Upfront fee	3.40%	Arranging/underwriting fee payable on financial close
Commitment fee (% of margin p.a.)	40%	Fee payable on undrawn facility balance prior to completion
Cost of LCs (%)		Fee payable to provider of LC by equity, if applicable
Facility drawdowns		Commence when the construction account is reduced to zero
Tranche B – project facility		
Facility use	580,000	Peak facility size
Type of facility	bullet	Senior bank project deed facility
Term (years)	7	Term post financial close
Interest only period (years)	7	Number of interest only years post finance close
Margin during construction		
If rated BBB	1.50%	Margin over bank bills/swaps charged prior to completion if project rated BBB
If rated BBB	1.50%	Margin over bank bills/swaps charged prior to completion if project rated BBB-
If unrated	1.50%	Margin over bank bills/swaps charged prior to completion if project unrated

Table 6.9   *Continued*

	Base Case Financial Model – Assumptions book	
	Value	**Description**
Margin during operations		
If rated BBB	1%	Margin over bank bills/swaps charged post completion if project rated BBB
If rated BBB	1.25%	Margin over bank bills/swaps charged post completion if project rated BBB-
Assumed margins		Margin grid over bank bills/swaps charged post completion if project unrated
30 June 2006	1.60%	Margin and period for which margin is applicable
30 June 2007	1.60%	Margin and period for which margin is applicable
30 June 2008	1%	Margin and period for which margin is applicable
Until first refinance	1%	Margin and period for which margin is applicable
Margin post 1st refinancing	1%	Margin applied to senior debt post first refinancing
Commitment fees	40%	Commitment fee payable on undrawn balances (% of margin)
Upfront fee	2%	Arranging/underwriting fee payable on financial close
Assumed rating	BBB	Assumed BBB at year 3, post Stage One completion
Number of refinancings	4	The number of times the facility is rolled don the above terms
Upfront fee payable on refinance	1%	Upfront fee payable for refinance
Refinance debt term	5	
Amortisation profile	credit fancier	Basis for principal amortised in each period
Amortisation term	11	Term over which the facility is amortised after the last refinancing
Tranche C – bond facility – standby facility		
Tranche C – bond facility amount – construction	25,000	Tranche C – bond facility amount during construction
Tranche C – bond facility amount – first 13 months of operations	5000	Tranche C – bond facility amount for 13 months post Stage Two completion
LC fee	1.50%	LC fee payable to tranche C – bond facility

Table 6.9 *Continued*

	Base Case Financial Model – Assumptions book	
	Value	Description
10. RPU		
Amount	72,613	Facility size
Underwriting fee	3%	Arranging/underwriting fee payable by equity underwriters out of equity underwriting fees at financial close
Base rate for coupon	5.50%	Seven year swap rate
Facility term	perpetual	Specified term of RPU
Bullet	yes	Repayment assumption for RPU
Margin-construction	n.a.	Margin during construction
Margin-operations	4%	Margin during operation
Refinance fee	1%	Refinance fee
Tax rate	30%	Corporate tax rate for the CCM

Source: Cross City Tunnel Parliamentary Notice, http://www.lee.greens.org.au/campaigns/crosscity.htm (accessed 7 October 2006).

Project equity came from CKI (50%), DB Capital Partners (30%) and Bilfinger Berger BOT GmbH (20%). The CCM consortium successfully completed the syndicated AU$580 million project loan transaction and initially 16 other banks joined the two mandated lead arrangers. Details of the loan were as follows:

- Borrower: CrossCity Motorway Finance Pty Ltd
- Guarantor: CrossCity Motorway Pty Ltd
- Type of loan: Term loan
- Amount: AU$580m
- Lifetime: 7 years
- Interest rate grid: benchmark BBSY, plus 150bp during 3-1/2 year construction period; 160bp post-construction; 170bp if no credit rating obtained by the borrower; thereafter 90bp if credit rating awarded of BBB+ or higher; 100bp with credit rating of BBB; 125bp if credit rating BBB- or less[9]
- Commitment fee: 40% of applicable margin
- Repayment: Bullet
- Estimated front-end fee: 1%

(International Financing Review, 2003)

---

[9] 1 basis point (bp) is equal to 0.01%; BBSY is the bank bill swap yield, a common AU$ benchmark domestic money market deposit rate.

6.9   Project outcomes

The CCT opened for traffic on 28 August 2005, less than two years and seven months after the commencement of construction. After opening, the CCT was often described as the 'ghost tunnel', due to its low volume of traffic. Statistics released by the CCM in February 2006 showed that around 30000 vehicles used the tunnel each day, well short of the projected 90000 per day. Many road users considered the toll too high for a short journey of 2.1 km. They would rather use a free alternative public road, even though it might take longer to reach the destination. Although many road users wanted to avoid the tunnel, a number of traffic calming measures – which many observers and media commentators suggested merely functioned to funnel traffic into the tunnel – were introduced. Instead of relieving congestion, however, such changes actually increased traffic congestion and resulted in numerous disruptions. For instance, there were lane reductions at William Street and changes in light phasing at various intersections across the city. Road users also complained about misleading signage indicating the tunnel was the only route to reach certain destinations when in fact alternative public roads were available.

The huge upfront consideration demanded by the NSW government to achieve its no-cost-to-government objective also sparked public outrage. It appeared to the community that, in return for such consideration, the government allowed inflated tolls to be set by the operator, while it might have given assurances to the operators that sufficient traffic throughput might be achieved for the tunnel through traffic calming measures. Some analysts were also sceptical about the traffic projections derived by the CCM consortium. It was suggested that the projection method used by the CCM employed a work back (or implied) process, in which the projections were derived from an internal rate of return promised to equity investors. In other words, the traffic projections and tolls might have been financially reverse-engineered to provide the required toll revenue and the CCM had failed to consider land use and transport interaction factors properly. To increase patronage and public acceptance, the CCM introduced a toll-free period from 24 October 2005 to late November 2005 and also announced a freeze on toll increases for one year and a waiver-of-fee for casual users. In all, however, in its first six months of operation the Sydney CCT was operating at roughly one-third of projected throughput, nowhere near projected income and rumoured to be near financial collapse. It was also a public relations disaster, attracting significant negative media and political attention, and was subject to strong general-user resistance. In response to public pressure, in October 2005 the government released some contract terms to the public. A few days after the release, the head of the RTA stepped down due to his failure to disclose an amendment to the contract.

In February 2006, there was media speculation that the Sydney CCT was for sale and rumours of a buy-out by the NSW government began and persisted throughout 2006, despite denials that the CCT was for sale and that NSW was not a potential buyer. The rumoured price was over AU$1 billion. However, despite denials by the government and the CCM, NSW Roads Minister Eric Roozendal said that '*any prospective buyers would be well advised to consider the cost of a trip through it*', while NSW opposition roads spokesman Andrew Stoner said '*taxpayers would be the ones to suffer most if the government was forced to pay compensation to the tunnel operator*' (Hassett, 2006). It was clear that, in addition to being financially problematic, the CCT was starting to attract significant political interest and had the potential to severely embarrass the NSW government that championed it.

The bottom line was looking unpleasant. With the poor operating performance of the CCT project, it seemed inevitable that CKI and other partners of the CCM might have to incur a significant impairment loss in the next financial quarter. It further seemed that the Sydney CCT investment was rapidly turning into a liability for CKI, a prospect that was not comforting to a company that had in the past been highly successful in Australia, whilst remaining relatively low key.

7

Financial Structure and Infrastructure Project Finance: the Hong Kong Western Harbour Crossing

The Hong Kong Western Harbour Crossing (WHC) is another model infrastructure project finance project that did not quite live up to expectations. Developed in the mid-1990s as part of the new road infrastructure developed to support the hugely ambititious (and very successful) new Hong Kong International Airport (Chep Lap Kok), it languished for the first ten years of its operation in losses – in fact, it turned its first profit in 2005–2006. This project has many parallels with the Channel Tunnel project, including a stellar cast of financiers. The WHC managed to turn operating profits and thus at least covered its costs but it also shows the importance of irreversibility in fixed infrastructure investment – one cannot turn a road tunnel into a shopping mall when things go wrong. This is the nature of asset-specificity. The project also provides a good case to assess project company debt capacity.

7.1 Refinancing the Western Harbour Crossing, Hong Kong[1]

Late one afternoon in November 2006, Paul Klein told Sophie Cheng:

---

[1] Mary Ho prepared this case under the supervision of the authors for class discussion. This case is not intended to show effective or ineffective handling of decision or business processes. Although the utmost care has been taken in preparing the history and fundamental background in this case, the investment decisions here are fictitious but typical. The opportunity discussed in this case does not represent an actual specific company opportunity. Ref. 07331C, Copyright © 2007 The University of Hong Kong. This material is used by permission of The Asia Case Research Centre at The University of Hong Kong (http://www.acrc.org.hk).

'The investors of the Western Harbour Crossing were under pressure to divest their interests in the tunnel and look for new opportunities for growth outside China. The time was ripe to propose a buy-out of the tunnel.'

Paul Klein was the Vice-President of GW Energy and Infrastructure Holdings Limited based in Stuttgart, Germany. He was in discussion with his investment advisor from Fragrant Harbour Bank Limited about the purchase of the Western Harbour Tunnel Company Limited in Hong Kong. With a turnover of €98 billion and assets of €292 billion in 2005, the GW Group developed and operated more than 700 km of toll roads, bridges and tunnels primarily in Germany, Eastern Europe and Australia. Its energy arm had also built up an impressive portfolio of energy assets, including water and electricity supply businesses in Europe and Australia. In mid-2006, the GW Group announced a bold strategy to diversify into transportation projects in the Asia Pacific region. Several project teams had been formed to evaluate the market opportunities in the region and recommend a course of action to the board of directors. Paul Klein had been assigned to a team charged with identifying toll road and tunnel investments in China. Contrary to the management's expectations, Paul's team had not come up with solid proposals to bid for greenfield infrastructure projects. Instead, the team had recommended the purchase of the WHC in Hong Kong, which had been in operation for almost 10 years. With just a few weeks left before the board of directors' meeting, Paul Klein came to Hong Kong and had a meeting with Sophie to finalise the buy-out proposal for the WHC. Paul hoped that as an independent advisor, Fragrant Harbour Bank could give him an objective and careful estimate of the proposed investment in the WHC, and provide the long-term debt financing required for the project when approval was granted by the board of directors.

7.2 History of the Western Harbour Crossing

7.2.1 *The need for a third harbour crossing*

Recognised as one of the deepest and busiest maritime ports in the world, the Victoria Harbour in Hong Kong has no bridge but three tunnels with the WHC completed in 1997 to provide the third underwater harbour crossing of Victoria Harbour (Figures 7.1, 7.2 and 7.3). The tunnel connected Sai Ying Pun on Hong Kong Island with the West Kowloon reclamation, where it carried traffic to the expressways leading to the Hong Kong International Airport at Chep Lap Kok. Distinct from other stand-alone infrastructure projects, the WHC was designed to form part of the Airport Core Programme organised by the Hong Kong Government during the 1990s. The programme involved the construction of the new Chep Lap Kok Airport to replace the Kai Tak International Airport in July 1998, and the development of supporting infrastructure including road works, bridge and tunnel links that facilitated access to the new

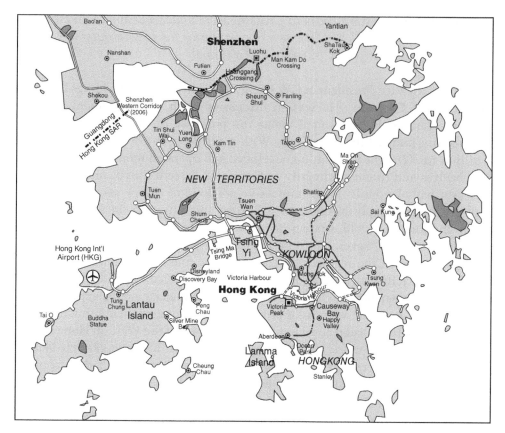

Figure 7.1   Map of Hong Kong.

airport. With a cost of HK$160.2 billion, the Airport Core Programme was one of the most expensive airport projects yet developed.

Apart from forming part of the key strategic networks linking the Chep Lap Kok Airport, one of the objectives of constructing the WHC was to alleviate congestion of the Cross-Harbour Tunnel and the Eastern Harbour Crossing, which had been in operation since 1972 and 1989 respectively. The Cross-Harbour Tunnel had always been the busiest tunnel in Hong Kong due to its central location. It ran between the main commercial district close to Causeway Bay on Hong Kong Island and Hunghom in Kowloon, and had an average daily throughput of 121 700 as of March 2005. As compared with the Cross-Harbour Tunnel, the location of the Eastern Harbour Crossing was less convenient. It connected Quarry Bay on Hong Kong Island and Cha Kwo Ling in Kowloon with an average daily traffic of 73 500 as of March 2005. Table 7.1 presents a summary of facts for the three Victoria Harbour crossings.

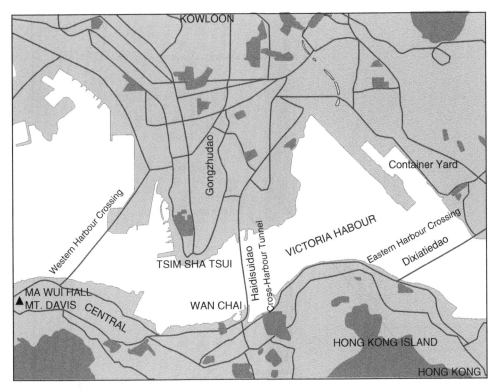

Figure 7.2 Map of Hong Kong (harbour crossings).

7.2.2 Structure of the Western Harbour Crossing project

With a length of approximately 2 km, the WHC is the longest cross-harbour crossing in Hong Kong. It has three lanes of traffic in each direction and a daily capacity of approximately 118 000 vehicles, which is 1.5 times the handling capability of the Cross-Harbour Tunnel or the Eastern Harbour Crossing (Figure 7.4). The extra capacity of the WHC was intended to support growing cross-harbour traffic levels in Hong Kong. The WHC was built by the Western Harbour Tunnel Company on a 30 year franchise build-operate-transfer (BOT) model. The Western Harbour Tunnel Company was a special project vehicle (SPV) established by China International Trust & Investment Corporation Hong Kong (Holdings) Limited (CITIC HK), CITIC Pacific (CITIC Pacific), Kerry Holdings (Kerry), the Cross Harbour Tunnel Company and China Merchant Holdings in 1992. Of the shares of the Western Harbour Tunnel Company Limited, 50% were held by Adwood Company, an SPV established by CITIC HK, CITIC Pacific, and Kerry for the purpose of investment holding (see Figure 7.5 for the holding structure). Under the general management agreement, the Cross Harbour Tunnel Company acted not only as an equity shareholder but also the general manager of the WHC project for the duration of the concession. By combining the investment

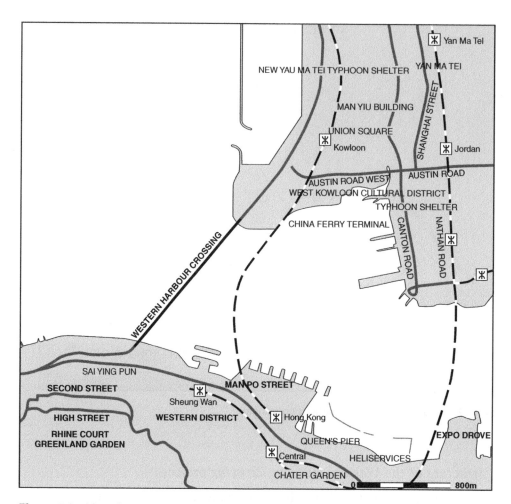

Figure 7.3 Map of Hong Kong: areas around Western Harbour Crossing.

in the WHC with an interest of 70.8% in the Eastern Harbour Crossing, CITIC Pacific was the largest investor in Hong Kong's harbour tunnels.

The BOT agreement for the WHC project comprised the Western Harbour Crossing Ordinance enacted in 1993 and the project agreement signed between the Hong Kong Government and the Western Harbour Tunnel Company. The Western Harbour Tunnel Company was granted the right to design, construct, operate and maintain the tunnel for a concession period of 30 years from the start of construction, i.e. from 1993 to August 2023. At the end of the concession period, the Western Harbour Tunnel Company had to transfer the tunnel back to the Hong Kong Government at no cost. The Hong Kong Government ensured land access for construction and was liable for compensation when there was a delay in availability. It was agreed that the Western Harbour Tunnel Company had to complete construction within 48 months

Table 7.1 Summary of Victoria Harbour crossings.

Tunnel	Opened	Length (km)	Franchise lasts until	Owner/operator	Cost for taxis/cars/minibuses/buses/lorries (HK$)	Vehicles daily (as at March 2005)	Capacity per day
Cross-Harbour Tunnel	1972	1.86	—	Hong Kong Government/Hong Kong Tunnels and Highways Management Co.	10/20/10/15/30	121,700	78,500
Eastern Harbour Crossing	1989	1.86	2016	New Hong Kong Tunnel Co.	25/25/38/75/38-75	73,500	78,500
Western Harbour Crossing	1997	2	2023	Western Harbour Tunnel Co.	35/40/50/100/55-110	39,200	118,000

Source: Hong Kong government.

The Build-Operate-Transfer (BOT) Franchise

Franchisee	: Western Harbour Tunnel Company Limited
Franchise Period	: 30 years including construction period
Construction Period	: August 1993 to April 1997 (45 months)
Construction Cost	: HK$5.7 billion
Total Project Cost	: HK$7 billion
Contractor	: Nishimatsu Kumagai Joint Venture

The Tunnel

No. of Traffic Lanes	: 6 (dual 3-lane tunnel)
Speed Limit	: 80 km/h (50km/h for one-tube-two-way operation)
Design Life of Tunnel Structure	: 120 years

Source: Western Harbour Tunnel Company Limited (2006), WHC Project Facts and Figures.

Figure 7.4 Key aspects of the Western Harbour Crossing.

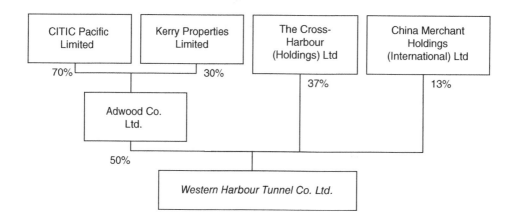

Source: Western Harbour Tunnel Company Limited (2006), Company Profile.

Figure 7.5 Shareholding structure of Western Harbour Tunnel Co. Ltd.

Table 7.2 Debt financing of the Western Harbour Crossing.

Facility	Purposes (to fund/support)
Tranche A	
Sub-tranche A	• Part of the design, construction, pre-operation and financing costs • The acquisition and financing costs of equipment
Sub-tranche B	• Until June 30th 1998: operating costs could not be funded out of toll revenues • Thereafter: as general working capital including the refinancing of third party loans
Sub-tranche C	• Until June 30th 1998, support HK$ denominated indebtedness incurred from export credit agencies, suppliers and other third party lenders • Thereafter, support other HK$ denominated third party loans
Tranche B	• Entry into interest rate swaps

Source: Lang, L., (1998) *Project Finance in Asia*, Elsevier: Amsterdam, p. 262.

from the beginning of construction, maintain a debt equity ratio not exceeding 69:31 on the date of commencement of operation, and keep the ownership structure intact in the first five years (Lang, 1998).

The design and construction of the WHC was conducted by a contractor consortium of Nishimatsu Construction and Kumagai Gumi, which would receive an agreed consideration of HK$5664 million from the Western Harbour Tunnel Company Limited. Besides the construction of the immersed tubes that formed the tunnel, the contract also covered the development of traffic interchanges, the construction of ventilation shafts or buildings, tunnel approach sections, an administration building complex and a toll plaza with twenty toll booths. Construction of the WHC began in September 1993 and was completed in April 1997, four months ahead of schedule. The total project cost was estimated to be HK$7534 billion. The shareholders of the Western Harbour Tunnel Company had agreed to pay for any cost overruns by way of equity share capital and/or subordinated debt.

Funds for the WHC project came from shareholders equity (32%) and a syndicated limited-recourse project loan (68%). HSBC was the advisor and arranger for the private sector consortium financing. The syndicated loan was used to fund various costs including contract price, operating costs and working capital. Tranche A of the bank facility was used to finance project costs, whilst tranche B included a swap facility to hedge the interest rate of 50% of the loan facilities from commencement of operation in 1997 for a period up to 5 years. Other than sub-tranche A, that had to be fully drawn prior to commencement of operation, the entire facility was available up to maturity, 15 years from the conclusion of the facility agreement. Prepayment was only permitted for sub-tranche A. Cancellation for tranche A was allowed in whole or in part but cancellation of tranche B was only permitted upon repayment or cancellation of tranche A (Lang, 1998). Interest would be paid at 1–1.5% above Hong Kong Interbank offered rate Hibor. The purposes of each tranche are summarised in Table 7.2. Figure 7.6 presents the balance sheet of the Western Harbour Tunnel Company at commencement of operation in 1997.

	HK$ million
Total contract costs	6,004
Contingency	328
Pre-operating costs	73
Interest during construction	1,007
Fees and commissions	122
Budgeted project costs	**7,534**

Shareholders' equity		
Adwood	1,216	
Cross Harbour Tunnel Company	900	
China Merchant Holdings	316	
Total equity financing	**2,432**	32% of total financing

Syndicated bank facility (from tranche A)		hibor + margins
Tranche A		(Up to HK$5,200 m)
Sub-tranche A (term loan)		(Up to HK$3,190 m)
Sub-tranche B (revolving credit facility)		(Up to HK$2,010 m)
Sub-tranche C (revolving letter of credit &		(Up to HK$2,010 m)
guarantee issuance facility)		
Tranche B (swap facility)		(Up to HK$2,600 m)
Total debt financing	**5,102**	68% of total financing
Total equity and debt	**7,534**	

Source: Lang, L., (1998) *Project Finance in Asia*, Elsevier: Amsterdam, p.262.

Figure 7.6   Balance sheet of the Western Harbour Tunnel Company at commencement of operation.

7.2.3   The toll adjustment mechanism

The Western Harbour Crossing Ordinance incorporated an innovative toll adjustment mechanism for the operation of the WHC. The mechanism was a profit control system designed with the objectives of generating sufficient cash flows for timely repayment of debt, delivering reasonable but not excessive returns to shareholders, and stabilising the toll rate for the community as much as possible. From the perspective of the government, the mechanism was an alternative to price cap regulation or scheme of control, and it worked in a more flexible way to protect public interest. From the investors' viewpoint, the mechanism provided better control to cover cost inflation

Dates on which the company is expected to be able to effect a toll increase	
1	1 January 2001
2	1 January 2005
3	1 January 2009
4	1 January 2013
5	1 January 2017
6	1 January 2021

Source: Western Harbour Crossing Ordinance, Schedule 4.

Figure 7.7   Schedule of toll adjustment arrangement.

or decline in patronage, thereby reducing the uncertainty of the income stream and return.

The toll adjustment mechanism involved six anticipated toll increases scheduled on specified dates, additional toll increases, and a toll stability fund (Figure 7.7). Where in any year the net revenue of the Western Harbour Tunnel Company Limited was greater than the upper estimated net revenue, but did not exceed the maximum estimated net revenue, the company had to pay into the fund 50% of the amount exceeding the upper estimated net revenue. In a case where the net revenue of the company was greater than the maximum estimated net revenue, the company had to pay into the fund the amount in excess of the maximum estimated net revenue and an amount equal to 50% of the difference between the upper estimated net revenue and the maximum estimated net revenue (Lang, 1998; Western Harbour Crossing Ordinance). Where the net revenue of the company is less than the upper estimated net revenue, the company may apply to increase the tolls by the amount of the appropriate anticipated toll increase (Western Harbour Crossing Ordinance).

Based on the above mechanism, the Company had to inject any excess annual net revenue into the toll stability fund. The net revenue was defined as the gross revenue less the interest less the operating costs. Depending on the cash flow derived and the amount of the fund, the timing of the toll increase could be deferred or advanced from the specified timetable. According to the Ordinance and Project agreement, the opening toll of the WHC was set at HK$30. The anticipated toll increases had been specified for every four years starting from 1 January 2001. The toll increase was limited to HK$10 and HK$15 for every increase up to 2010 and thereafter respectively.

The toll adjustment mechanism served to minimise the problems of disputes over timing and the level of toll increase. Unlike the control system scheme that was designed based on a permitted return on fixed assets, the toll adjustment mechanism was operated on a net revenue basis, which had the advantage of eliminating the tendency for the tunnel company to over-capitalise its assets. However, the toll adjustment mechanism provided fewer incentives for the tunnel company to reduce costs and undertake renovations.

7.2.4 Historical financial performance

Since its commencement of operation in 1997, the WHC has been consistently under-utilised and its traffic throughput well below expectations. For the year ended 31 July 2006, the total daily traffic throughput of the WHC was 42 995 (Western Harbour Tunnel Company Limited, 2006), which represented less than half of its designed daily capacity of 118 000. To stimulate usage of the tunnel, the Western Harbour Tunnel Company introduced a number of promotion schemes such as the: Midnight Empty Taxi promotion, Midnight Goods Vehicle promotion, Fantastic Holidays and other promotion programmes. The Company also explored ways to generate additional non-toll revenues from outdoor advertising and the hosting of telecommunications network equipment. A refinancing by the company in 2002 had extended the loan period from 2007 to 2011. To minimise the fluctuations of interest expense arising from the volatility of interest rates, the company had entered into interest rate swap agreements. The above measures had helped to strengthen the financial position of the company. During the year ended 31 July 2006, the Western Harbour Tunnel Company recorded a profit after tax of HK$299 million, representing a growth of 19% as compared with the figure in the preceding year (see Tables 7.3–7.7 and Figures 7.8 and 7.9 for the historical financial statements of the company) (Western Harbour Tunnel Company Limited, 2006). Market share of the WHC had increased from 17% from 2004/05 to 19% in 2005/06. Despite gradual improvement in financial performance, the company was still unable to meet the targeted minimum net revenue that was stipulated in the Western Harbour Crossing Ordinance (Tables 7.8 and 7.9). As of July 2006, the IRR of the Company was negative 32.48% (Table 7.10) (Western Harbour Tunnel Company Limited, 2006).

As the net revenue fell short of the estimated statutory minimum, the Western Harbour Tunnel Company had the right to raise its tolls in accordance with the toll adjustment mechanism. The company had raised its gazetted tolls several times and offered concessionary rates each time to keep the actual tolls down (Table 7.11).

7.2.5 The competitive landscape

It was obvious that the WHC had failed to meet the government's objective of easing congestion at the Cross-Harbour Tunnel, which remained over-utilised despite its smaller capacity. The main reasons for the under-utilisation of the WHC were that it charged the highest toll but did not have the best location for motorists compared to its competitors. For historical reasons, the WHC was unable to compete on a level playing field because the three cross-harbour tunnels were constructed at different times by different operators, who set their tolls based on different costs and traffic projections (Lau, 2001). As the oldest and most heavily utilised tunnel, the Cross-Harbour Tunnel could afford to charge the lowest toll because it had already been reverted back to government ownership and could enjoy toll price privileges that were not available to the other privately owned tunnels. Since its commencement of operation in 1972, the tolls of the Cross-Harbour Tunnel had not kept up with inflation

Table 7.3 Daily traffic for the Western Harbour Crossing (WHC).

Year ended July	2006	2005	2004	2003	2002	2001	2000	1999*	1998*	1997*(5–7/1997)
Motorcycles	412	385	470	461	520	566	655	445	375	300
Private cars	22,522	21,232	21,986	21,808	23,778	24,377	25,343	25,858	20,626	15,860
Taxis	8,522	7,044	6,007	4,679	4,824	5,107	4,685	—	—	—
PC sub-total	**31,456**	**28,661**	**28,463**	**26,948**	**29,122**	**30,050**	**30,683**	**26,303**	**21,001**	**16,160**
Light goods vehicles	3,886	3,434	3,500	3,499	3,889	3,954	4,245	5,117	2,914	2,209
Medium goods vehicles	740	658	633	641	728	749	781	870	453	314
Heavy goods vehicles	96	77	46	64	81	85	71	123	21	20
GV sub-total	**4,722**	**4,169**	**4,179**	**4,204**	**4,698**	**4,788**	**5,097**	**6,110**	**3,388**	**2,543**
Public light buses	2,414	2,414	2,460	2,477	2,291	2,208	2,061	1,357	554	162
Single-decked buses	1,221	1,128	1,069	943	962	933	913	821	309	131
Double-decked buses	3,182	3,168	3,127	3,048	3,063	3,009	3,001	3,034	1,980	1,223
Bus sub-total	**6,817**	**6,710**	**6,656**	**6,468**	**6,316**	**6,150**	**5,975**	**5,212**	**2,843**	**1,516**
Total	**42,995**	**39,540**	**39,298**	**37,620**	**40,136**	**40,988**	**41,755**	**37,625**	**27,232**	**20,219**

*Remarks: taxis were previously grouped under Private Cars

Traffic mix for the WHC										
PC	73.2%	72.5%	72.4%	71.6%	72.6%	73.3%	73.5%	69.9%	77.1%	79.9%
GV	11.0%	10.5%	10.6%	11.2%	11.7%	11.7%	12.2%	16.2%	12.4%	12.6%
Bus	15.8%	17.0%	17.0%	17.2%	15.7%	15.0%	14.3%	13.9%	10.5%	7.5%
	100.0%	100.0%	100.0%	100.0%	100.0%	100.0%	100.0%	100.0%	100.0%	100.0%

Source: Western Harbour Tunnel Company Limited (2006) Annual Report.

Table 7.4 Total cross-harbour traffic and Western Harbour Crossing's market share (year ended July).

Daily traffic for total market	2006	2005	2004	2003	2002	2001	2000	1999*	1998*	1997* (5-7/97)
Motorcycles	7,976	8,398	8,433	8,102	7,886	7,260	6,847	6,798	6,630	6,521
Private cars	99,797	105,019	108,760	109,387	113,063	113,536	115,656	152,193	158,284	157,324
Taxis	51,110	49,868	48,411	44,188	46,041	47,290	40,978	—	—	—
PC sub-total	**158,883**	**163,285**	**165,604**	**161,677**	**166,990**	**168,086**	**163,481**	**158,991**	**164,914**	**163,845**
Light goods vehicles	35,941	36,259	36,058	35,425	36,382	36,543	36,728	35,702	36,405	38,061
Medium goods vehicles	6,912	6,904	6,829	6,908	7,036	7,060	7,486	7,787	7,593	7,675
Heavy goods vehicles	1,099	1,236	1,207	1,512	1,559	1,525	1,565	1,430	1,203	1,132
GV sub-total	**43,952**	**44,399**	**44,094**	**43,845**	**44,977**	**45,128**	**45,779**	**44,919**	**45,201**	**46,868**
Public light buses	6,806	6,786	6,652	6,294	6,216	6,088	5,721	4,639	3,803	3,495
Single-decked buses	5,849	5,564	5,221	4,353	4,711	4,450	4,418	4,119	4,066	4,279
Double-decked buses	11,465	11,679	11,869	11,793	11,922	11,746	11,563	10,979	9,043	7,964
Bus sub-total	**24,120**	**24,029**	**23,742**	**22,440**	**22,849**	**22,284**	**21,702**	**19,737**	**16,912**	**15,738**
Total	**226,955**	**231,713**	**233,440**	**227,962**	**234,816**	**235,498**	**230,962**	**223,647**	**227,027**	**226,451**

*Remarks: taxis were previously grouped under Private Cars

Market shares for the WHC by vehicle category

	2006	2005	2004	2003	2002	2001	2000	1999*	1998*	1997* (5-7/97)
PC	19.8%	17.6%	17.2%	16.7%	17.4%	17.9%	18.7%	16.5%	12.7%	9.9%
GV	10.7%	9.4%	9.5%	9.6%	10.5%	10.6%	11.1%	13.6%	7.5%	5.4%
Bus	28.3%	27.9%	28.0%	28.8%	27.6%	27.6%	27.6%	26.4%	16.8%	9.6%
Total	**18.9%**	**17.1%**	**16.8%**	**16.5%**	**17.1%**	**17.4%**	**18.1%**	**16.8%**	**12.0%**	**8.9%**

Source: Western Harbour Tunnel Company Limited (2006) Annual Report.

Table 7.5 Profit and loss account for the year ended 31 July yearly.

HK$'000	2006	2005	2004	2003	2002	2001	2000	1999	1998	1997
Turnover	766,524	704,709	636,266	600,830	642,932	621,372	555,611	477,104	327,450	55,960
Operating and administrative expenses*	(261,845)	(255,918)	(257,277)	(246,671)	(226,516)	(215,329)	(202,808)	(186,015)	(191,676)	(49,450)
Operating profit	504,679	448,791	378,989	354,159	416,416	406,043	352,803	291,089	135,774	6,510
Finance costs										
– Interest on bank loans	(113,712)	(114,704)	(124,980)	(165,513)	(245,280)	(355,802)	(408,347)	(449,831)	(441,070)	(91,149)
– Interest on shareholders' loans	(28,426)	(29,251)	(30,238)	(29,744)	(25,963)	(23,513)	(22,150)	(20,041)	(119,624)	(44,950)
	(142,138)	(143,955)	(155,218)	(195,257)	(271,243)	(379,315)	(430,497)	(469,872)	(560,694)	(136,099)
Profit/(loss) before taxation	362,541	304,836	223,771	158,902	145,173	26,728	(77,694)	(178,783)	(424,920)	(129,589)
Deferred tax provision	(63,458)	(53,625)	(39,833)	(5,484)	—	—	—	—	—	—
Profit/(loss) after taxation	299,083	251,211	183,938	153,418	145,173	26,728	(77,694)	(178,783)	(424,920)	(129,589)
Accumulated losses brought forward	37,952	(213,259)	(397,197)	(639,085)	(784,258)	(810,986)	(733,292)	(554,509)	(129,589)	—
Prior year adjustment on deferred taxation	—	—	—	88,470	—	—	—	—	—	—

Table 7.5 *Continued*

HK$'000	2006	2005	2004	2003	2002	2001	2000	1999	1998	1997
Retained profits/ (accumulated losses) carried forward	337,035	37,952	(213,259)	(397,197)	(639,085)	(784,258)	(810,986)	(733,292)	(554,509)	(129,589)
***Breakdown of operating & administrative expenses**										
Staff costs	48,288	46,411	46,538	46,590	46,845	46,336	46,221	44,361	48,228	14,116
Maintenance costs	8,598	8,513	10,319	9,862	6,254	7,590	6,067	4,783	3,386	1,177
Utility costs	6,767	6,582	6,673	6,795	6,676	6,679	6,475	7,017	8,341	2,339
Rates	26,450	24,318	23,082	22,839	21,247	19,340	13,912	6,096	7,260	1,660
Insurance	19,236	18,390	17,565	15,835	6,981	3,714	3,441	3,276	3,925	871
PR & promotion, professional fees, administrative expenses and others	7,143	7,744	6,524	7,708	8,711	8,242	10,737	12,034	19,317	4,991
Depreciation	145,363	143,960	146,576	137,042	129,802	123,428	115,955	108,448	101,219	24,296
	261,845	255,918	257,277	246,671	226,516	215,329	202,808	186,015	191,676	49,450

Source: Western Harbour Tunnel Company Limited (2006) Annual Report.

Table 7.6 Balance sheet as at 31 July yearly.

HK$'000	2006	2005	2004	2003	2002	2001	2000	1999	1998	1997
Non-current assets										
Fixed assets	5,606,227	5,750,394	5,935,070	6,080,760	6,217,272	6,345,746	6,468,422	6,583,475	6,690,707	6,789,066
Deferred taxation assets	—	—	43,153	82,986	—	—	—	—	—	—
Derivative financial instruments	21,386	—	—	—	—	—	—	—	—	—
	5,627,613	**5,750,394**	**5,978,223**	**6,163,746**	**6,217,272**	**6,345,746**	**6,468,422**	**6,583,475**	**6,690,707**	**6,789,066**
Current assets										
Inventories	1,519	1,296	1,291	1,306	1,255	1,225	1,296	1,345	1,376	965
Accounts receivable and prepayments	39,071	36,909	32,234	31,316	27,805	10,167	12,225	6,331	5,152	9,398
Derivative financial instruments	12	—	—	—	—	—	—	—	—	—
Cash and bank balances	6,378	7,433	7,287	4,266	4,654	3,279	5,441	4,982	11,926	12,029
	46,980	**45,638**	**40,812**	**36,888**	**33,714**	**14,671**	**18,962**	**12,658**	**18,454**	**22,392**
Current liabilities										
Accounts payable and accruals	86,521	105,300	149,379	144,318	124,240	121,807	111,015	110,220	151,506	167,507
Current portion of long-term bank loans	263,110	234,566	210,907	185,907	578,000	514,000	351,000	209,000	143,000	51,000
Derivative financial instruments	1,319	—	—	—	—	—	—	—	—	—
	350,950	**339,866**	**360,286**	**330,225**	**702,240**	**635,807**	**462,015**	**319,220**	**294,506**	**218,507**
Net current liabilities	(303,970)	(294,228)	(319,474)	(293,337)	(668,526)	(621,136)	(443,053)	(306,562)	(276,052)	(196,115)
Net assets	**5,323,643**	**5,456,166**	**5,658,749**	**5,870,409**	**5,548,746**	**5,724,610**	**6,025,369**	**6,276,913**	**6,414,655**	**6,592,951**
Share capital	400,000	400,000	400,000	400,000	400,000	400,000	400,000	400,000	400,000	400,000
Retained profits/ (accumulated losses) brought forward	37,952	(213,259)	(397,197)	(639,085)	(784,258)	(810,986)	(733,292)	(554,509)	(129,589)	—
Prior year adjustment on deferred taxation	—	—	—	88,470	—	—	—	—		—
Retained profits/ (accumulated losses) after reinstatement										

Table 7.6 *Continued*

HK$'000	2006	2005	2004	2003	2002	2001	2000	1999	1998	1997
Profit/(loss) after taxation	37,952	(213,259)	(397,197)	(550,615)	(784,258)	(810.986)	(733,292)	(554,509)	(129,589)	—
Retained profits/(accumulated losses) carried forward	299,083	251,211	183,938	153,418	145,173	26,728	(77,694)	(178,783)	(424,920)	(129,589)
	337,035	**37,952**	**(213,259)**	**(397,197)**	**(639,085)**	**(784,258)**	**(810,986)**	**(733,292)**	**(554,509)**	**(129,589)**
Hedging reserve	16,565	—								
Shareholders' equity/(capital deficiency)	753,600	437,952	186,741	2,803	(239,085)	(384,258)	(410,986)	(333,292)	(154,509)	270,411
Shareholders' loan	2,863,490	2,945,064	3,045,813	3,015,575	2,795,831	2,519,868	2,336,355	2,142,205	1,984,164	1,864,540
Shareholders' funds	3,617,090	3,383,016	3,232,554	3,018,378	2,556,746	2,135,610	1,925,369	1,808,913	1,829,655	2,134,951
Long-term bank loans	1,629,109	2,062,678	2,426,195	2,852,031	2,992,000	3,589,000	4,100,000	4,468,000	4,585,000	4,458,000
Deferred taxation liabilities	77,444	10,472	—	—	—	—	—	—	—	—
	5,323,643	**5,456,166**	**5,658,749**	**5,870,409**	**5,548,746**	**5,724,610**	**6,025,369**	**6,276,913**	**6,414,655**	**6,592,951**
Long-term bank loans										
Secured bank loans	1,639,610	2,081,390	2,449,000	2,879,000	2,992,000	3,589,000	4,100,000	4,468,000	4,585,000	4,458,000
Less: Unamortised deferred expenditure	(10,501)	(18,712)	(22,805)	(26,969)	—	—	—	—	—	—
	1,629,109	**2,062,678**	**2,426,195**	**2,852,031**	**2,992,000**	**3,589,000**	**4,100,000**	**4,468,000**	**4,585,000**	**4,458,000**
Current portion of long-term bank loans										
Secured bank loans	268,856	238,659	215,000	190,000	578,000	514,000	351,000	209,000	143,000	51,000
Less: Unamortised deferred expenditure	(5,746)	(4,093)	(4,093)	(4,093)	—	—	—	—	—	—
	263,110	234,566	210,907	185,907	578,000	514,000	351,000	209,000	143,000	51,000
Total deferred expenditure	**(16,247)**	**(22,805)**	**(26,898)**	**(31,062)**	—	—	—	—	—	—

Source: Western Harbour Tunnel Company Limited (2006) Annual Report.

Table 7.7 Cash flow statement for the year ended 31 July yearly.

HK$'000	2006	2005	2004	2003	2002	2001	2000	1999	1998	1997
Net cash surplus/(deficit) from operating activities	523,083	476,388	409,876	346,650	286,047	146,925	56,175	(90,112)	(150,384)	(11,099)
Cash flow from investing activities										
Net receipt/(payment) of tunnel costs and fixed assets	(2,555)	(2,291)	(1,855)	(1,172)	(1,672)	38,913	(1,716)	(3,832)	(68,719)	(1,502,928)
Deferred expenditure incurred	—	—	—	(34,866)	—	—	—	—	—	—
Net cash used in investing activities	**(2,555)**	**(2,291)**	**(1,855)**	**(36,038)**	**(1,672)**	**38,913**	**(1,716)**	**(3,832)**	**(68,719)**	**(1,502,928)**
Cash flow from financing activities										
Net loans from/(repayment to) shareholders	(110,000)	(130,000)	—	190,000	250,000	160,000	172,000	138,000	—	—
New secured bank loans drawn down	—	—	—	3,350,000	—	—	—	—	—	—
New repayment of secured bank loans	(411,583)	(343,951)	(405,000)	(3,851,000)	(533,000)	(348,000)	(226,000)	(51,000)	219,000	1,522,900
Net cash inflow/(outflow) from financing	**(521,583)**	**(473,951)**	**(405,000)**	**(311,000)**	**(283,000)**	**(188,000)**	**(54,000)**	**87,000**	**219,000**	**1,522,900**
Net increase/(decrease) in cash and bank balance	(1,055)	146	3,021	(388)	1,375	(2,162)	459	(6,944)	(103)	8,873
Cash and bank balances at beginning of year	7,433	7,287	4,266	4,654	3,279	5,441	4,982	11,926	12,029	3,156
Cash and bank balances at end of year	**6,378**	**7,433**	**7,287**	**4,266**	**4,654**	**3,279**	**5,441**	**4,982**	**11,926**	**12,029**

Source: Western Harbour Tunnel Company Limited (2006) Annual Report.

Table 7.7 *Continued*

Cash flow since construction to 31 July 2006	HK$'000
Inflow	
Fund from shareholders	3,110,000
Bank loans	4,776,500
Revenue	5,388,758
	13,275,258
Outflow	
Capital expenditure	6,730,771
Operating expenses (excluding depreciation)	917,416
Finance charge on bank loan	2,510,388
Bank loan repayments	2,868,034
Loan repayment to shareholders	240,000
Working capitals	2,271
	13,268,880
Net Cash Balance	**6,378**

Source: Western Harbour Tunnel Company Limited (2006) Annual Report.

Table 7.8 Annual net revenue.

Year ended (HK$m)	Jul-06	Jul-05	Jul-04	Jul-03	Jul-02	Jul-01	Jul-00	Jul-99	Jun-98	Cumulative
Minimum net revenue in the ordinance	1,455	1,190	880	794	713	506	253	201	154	6,146
Annual net revenue/deficit	567	492	400	325	299	172	59	(52)	(208)	2,054
Shortfall	**888**	**698**	**480**	**469**	**414**	**334**	**194**	**253**	**362**	**4,092**

Source: Western Harbour Tunnel Company Limited (2006) Annual Report.

Table 7.9 Estimated net revenue, Western Harbour Crossing Ordinance.

Year ended 31 July in	Minimum estimated net revenue	Upper estimated net revenue (HK$ million)	Maximum estimated net revenue
1998	154	336	403
1999	201	399	471
2000	253	461	538
2001	506	768	865
2002	713	1,016	1,128
2003	794	1,106	1,221
2004	880	1,202	1,321
2005	1,190	1,570	1,711
2006	1,455	1,881	2,039
2007	1,549	1,983	2,143
2008	1,623	2,061	2,223
2009	1,876	2,369	2,551
2010	2,028	2,562	2,760
2011	1,892	2,405	2,594
2012	1,821	2,326	2,513
2013	2,212	2,815	3,038
2014	2,573	3,267	3,524
2015	2,733	3,474	3,749
2016	2,891	3,682	3,974
2017	3,507	4,449	4,797
2018	4,018	5,090	5,486
2019	4,220	5,355	5,775
2020	4,422	5,621	6,064
2021	5,192	6,583	7,098
2022	5,747	7,285	7,855
2023	5,726	7,286	7,864

Source: Western Harbour Crossing Ordinance, Schedule 5.

(Wilson, 2004). However, tolls at the WHC had been set at a high level because it was more costly to build and the private investors had to obtain a fair return pursuant to terms agreed with the government.

The big gap in tolls among the three cross-harbour tunnels explained to a larger extent the relatively low traffic volume of the WHC and the Eastern Harbour Crossing. Many motorists benefited from the artificially low tolls set by the Cross-Harbour Tunnel and were unwilling to shift to use other tunnels. Whenever there was a toll increase for the WHC or the Eastern Harbour Crossing, the Cross-Harbour Tunnel noted an increase in traffic, revenue and market share because of traffic divergence. Apart from the high tolls, some motorists approaching from Wanchai (or east Hong Kong) were deterred from using the WHC as it would take more time to pass through the bottlenecks in Central district to reach the tunnel. However, the congestion problem in Central was expected to be resolved when the Central-Wanchai Bypass was completed by 2011 according to the Hong Kong Government's plan. Other traffic

Table 7.10 Western Harbour Crossing's IRR up to July 2006.

Period	Year ended	Shareholders' investment/return	#Discount factor b	NPV of 1994 cash flow × discount factor
		HK$'000 a IRR = −32.48%	−	HK$'000 a × b
0	Jul-94	(555,000)	1.0000	(555,000)
1	Jul-95	(515,200)	1.4810	(762,989)
2	Jul-96	(537,500)	2.1932	(1,178,863)
3	Jul-97	(592,300)	3.2481	(1,923,841)
4	Jul-98	—	4.8103	—
5	Jul-99	(138,000)	7.1238	(983,086)
6	Jul-00	(172,000)	10.5501	(1,814,610)
7	Jul-01	(160,000)	15.6242	(2,499,869)
8	Jul-02	(250,000)	23.1387	(5,784,685)
9	Jul-03	(190,000)	34.2675	(6,510,821)
10	Jul-04	—	50.7487	—
11	Jul-05	130,000	75.1566	9,770,358
12	Jul-06	110,000	111.3037	12,243,406
	Total	**(2,870,000)**		**0**

\# Discount factor for period 0 = 1
Discount factor for period n = $1/(1 + IRR)^n$

Source: Western Harbour Tunnel Company Limited (2006) Annual Report.

Table 7.11 Toll level of the Western Harbour Crossing.

HK$	Gazetted tolls effective 31 July 06	Actual tolls* (since 4/7/2004)	Actual tolls* (16/2/2003– 3/7/2004)	Actual tolls* 3/12/2000– 15/2/2003)	Actual tolls* (1/5/1997– 2/12/2000)
Motorcycles	40	22	20	20	15
Private cars	80	40	37	35	30
Taxis	80	35	35	35	30
Light goods vehicles	120	55	50	50	45
Medium goods vehicles	165	80	70	70	65
Heavy goods vehicles	245	110	100	100	95
Light buses	90	50	47	45	40
Single-decked buses	90	70	60	50	40
Double-decked buses	130	100	85	70	55
Each extra axle (for GV only)	80	30	30	30	30

* Actual toll level before specific promotion.
Source: Western Harbour Tunnel Company Limited (2006) Annual Report and Western Harbour Crossing Toll Schedule.

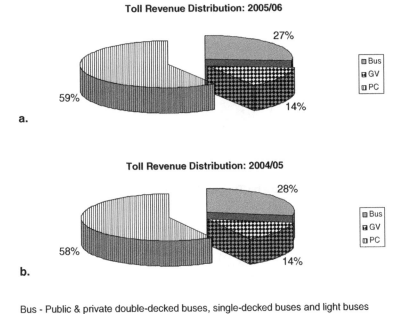

Bus - Public & private double-decked buses, single-decked buses and light buses
GV - Goods vehicles
PC - Private cars, taxis and motorcycles

Source: Western Harbour Tunnel Company Limited (2006) Annual Report.

Figure 7.8   Toll revenue distribution.

improvement schemes, including the Route 8 project between Shatin and Cheung Sha Wan and the proposal to revert traffic direction of Woosung Street and Parkes Streets, were expected to benefit the WHC in the long run.

Cyclical economic factors and government policy functioned to slow traffic flow through WHC over its existence. In 1997–98 the Asian Financial Crisis was a trigger for a prolonged downturn in the Hong Kong economy, with the property sector hit particularly hard. Residential development on the West Kowloon reclamation, part of the original mix of factors expected to create demand for the WHC, was limited and delayed by government and public concerns, while growth in car ownership in Hong Kong was significantly below early 1990s forecasts (Table 7.12). The delay in the development of the West Kowloon reclamation into an integrated arts, culture, entertainment and residential district had also been cited as a factor underlying the lower-than-expected traffic demand for the WHC. The West Kowloon Cultural District Project, which was first announced in 1998 and defined in the Invitation for Proposals in 2003, was temporarily put on hold by the government in 2006 for re-examination after serious public debate. Under the original proposal, the project comprised the development of a number of arts and cultural facilities, including three theatres, four museums, an art exhibition centre, a performance venue and a water amphitheatre (Hong Kong SAR Government Press Release, 2006). As of late 2006, the government was still gauging views from the public on how to make the best use of the land and financing options before taking the project forward.

Market Share by Tunnel: 2005/06

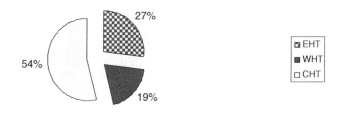

a.

Market Share by Tunnel: 2004/05

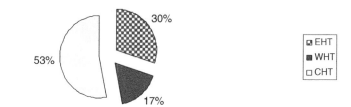

b.

WHT - Western Harbour Crossing
EHT - Eastern Harbour Crossing
CHT – Cross-Harbour Tunnel

Source: Western Harbour Tunnel Company Limited (2006) Annual Report.

Figure 7.9   Market share by tunnel.

7.3   Valuation of the project

After the meeting with Paul Klein, Sophie Cheng worked with her assistant Eddie Lee to conduct a valuation for the WHC. Sophie had to be prepared to explain the economic viability of the WHC venture in two weeks and also answer Paul's questions about financing for implementation. In order to get a full understanding of the future prospects of the WHC, Eddie researched articles to probe the latest news about the WHC and its competitors. He read the following article about the imbalance of tolls:

> *'The WHC had fallen victim to a toll imbalance. It was likely that this problem would drag on in the short-term as there was no mechanism in place to close the difference in toll levels and even out the discrepancies in traffic flow among the three cross harbour tunnels'.*[2]

---

[2] Quotation prepared by case writer.

Table 7.12 Vehicle registration 1995–2005.

Vehicles Registered		End of Year										
		1995	**1996**	**1997**	**1998**	**1999**	**2000**	**2001**	**2002**	**2003**	**2004**	**2005**
A. Motorised vehicles												
Motor Cycles		29,073	30,164	31,212	32,004	33,079	34,085	36,191	38,678	41,128	43,619	45,941
Motor Tricycles		23	2	—	—	—	—	—	—	—	1	2
Private Cars		318,233	325,131	348,450	359,694	365,533	374,013	381,757	384,864	382,880	385,028	388,311
H.K. & Kln. Taxis		15,321	15,249	15,249	15,250	15,250	15,250	15,250	15,250	15,250	15,250	15,250
N.T. Taxis		2,829	2,837	2,837	2,838	2,838	2,838	2,838	2,838	2,838	2,838	2,838
Lantau Taxis		40	40	50	50	50	50	50	50	50	50	50
Public buses:												
CMB		881	855	823	—	—	—	—	—	—	—	—
	Single Deck	31	29	29	—	—	—	—	—	—	—	—
	Double Deck	850	826	794	—	—	—	—	—	—	—	—
NWFB		—	—	—	841	946	891	780	835	775	731	706
	Single Deck	—	—	—	62	96	95	66	66	46	36	36
	Double Deck	—	—	—	779	850	796	714	769	729	695	670
KMB		3,506	3,594	3,839	3,990	4,078	4,251	4,386	4,438	4,295	4,150	4,029
	Single Deck	308	285	278	275	265	250	249	243	237	175	156
	Double Deck	3,198	3,309	3,561	3,715	3,813	4,001	4,137	4,195	4,058	3,975	3,873
NLB		79	78	80	87	86	89	75	78	80	86	86
	Single Deck	74	74	80	87	86	89	75	78	80	86	86
	Double Deck	5	4	—	—	—	—	—	—	—	—	—
City Bus		368	451	590	756	794	794	794	793	779	764	759
	Single Deck	22	32	57	137	103	103	95	95	87	81	82
	Double Deck	346	419	533	619	691	691	699	698	692	683	677
Long Win Bus		—	—	22	163	160	160	160	145	145	145	148
	Single Deck	—	—	—	3	10	10	10	9	9	9	9
	Double Deck	—	—	22	160	150	150	150	136	136	136	139
City Bus 2		—	—	—	181	166	167	164	164	164	164	160
	Single Deck	—	—	—	20	25	25	22	22	22	22	12
	Double Deck	—	—	—	161	141	142	142	142	142	142	148
KCRC		—	—	84	93	59	81	85	91	91	91	107
	Single Deck	—	—	14	21	20	20	24	24	24	24	26
	Double Deck	—	—	70	72	39	61	61	67	67	67	81
Others		4,765	5,287	5,546	5,775	5,884	6,065	6,368	6,728	7,115	7,121	7,059
	Single Deck	4,487	5,037	5,369	5,597	5,691	5,875	6,188	6,555	6,936	6,987	6,932
	Double Deck	278	250	177	178	193	190	180	173	179	134	127

Table 7.12 *Continued*

Vehicles Registered	1995	1996	1997	1998	1999	End of Year 2000	2001	2002	2003	2004	2005
Private Buses	333	383	423	437	447	451	485	489	490	492	493
Public Light Buses	4,350	4,348	4,350	4,350	4,350	4,350	4,350	4,350	4,350	4,350	4,350
Private Light Buses	2,585	2,481	2,392	2,297	2,228	2,158	2,098	2,042	1,979	1,935	1,897
Light Goods Vehicles	96,070	92,900	91,823	89,608	86,735	83,771	80,889	78,332	75,987	75,200	75,522
Medium Goods Vehicles	38,415	39,375	40,492	40,442	40,776	41,817	42,036	42,319	42,059	42,504	42,794
Heavy Goods Vehicles	1,831	2,144	2,546	2,735	2,863	3,068	3,308	3,485	3,541	3,585	3,500
Special Purpose Vehicles	311	345	406	457	503	550	617	759	851	942	1,095
Government vehicles:											
Motor Cycles	1,322	1,207	1,308	1,238	1,264	1,201	1,240	1,243	1,204	1,152	1,194
Others	5,961	6,075	6,381	6,125	6,104	6,041	5,887	5,580	5,450	5,346	5,200
Total (Motor Vehicles)	526,296	532,946	558,903	569,411	574,193	582,141	589,808	593,551	591,501	595,544	601,491
B. Non-motorised vehicles											
Trailers	24,937	26,024	27,388	26,479	24,474	23,761	23,205	23,220	22,566	21,930	21,171
Rickshaws	—	—	—	—	—	—	—	—	—	—	—

Source: Transport Department, Annual Transport Digest (2006).

Table 7.13 GDP in Hong Kong.

Year	Quarter	GDP			
		At current market prices		At constant (2000) market prices	
		HK$ million	Year-on-year % change	HK$ million	Year-on-year % change
1990		598,950	11.7	845,515	4.0
1991		690,324	15.3	893,625	5.7
1992		805,082	16.6	951,270	6.5
1993		927,996	15.3	1,011,492	6.3
1994		1,047,470	12.9	1,068,193	5.6
1995		1,115,739	6.5	1,110,086	3.9
1996		1,229,481	10.2	1,156,923	4.2
1997		1,365,024	11.0	1,216,102	5.1
1998		1,292,764	−5.3	1,149,662	−5.5
1999		1,266,702	−2.0	1,195,624	4.0
2000		1,314,789	3.8	1,314,789	10.0
2001		1,298,813	−1.2	1,323,167	0.6
2002		1,276,757	−1.7	1,347,495	1.8
2003		1,233,983	−3.4	1,390,610	3.2
2004#		1,291,568	4.7	1,510,182	8.6
2005#		1,382,052	7.0	1,619,984	7.3
2006#	Q1	346,710	7.9	407,793	8.0
	Q2	349,330	5.4	409,078	5.5
	Q3	379,349	6.6	447,784	6.8

Source: Hong Kong Census and Statistics Department (2006), "Gross Domestic Product".
#: Subject to revision as new information becomes available.

Eddie also conducted some research on the economic environment in Hong Kong. In the third quarter of 2006, the Gross Domestic Product (GDP) in Hong Kong rose by 6.8% in real terms over the preceding year, compared with the 5.5% increase in the second quarter (Hong Kong Census and Statistics Department, 2006a). This signified the twelfth consecutive quarter of distinctly above-trend growth since the upturn began in mid-2003 (Hong Kong SAR Government, 2006). In the light of the strong GDP growth for the first three quarters of 2006, the Hong Kong Government had revised forecast GDP growth for 2006 as a whole from 4–5% to 6.5%. Overall consumer prices increased by 2% in October 2006 over a year earlier, which was slightly lower than the increase of 2.1% in September 2006 (Hong Kong Census and Statistics Department, 2006b) (see Tables 7.13–7.15: economic and monetary statistics in Hong

Table 7.14　Consumer price index in Hong Kong.

Year	Month	Composite consumer price index		Consumer price index (A)	
		Index	Year-on-year % change	Index	Year-on-year % change
1990		63.8	+10.2	63.5	+9.8
1991		71.0	+11.6	70.8	+12.0
1992		77.8	+9.6	77.4	+9.4
1993		84.6	+8.8	84.0	+8.5
1994		92.1	+8.8	90.8	+8.1
1995		100.4	+9.1	98.7	+8.7
1996		106.7	+6.3	104.6	+6.0
1997		113.0	+5.8	110.6	+5.7
1998		116.2	+2.8	113.5	+2.6
1999		111.6	−4.0	109.8	−3.3
2000		107.4	−3.8	106.6	−3.0
2001		105.7	−1.6	104.8	−1.7
2002		102.4	−3.0	101.4	−3.2
2003		99.8	−2.6	99.3	−2.1
2004		99.4	−0.4	99.3	0.0
2005		100.3	+1.0	100.3	+1.1
2006	Jan	101.3	+1.9	101.1	+1.6
	Feb	101.1	+1.2	101.1	+1.1
	Mar	101.5	+1.6	101.4	+1.4
	Apr	102.0	+1.9	101.7	+1.7
	May	102.1	+2.1	101.8	+1.8
	Jun	102.4	+2.2	102.3	+1.9
	Jul	102.7	+2.3	102.4	+2.0
	Aug	102.7	+2.5	102.5	+2.3
	Sep	102.8	+2.1	102.5	+1.9
	Oct	102.9	+2.0	102.5	+1.7

Source: Hong Kong Census and Statistics Department (2006), Consumer Price Index.

Kong). The profits tax rate in Hong Kong for the year of assessment 2006/07 was 17.5%.

To finalise the valuation report, Sophie and Eddie had to complete the following tasks:

- construct a financial model for the WHC venture, and estimate the value of the venture as a going concern using multiple scenarios
- estimate the debt capacity of the WHC and recommend a loan structure for the potential purchase
- comment on the nature and desirability of the WHC as a portfolio investment.

Table 7.15 Hong Kong interbank offered rates.

As at end of		1-month	12-month
		(% per annum)	
2004	Dec	0.22	0.84
2005	Jan	0.50	1.34
	Feb	1.63	2.28
	Mar	2.47	3.09
	Apr	1.84	2.38
	May	3.41	3.34
	Jun	3.34	3.38
	Jul	3.34	3.63
	Aug	3.56	4.00
	Sep	4.09	4.22
	Oct	4.16	4.53
	Nov	4.00	4.47
	Dec	4.06	4.38
2006	Jan	3.66	4.19
	Feb	4.03	4.38
	Mar	4.13	4.63
	Apr	4.56	4.63
	May	4.44	4.69
	Jun	4.12	4.88
	Jul	3.83	4.63
	Aug	4.04	4.38
	Sep	4.17	4.20
	Oct	3.84	4.13

Source: Hong Kong Monetary Authority (2006), "Hong Kong Interbank Offered Rates".

Sophie was aware that the valuation of an infrastructure project that had been in full operation was quite different from that of a green-field project. She was concerned that the sunk costs associated with the WHC venture could seriously weaken the attractiveness of the project to the GW Group.

8

Institutional Risks and Infrastructure Project Finance: the Dabhol Power Project

This chapter's case study is likely to be only one of a hundred possible narratives of a project that epitomises how complication and controversy can become embroiled in commercial project finance and development. Anyone with insight or experience of the former Enron Corporation, its Dabhol Power Company vehicle, or the international and Indian financial intermediaries involved at different stages of the project almost certainly now holds a distinct view as to the merits and fate of the scheme. Observers and participants in the layers of contemporary Indian politics might easily have other views to add as to the roles of powerful foreign commercial and financial interests in an important part of the domestic Indian economy, which was itself beginning a period of major reform. The Dabhol Power Project has come to represent both victim and aggressor, and none of those involved in its creation and demise can be wholly pleased with the outcome.

Different as they may be, those many narratives could easily be accurate. Truth is personal, and the full facts of such a labyrinthine and protracted case can be known to no-one. So in preparing this chapter, we have looked at Dabhol as a means to illustrate certain aspects of project finance that are most important to our treatment of the subject. This means that the study dwells on the institutional concerns that are presented in Chapter 2, and the ways that important factors of risk discussed in Chapter 4 – by which the entire rationale for non-recourse finance stands or falls – were fully or partly anticipated in the project's financing structure, or failed to be fully ameliorated. Dabhol may have been wrongly conceived, wrongly analysed and structured, or a foreign-inspired device to allow vehement debates on Indian nationalism to become tangible in a commercial setting, but it appears to have powerful lessons for project sponsors and their financiers.

8.1   Dabhol Power Project[1]

In 1992, Indian public sector authorities entered an agreement with the Houston-based Enron Corporation to build a gas-fired power plant in Dabhol in Maharashtra state. This was intended as the world's largest independent power project and the largest single foreign investment in India.

8.1.1   India

Since gaining independence from the UK in 1947, India's economy was managed with socialist-oriented developmental state policies and emphasis on central planning and intricate regulation of the private sector. Influenced by Mahatma Gandhi's doctrine of *swadeshi*, or self-reliance from foreign influence, successive Congress Party governments – all tracing their roots to Gandhi's independence campaigns – restricted foreign investment and imposed high import tariff barriers so as to protect domestic industries.

By the early 1990s, India faced both a severe external payments crisis and a debilitating federal budget deficit. At the same time, it was losing a strategic ally and economic partner with the collapse of the Soviet Union. Faced with apparently hostile neighbours and an uncertain post-Cold War balance of power, India began to look for new international partners, and for the first time considered allowing foreign commercial interests to participate in key sectors of the domestic economy. Thus in 1991, the Congress Party-led national government under Prime Minister P. V. Narasimha Rao began a programme of liberalisation with two major policy changes: relaxing control of industrial licensing and opening the economy to foreign investments.

8.1.2   India's power sector

In October 1991, India's government opened the energy sector for private investors to build and operate power generation plants without restriction on foreign ownership (Edwards and Shukla, 1995). With a steadily growing population, mass urban migration and agglomeration, and the beginning of a phase of above-trend growth in industrialisation, the power sector was critical to India's economic development. National power demand was growing at an annual rate of almost 8% and was projected to increase to some 140 000 MW by 2005 but total installed capacity was only around 80 000 MW in the early 1990s (Hill, 2005).

---

[1] Grace Loo prepared this case under the supervision of the authors for class discussion. This case is not intended to show effective or ineffective handling of decision or business processes. The utmost care has been taken in preparing the history and fundamental background in this case. Ref. 07/335 © 2007 by The Asia Case Research Centre, The University of Hong Kong (http://www.acrc.org.hk). No part of this publication may be reproduced or transmitted in any form or by any means – electronic, mechanical, photocopying, recording, or otherwise (including the Internet) – without the permission of The University of Hong Kong.

The power generation system was both inefficient and incapable of meeting demand on this scale. Power plants were fuelled mostly by coal mined in the eastern state of Bihar and transported to the rest of the nation by an ageing and inefficient railway network, a process involving significant pilferage. The average mid-1990s plant load factor, that is, the plant capacity used in production, was 57%, falling short of peak load demand by between 20% and 30% (Nicholson, 1995). In addition, up to 20% of the power generated was lost through transmission over poorly maintained and under-invested low-voltage distribution networks. Pilferage of electricity was common, with many consumers tapping illegally into the grid for their own use or wiring around electricity meters to avoid payment. As much as 25% of the power connections in New Delhi were estimated to be illegal at this time (Choukroun, 2002).

The power sector operated a two-tier system, with responsibility shared between the federal centre and each state, all of which maintained a State Electricity Board (SEB) involved in the generation, transmission and distribution of electricity. While some central government utilities also generated electrical power, the functions of transmission and distribution were largely left to the SEBs. In the mid-1990s, about 70% of national power was generated and distributed by SEBs (Hill, 2005). As India struggled with power shortages, the SEBs were prone to subsidise the tariffs of agricultural consumers while industrial consumers bribed state officials for lower power bills. With losses accumulating at up to US$2 billion annually,[2] the SEBs' ability to invest in new capacity was severely limited.

Given the importance of the power sector to India's economic development, the government decided to fast-track eight projects to induce foreign investment in this sector as part of a wider programme of economic liberalisation. These projects were to be initiated through individual negotiation rather than public tendering and also enjoyed certain support from central government, including certain guarantees (Paterson, 2006). Phase I of the Dabhol power plant was the first such project to be launched and was widely seen as the flagship initiative marking the opening of India's domestic energy sector to foreign interests. At the time of its instigation, Dabhol was the world's largest independent power project and the largest foreign-sourced investment project in India.

8.1.3 Enron

The Enron Corporation began in Texas in 1985 as a pipeline distribution company delivering gas to power utilities and other commercial businesses. With the landmark deregulation of the US electrical power markets, Enron expanded activities to include trading in electricity, collecting substantial margins from differences in inter-state wholesale and resale prices. Enron began to devise complex supply contracts that sought to protect its customers from certain risks, such as changes in interest rates

---

[2] US$1 = Rs33.18 as at the end of 1995.

or an inability to pay, but also entered into a far higher number of such contracts than its supply contracts would ever need for risk management purposes. The company became authorised by the Federal Securities and Exchange Commission to adopt accounting practices customarily associated with securities or commodity traders, and so marked-to-market portions of its balance sheet assets and other receivables regardless of their liquidity or likelihood of being realised. The result was earnings generation that could best be termed mythical.

This aggressive corporate approach became further manifested in the falsification of revenue and income to support the group's share price. This grew increasingly complex, in particular involving a series of transactions to conceal overall funded liabilities and present a less highly leveraged appearance to shareholders and financial analysts. Doubts about Enron's reporting and true performance turned from Wall Street rumour to widespread concern in 2001, and towards the end of that year Enron admitted to having overstated group earnings over four prior years by US$586 million and of concealing liabilities of around US$3 billion. The ensuing loss of confidence and collapse was precipitate: in November 2001, Enron's debt was downgraded to junk-bond status and the company filed for bankruptcy within weeks. Enron then employed some 21 000 employees worldwide.

In the early 1990s, Enron began to expand downstream activities, for example in developing power generation projects to serve as buyers of natural gas (Allison, 2001). It also sought to move into newly privatised non-US markets (Fidler, 2002). A subsidiary, Enron International, was set up in 1990 with the mandate of running power generation projects in developing markets. Rebecca Mark, a protégé of the Enron chairman Kenneth Lay, was appointed its chief executive. Mark was considered ambitious and influential, and became one of Enron's more high-profile figures (Anderson, 2001). Mark saw Enron becoming the largest distributor of liquefied natural gas (LNG) in India, requiring capital investment of up to US$20 billion by 2010. In this vision, Enron would set up two LNG terminals and re-gasification units, the first being at Dabhol in Maharashtra. LNG would be imported to meet projected national needs through a terminal facility that Enron was building in Qatar, a gas-rich Gulf state. The two Indian LNG terminals would also help Enron to meet the 15% annual return that it had promised to investors in the Qatar facility (Hill, 2005).

Thus Enron opened discussions in 1992 with the Indian authorities over the Dabhol power plant, to be built, owned and operated by Dabhol Power Company (DPC), an Indian private company unusually owned entirely by foreign interests. Enron owned 80% of the company, with General Electric Corporation (GE) and Bechtel Enterprises Inc. (Bechtel) each holding 10% interests. GE would supply gas turbines for Dabhol and Bechtel serve as the turnkey engineering procurement and construction contractor. The three shareholders controlled DPC through a series of companies registered in Mauritius, which had favourable double taxation agreements with India that might help the owners lessen withholding taxes in the event of the project producing surpluses for distribution (Allison, 2001). A further Enron subsidiary was made turnkey contractor for the re-gasification plant.

8.2 Development of the Dabhol Power Project

The Dabhol plant was controversial from inception. Following the decision to open the power sector to foreign direct investment, Indian officials visited the United States in May 1992 to seek investment and technological resources for the sector. The delegation met Enron officials, including Rebecca Mark, and indicated their interest in commissioning a private power plant. Immediately upon the delegation's return, the Secretary of Power informed the state-controlled power supplier, Maharashtra State Electricity Board (MSEB), that Enron would visit the state in June to inspect potential sites along the state's coastal areas for a proposed power project.

Maharashtra is one of India's most prosperous and developed states and state capital Mumbai (formerly Bombay) is the country's main financial and commercial hub. When Enron and GE executives arrived in India, they first met central government officials in Delhi and then travelled to Mumbai to inspect sites and meet state representatives. Soon afterwards on 20 June 1992, Enron and the Maharashtra state government signed a memorandum of understanding (MoU) to develop a 2000 MW LNG-fired power plant at Dabhol, 180 km south of Mumbai. Agreement in principle over the US$3.1 billion project was made unusually quickly (Callari, 2002). Neither central nor state government engaged independent technical assistance or conducted a financial appraisal of the project and the main contract was concluded without competitive bidding of any kind (Allison, 2001).

A state government committee reviewed the project in 1995 and commented that:

> *'in a matter of less than three days, a MoU was signed between Enron and MSEB in a matter involving a project of the value of over 10000 crore rupees [US$2.5 billion] at the time, with entirely imported fuel and largely imported equipment, in which, admittedly, no one in the government had expertise or experience. In fact, the file [on the Dabhol project] did not even show what Enron was – what its history is, business or accomplishment. [...] It looked more like an ad hoc decision rather a considered decision on a durable arrangement with a party after obtaining adequate and reliable information. Neither the balance sheet and annual accounts of Enron, nor any information about its activities, area of operation, its associates, etc, was obtained by the government then or even later'* (Report of the Maharashtra State Cabinet, 1995).

8.2.1 Project structure

Two months after signing the MoU, Enron submitted detailed implementation proposals to India's Foreign Investment Promotion Board (Allison, 2001) and at its recommendation split the project into two separate phases. MSEB had earlier suggested that the project be divided in such a way but Enron disagreed on the grounds that it would adversely affect Dabhol's economies of scale in production. MSEB's motivation is unclear but splitting the project into two phases might have made it easier for the board as Dabhol's sole power buyer to cope with the financial consequences of the project, which in every sense was transforming for the state in terms of capacity, output and costs.

Phase I of the project involved the construction of a 695 MW gas-fired power station that would run on imported distillate oil, scheduled to commence production in December 1997. Phase II would expand the capacity of the plant to 2015 MW and involved the construction of a 1320 MW gas-fired plant, a re-gasification facility, a LNG carrier plus corresponding port facilities including fuel jetty, navigation channel and breakwater (Project Finance, 2000). The second phase was scheduled for commissioning at the end of 2001 and upon the completion of Phase II, the entire plant would switch to using LNG for fuel.

With the completion of Phase II, Dabhol would run on Qatar-sourced LNG, where Enron had extensive interests, including the US$4 billion development of a new gas field (Brauchli, 1995) and other pipeline construction. Dabhol's total cost would exceed US$2.8 billion. The scheme was designed as a gas-fired baseload station that generates electricity constantly, compared to peakload stations that generate power only during peak hours. Under a 20 year contract, MSEB would buy a minimum amount of electricity at a plant load factor of 90%. The project's debt-to-equity ratio was intended not to exceed 70:30. The capital cost of Phase I was US$920 million (Sant *et al.*, 1995), and that of Phase II would be about US$1.9 billion.

8.2.2 Objections

Both Enron and the central government sought financing from the World Bank for Dabhol. This was a strategic decision, for not only did the World Bank offer loans on favourable terms for many such power projects but to raise funds in this way would represent endorsement of the project so as to make commercial funding easier to obtain. However, World Bank analysis showed that the project would produce too much power at too high a cost for Maharashtra. New unwanted capacity would compel MSEB to replace cheaper coal-fired power output with a source that was as much as five times more costly, for MSEB was contracted to buy a minimum amount from Dabhol even if the plant's output was unwanted (Allison, 2001). As transmission network capacity between India's five regional grids was limited and most state electricity boards were in poor financial condition, selling power to other states was scarcely viable.

The World Bank also noted that the Dabhol plant was planned as a baseload power station when Maharashtra faced electricity shortage only during peak hours, and the use of LNG was dubious given the considerably lower cost of domestic coal. It concluded that the project did not meet a least-cost test, lacked an overall justification, and the Maharashtra-Enron MoU was unduly favourable to Enron. In April 1993, the

World Bank turned down the Maharashtra government's loan application on the grounds that the project was 'not economically viable'.

India's federal authority supervising the power sector, the Central Electricity Authority (CEA), studied the project independently and also found that 'the entire MoU was one-sided' in favour of Enron. The CEA specifically noted that (Allison, 2001):

- specific details about project costs as required under Indian law were absent
- the price of electricity generated by Dabhol was high
- the MoU failed to specify when the 20 year contract would begin, and lacked a specification as to when electricity supply or payment would commence
- the structure of payments did not conform to earlier guidelines issued by the central government in 1992
- no provision was made for auditing the project to ensure that the price of electricity was commensurate with its cost
- MSEB guaranteed a minimum fuel purchase but the fuel supplier was not bound to provide a minimum quantity of fuel
- MSEB had not verified whether the price of fuel was economical.

8.2.3 Deal closed

CEA withheld clearance for the project on the grounds that the tariff was overly high. Nonetheless, the Maharashtra government urged the central government to give clearance to the project and the central government succumbed, despite CEA clearance for such projects being mandated by law (Dutta, 2002). In December 1993, Maharashtra signed a 20 year power purchase agreement with DPC, putting the final seal on the project (Allison, 2001). To allow Dabhol to proceed, Enron needed to secure some 150 federal and state approvals, resolve many legal issues, and deal in principle with complicated federal and state taxes (Hill, 2005). Enron, with its aggressive lobbying efforts, overcame the notoriously conservative Indian bureaucracy with exceptional speed.

8.3 Power purchase agreement

The power purchase agreement between MSEB and Enron was a take-or-pay contract under which MSEB committed to purchase an agreed amount of power capacity without heed to the amount of energy it used. MSEB would buy power from Dabhol for 20 years, regardless of demand or whether cheaper sources of fuel were available. While the tariff for electricity for most state power projects increased or decreased marginally over time, the tariff for Dabhol was structured such that it was expected to increase steadily over the project's life (Sant *et al.*, 1995).

Each unit of electricity purchased by MSEB comprised a capacity charge and an energy charge. The capacity charge covered the recovery of capital invested in the project and was calculated predominantly in US dollars with MSEB bearing all currency risks. The capacity charge would be applied in full whenever a plant load factor of 90% was achieved. The energy charge was a variable component of the tariff based upon the volume of fuel consumed, variable operations and maintenance charges, take-or-pay charges for fuel supplies and special operations fees. The final price of Dabhol power depended largely on the US$–rupee exchange rate, oil price levels, and the plant's actual load factor.

The power purchase agreement assured DPC of an internal rate of return of 16%, which was less than many analysts deemed adequate to attract foreign capital, taking into account foreign investors' perception of the risks involved in investing in India. Nonetheless, industry observers calculated DPC's real post-tax internal rate of return to be between 26% and 32%, which amounted to annual excess payments of US$15.9–US$20.4 million from MSEB (Sant *et al.*, 1995).

8.3.1   *Phase I financing*

Financing for Phase I of the project was arranged by March 1995 (Euroweek, 1995). The three foreign sponsors together contributed equity investment of US$276 million. Debt financing came in four main parts (International Financing Review, 1995):

- US$150 million through a syndication of 12 banks, arranged by Bank of America and ABN Amro Bank NV
- about US$298 million in commercial export credit loans guaranteed by the US Export-Import Bank
- US$100 million loan from the Overseas Private Investment Corporation (OPIC), a US agency that provides political risk insurance
- US$98 million in long-term loans through Industrial Development Bank of India (IDBI), a state-sector lender.

Credit support came in the form of a letter of credit, guarantee from the state government, counter-indemnities from central government, and an escrow account over certain of MSEB's payments. Dabhol was the first of the eight fast-tracked energy projects for which central government offered guarantees in order to attract private sector involvement.

8.3.2   *The agreement revised*

In March 1995, prior to the commencement of construction of Phase I (Nainan, 1995), a new coalition of the nationalist Bharatiya Janata Party (BJP) and the Hindu nationalist Shiv Sena replaced the Congress state government that had been responsible for entering the Dabhol contract. Both BJP and Shiv Sena campaigned for election on an anti-Dabhol platform, fanning the widespread perception that dealings with foreigners were likely to be to India's disadvantage (Edwards and Shukla, 1995).

Antagonism to Dabhol was accentuated by a statement before a US congressional hearing by a senior company official that Enron had spent US$20 million in education in India to show the benefits of private power projects, a sum which many in India interpreted as bribe money. The new Maharashtra coalition government quickly formed a committee to review the project led by Gopinath Munde, deputy chief minister and state president of the BJP, who during the campaign had visited Dabhol and pledged to *'throw Enron into the Arabian Sea'* (Dutta, 2002). The Munde Committee studied the project and concluded the following.

- There was no reason not to include competition in bidding for the project. It was an act of impropriety for the central government to negotiate solely with Enron.
- The capital costs of the Dabhol project were artificially inflated.
- The foreign currency denomination of tariff payments would lead to unjustifiably high rates for consumers.
- The high cost of power generated by Dabhol would adversely affect the economic development of Maharashtra.

The report was never published although sections were leaked to the Indian press. In response, Enron's Mark indicated that the company had taken into account that duties on imported equipment would be subject to the whim of Indian customs, implying that Enron had indeed fudged the costs (Hill, 2005).

The state government declared the Dabhol project cancelled in August 1995 at the recommendation of the Munde Committee. State chief minister Manohar Joshi called a halt to construction work of Phase I until the Dabhol contract could be rescinded (Fernandes, 1995). *'This deal is against the interests of Maharashtra,'* Joshi said before the state legislature, while *'accepting this deal would indicate an absolute lack of self respect and would amount to betraying the trust of the people'* (Brauchli, 1995). Nonetheless, Joshi also hinted that the state government would be open to renegotiation if Enron took the appropriate initiative.

The state government's decision to declare the Dabhol project cancelled prompted Enron to issue an arbitration notice and demand compensation for US$300 million that DPC had by then injected into the project. The company ran advertisements in prominent Indian newspapers publicising the benefits of the project (Hill, 2005). By September 1995, polls showed that 80% of the state and 60% of the nation wanted the project to resume. In the hope of saving its investment, Mark suggested renegotiation to Joshi, proposing a tariff revision to take account of Dabhol's special infrastructure and tax requirements and to match the most favourable offer delivered under similar private power projects recently approved in Maharashtra (Nainan, 1995).

Enron also suggested switching from distillate fuel to naphtha or LNG from domestic suppliers, and offered MSEB a 30% share in DPC. The state government began negotiations with Enron and in February 1996 the two parties agreed to revise the Dabhol agreement. Under the new terms (India Business Intelligence, 1995):

- a new turbine design with a fall in price of generation equipment would reduce capital costs by US$300 million, while increasing total capacity from 2015 MW to 2184 MW
- the re-gasification plant would be devolved into a separate venture
- Phase I of the project would use domestic naphtha instead of distillate oil
- separation of the re-gasification plant and the fuel switch would reduce costs by US$450 million and lower the unit rate of power by cutting the fixed cost component of the tariff. This would lessen Dabhol's average unit rate of power[3] from US$0.055 for Phase I to US$.043 for the entire project, making Dabhol the cheapest large-scale power project under negotiation or implementation in India
- MSEB was given the option to acquire 30% share of DPC, which would reduce the project's annual cash outflow by US$150-US$170 million (Inkpen, 2002)
- the sponsors could sell re-gasified LNG to customers other than MSEB (Naik, 2003)
- Phase I and II would be implemented simultaneously to cover four months' lost construction time (Economist Intelligence Unit, 1995).

A new power purchase agreement was signed in August 1996 and construction work on Phase I resumed by year end (Naik, 2003). The revised agreement was given a counter-guarantee by the central government and Phase I of Dabhol went online in May 1999 (Sarkar and Sharma, 2006). Under the revised agreement, DPC's annual return on the entire project was expected to be around 20% (Economist Intelligence Unit, 1995).

The state government's attempt to cancel Dabhol was a serious setback to the central government's effort to attract foreign investment, especially in India's energy sector (Nicholson, 1995). Dabhol was intended as a trailblazer to lure foreign investment, but the state's behaviour was chilling to certain foreign investors, heightening their perception of India being relatively risky for cross-border investment. To many observers, the state government's decision was a political ploy by the BJP to discredit the Congress Party central government prior to a national election due in 1997 (Fernandes, 1995). Critics suggested that the new agreement failed to address the underlying cost issues of the project and worsened MSEB's position (Economist Intelligence Unit, 1995), by removing the optional nature of Phase II of the project.

---

[3] Average unit rate of power refers to the unit rate of power over the lifetime of the project.

8.3.3   Phase II financing

More than 40 lenders became financially committed to aspects of the project but over 12 months were needed before a bank syndicate could be formed to fund the foreign currency debt requirements of Phase II. Financing for Phase II was completed in May 1999, with the loan described in the specialist financial press as among the most successful international project financings. The projected capital cost of Phase II, including re-gasification facilities, was US$1.9 billion with US$1.414 billion in loans and US$452 million in commitments from the original foreign sponsors. US$1.082 billion comprised foreign debt, representing the most sizeable foreign borrowing sanctioned by India (Allison, 2001). The debt package was obtained from a number of sources (International Financing Review, 1999):

- US$333 million equivalent in a local currency loan was provided by Indian banks with IDBI acting as arranger
- US$497 million syndicated loan was provided by groups of both domestic and off-shore lenders
- OPIC provided US$60 million in project finance loans
- US$433 million export credit loan of which US$258 million was provided by the Export-Import Bank of Japan (Jexim) and guaranteed by Indian financial institutions and commercial banks. The remaining US$175 million was provided by commercial banks with a guarantee from Japan's Ministry of International Trade and Industry
- a second US$90.8 million export credit loan was provided by commercial banks with Belgium's Office Nationale du Ducroire providing a 95% guarantee and the other 5% given by long-term Indian policy banks.

The proceeds were to be applied towards the construction of an additional 1320 MW of generation capacity, a LNG re-gasification facility and port facilities. Enron had by now secured 20 year contracts to buy annually 1.6 million tonnes of LNG from Oman LNG LLC and 460 000 tonnes from Abu Dhabi Liquefaction Gas Co. Ltd (Project Finance, 2000). Unlike Phase I, financiers of Phase II were given no central government counter-indemnities but received support from the state government.

8.3.4   Godbole committee

During the 1999 Maharashtra elections, the Congress Party, now projecting Dabhol as a symbol of its opponents, campaigned against the project and defeated the ruling coalition. The new administration set up a committee to review the second phase of the project. Many saw the move as a ploy by the incoming administration to undermine the BJP, which at the time was leading the federal government. The committee, led by Madhav Godbole, released its findings on 12 April 2002 after studying the projects for two months and recommended a renegotiation of the PPA with a reduction and re-denomination of the tariff.

> '*DPC has emphasized the sanctity of the contracts entered into with it. However, it is well known that many commercial contracts are routinely renegotiated with major changes. In a sense, economic reality dominates technical legality in the commercial world*,'
> the report said (Report of the Energy Review Committee, 2001).

With 90% of the construction cost of Phase II already committed, Dabhol's sponsors strongly opposed the recommendations of the committee but negotiations between the state and Enron broke down for lack of an agreement.

8.3.5   Cash crises

Phase I of the Dabhol plant began operations in May 1999 (The Economist, 2001a) and within one month MSEB raised concern about its ability to pay a US$20 million monthly bill. Although MSEB had a relatively strong financial position among India's state electricity boards, it faced common longstanding financial problems, in part due to receiving payment for only a portion of the electricity that it supplied (Paterson, 2006). Much of its supplied power was lost through transmission inefficiency and pilferage, while MSEB also subsidised tariffs to the state's most sensitive electorate, especially in the agricultural sector (The Economist, 1994). Phase I of Dabhol consumed naphtha, a comparatively expensive fuel oil derivative, and crude oil prices had risen sharply in 2000. This was aggravated by a falling rupee, given that MSEB was bound to a US dollar-denominated price (Inkpen, 2002). While MSEB was buying only 33–60% of Dabhol's output, its take-or-pay commitment of a 90% capacity take-up converted its low usage into a higher average purchase price (Inkpen, 2002; Pesta, 2000). By end-2000, MSEB was paying almost five times more at US$0.161 for each kilowatt-hour of electricity compared with US$0.034 in most other states (Pesta, 2000). Faced with financial difficulties, MSEB sold half of its 30% stake in DPC acquired after the 1996 renegotiations, increasing Enron's shareholding in DPC to 65%.

With its single customer unable to pay for power, DPC invoked the central government counter-indemnity in February 2001 to recover US$17 million owed by MSEB since November 2000. The state government eventually paid the amount (New York Times, 2005). In the same month, MSEB accused DPC of misrepresenting Dabhol's technical capability and made a claim for compensation of around US$86 million. This was disputed by DPC as an attempt by MSEB to evade contractual payments. While power generated from Dabhol would become cheaper when Phase II went into operation at year-end and natural gas was substantially cheaper than naphtha, MSEB's overall costs would increase as it was bound to buy LNG under a 20 year supply contract (The Economist, 2001a).

DPC had few options if MSEB was unable or unwilling to pay or stopped taking power from Dabhol. Although electricity was in shortage throughout India, cross-state sales were hampered by poor infrastructure. Independent power projects wishing to sell outside their home state needed approval from the state of domicile as well as approval to the tariff by the consuming state's regulatory authority (Joshi, 2002). Since 1999, Enron had tried without success to persuade the federal and Maharashtra governments to relax these rules so as to enable independent power projects to sell power to more than one state (Joshi, 2002).

Foreign creditor banks were now also concerned about MSEB's payment problems, not least since the central government counter-indemnity applied only partially to Phase II and funds for its construction had ceased to be made available in March 2001 when construction was about 90% finished (Joshi, 2002). Indian creditors, on the other hand, reduced interest rates on DPC's local currency loans to help avoid statutory provisioning for their loans (Rao, 2001). In April, DPC commenced arbitration proceedings in London and served notice of political *force majeure* (The Economist, 2001b). With payments outstanding from four months earlier, DPC threatened to cease supply even though MSEB had paid a later invoice.

Neither state nor central government would meet MSEB's payments on the grounds that the bills were in dispute (Paterson, 2006). DPC tried to recover payment through the project escrow account but MSEB sought injunctions to block the attempt (Naik, 2003). In May 2001, DPC stopped Phase I operations, sought international arbitration and issued a preliminary termination notice to MSEB, a prescribed step towards ending the supply contract that also triggered a six-month cooling-off period. MSEB disputed the process and asked the Maharashtra Electricity Regulatory Commission (MERC) to adjudicate the disputed resolution process, claiming that MERC held quasi-judicial powers to adjudicate all state power disputes. In May 2001, MSEB rescinded the power purchase agreement (PPA), claiming that DPC had misrepresented the power plant's ramp up capability,[4] and declared the PPA null and void (Naik, 2003). Construction of Phase II was halted in June.

8.3.6 *Claims and settlements*

In September 2001, Enron's chairman Kenneth Lay announced that the corporation planned to divest assets of US$4–US$5 billion (McLean *et al.*, 2001) and in the same month offered to sell Enron's 65% DPC stake to the Indian government for US$2.3 billion, with US$1.1 billion covering Enron's investment and US$1.1 billion to acquire the claims of offshore lenders. The two parties disagreed with the terms and no settle-

---

[4] A measure of the time taken for the plant to go from a cold start to full output.

ment was reached. Enron then invoked clauses in the Dabhol contract to claim US$5 billion from the central and state governments on the grounds that MSEB had violated the PPA (McLean, 2001; Weil, 2000). By end-October 2001, DPC was preparing to issue a final termination notice to MSEB (Naik, 2003).

However at home, Enron's fraudulent financial and accounting strategies were being exposed, and it sought bankruptcy protection in December 2001 (McLean, 2001; Weil, 2000). DPC creditors in India asked the Mumbai courts to place Dabhol in receivership to prevent it from being absorbed in Enron's bankruptcy proceedings.

Dabhol then lay idle for several years, as sponsors, domestic and foreign financial creditors, national and state governments spun a web of settlement, damages and counter-claims. In March 2004, Indian receivers took control of the Dabhol plant (BBC, 2002) and one month later foreign commercial lenders terminated the PPA. At the same time, Bechtel and GE acquired Enron's 65% stake in DPC for US$23 million through an American bankruptcy court and initiated arbitrations to recover their losses.

8.4 Epilogue

In 2004, India's Congress-led United Progressive Alliance took over the central government and set out to clear the foreign debts that arose from Dabhol and restart operations at the plant. By July 2005, the government had concluded the arbitration and legal cases arising from Dabhol after settling with Bechtel, which received US$160 million in settlement, and GE, which accepted US$145 million (Paterson, 2006). Both sought to minimise their losses and, unlike DPC's financial creditors, were able to use the leverage in bargaining that their technical resources provided, without which a restart was difficult to achieve (The Economist, 2004).

Amounts overdue to 19 overseas lenders were settled for US$230 million at a 20% discount (Sridhar, 2005) while OPIC received a settlement of US$228 million from India's Gas and Power Investment Company Limited (GPICL) funded by domestic lenders including IDBI, ICICI Bank, IFCI and Canara Bank, and the sale of domestic bonds (Kapoor, 2005). The settlement with foreign debtors enabled Dabhol to be revived, and the plant was restarted in April 2006.

8.5 Analysis

The only beneficiaries of Dabhol's collapse may have been parties without prior involvement that became able to acquire interests in the project or its assets at a discount to their costs or accounting value. The plant's technical sophistication is generally undoubted, and at the time of writing it appears to be a potential resource, but without further funding and high prevailing energy prices the plant's future and usefulness must remain economically uncertain. The forces that led Dabhol to disaster were many, and not all aspects of its fate were intrinsic to the project or its conceptual design. Many of Dabhol's most vociferous supporters can no longer be questioned as to what went wrong, or what might have happened. But this brief study exposes structural project weaknesses that were always likely to be tested by profound changes in conditions. Some of these were genuinely exogenous to Dabhol, but were unguarded against and – worse to contemplate – others were matters which Dabhol and its sponsors helped bring about, that is, events that were partly endogenous, and must be seen as likely to have damaged the project's performance in all circumstances.

We make no assessment in this chapter of the course of either state or federal politics in India. The institutional problem is that Enron and others effectively did just as little in this regard. Dabhol failed to anticipate political change at either local or national level, or the tensions that could result. In its promotional work, partly to secure government approvals, Enron invested the project with such notoriety that the political leaders of the time could not refuse to play the project as a pawn in their wider dealings. We do not deal here with commercial matters of political risk insurances that may in reality have been deployed by some parties to mitigate specific risks, but in a wider sense the project's structure left it highly vulnerable to political change, tension and dispute. Even if Dabhol could not be isolated from all these factors, certain institutional elements in the agreements between the Indian authorities and the sponsors would have been of some protection but were missing or defective. These would have included explicit time-sensitive federal and state commercial indemnities or third party financial guarantees.

Political change was thus neglected in the project structure. This is crucial for a related reason, given that the cost of Dabhol's output was relatively high, and the plant was effectively exposed to a monopsony buyer with a poor credit standing. Enron and others sought or obtained inadequate institutional support for their project's risk exposure to MSEB, and no real contingent provision appears to have been made at a sufficiently early stage for Dabhol output to be used elsewhere. In due course, all these factors were gradually cemented into the project, because the peculiar financial interests of Enron were so entrenched that the company's ability to negotiate changes to the project, least of all to command new financial resources, were increasingly strained. Enron International wanted a display trophy, a high-profile, high-technology project in a new overseas market, to show success to future partners elsewhere, and may have been hampered in being unable to change course sufficiently radically with Dabhol, or abandon the project at an earlier stage. It was also driven by the group's need to conserve capital and maintain its reported earnings performance. Thus while it would be incorrect to see in Dabhol a critical reason for Enron's later collapse, nor to suggest that Enron was in financial distress throughout its involvement with Dabhol, it remains true that the marks of Enron's approach wholly infected Dabhol. This includes concerns for earnings management, capital conservation, exorbitant promotion, and highly visible performance. The promoters were thus able to accept vulnerable transaction economics. In effect the greatest controversy in Dabhol was not the role of foreign commercial interests in a sensitive sector, but the result of

Dabhol's promoters needing to establish economic rents in the project's operations, in order to make their cash flow numbers satisfactory and satisfy their wider objectives. At a level of institutional detail, it may also be the case that the weighted shareholdings in Dabhol left Enron with greater freedom to control the project's direction than was optimal, and left the two other original shareholders without the same degree of interest in the project's ultimate fate in their role as shareholders. All this left many potential financiers with a good excuse to avoid the project.

9

Extreme Complexity in Transacting: public private partnerships at work in the London Underground*

This chapter deals with what has been described as one of the most complex contracts ever to be seen in the United Kingdom, namely the public private partnerships (PPP) entered into between London Underground Limited (LUL) and two infrastructure companies, Tubes Lines and Metronet, to refurbish, upgrade, maintain and operate the London Underground metropolitan rail system. Given the scale of the transaction and its term of 30 years, it was notable that such a PPP was even concluded, and an indication of the commitment of certain political leaders to the principles of public private partnerships and private finance initiatives.

However, with these transactions presently into their fifth year of operation, all does not seem entirely under control, both for LUL and the infrastructure companies. The LUL PPP case outlines elements of the relevant history of LUL and the PPPs, and presents a hypothetical scenario: would it be wise for one of the LUL infrastructure companies, Metronet Rail, to offer a shareholding to the Hong Kong Mass Transit Railway Corporation (MTRC), one of the world's most efficient urban railway operators; and, indeed, would it be wise for MTRC to consider such an offer seriously?

9.1 Public private partnership: London Underground[1]

In 2007, the Board of Directors of the Hong Kong MTR Corporation Limited (MTRC) was considering buying into the problematic Metronet Rail, one of the two specially

---

* See endnote to this chapter.
[1] Alison Bate prepared this case under the supervision of the authors for class discussion. This case is fictitious and not intended to show effective or ineffective handling of decisions or business

created private sector engineering consortia contracted to run and maintain the London Underground rail network under 30 year PPP agreements with LUL. The LUL PPPs were among the largest and most complex private finance arrangements in the (UK) public sector. The UK government had intended the PPPs to overcome decades of critical under-funding of the London Underground system by providing sufficient long-term stable investment to restore the system to a good state of repair and deliver benefits to customers through improved reliability, increased capacity and an enhanced travelling environment (Transport for London, 2007).

Acquiring a significant interest in Metronet would allow the MTRC to obtain a foothold in the operation of urban rail systems in Europe. With the LUL franchise already one-tenth of the way into its contracted commitment to fix the infrastructure, progress had been disappointing and public dissatisfaction was tangible. With five years to go until the 2012 London Olympics, the window of opportunity for the MTRC was fast closing. While aware that this would be an unusual decision, the longer the company debated the investment, the less time it was giving itself for preparations, if it eventually opted to go ahead. Clearly not all goals of the LUL PPP would be completed by 2012, but it would be a national public relations disaster for London and the United Kingdom if the expected improvements to the system did not transpire by the time of the Olympic's. On the other hand, it would be an unprecedented high-profile coup for the MTRC if its timely involvement could effect a major improvement in the consortium's performance.

9.2   Two countries, two systems

9.2.1   The London Underground

'The Underground is the core of London's public transport system. There can be no sustained, long-term improvement in transport in the capital without tackling the legacy of years of under-investment in the Tube. The consequences are evident in the unreliability of the service, the constant breakdowns of escalators and many other manifestations of a deteriorating system functioning far below its potential.'
Ken Livingstone, Mayor of London (Greater London Authority, 2001).

LUL was established in 1985 but its antecedants date back to 1863 when the world's first underground railway system came into operation in London. The Underground was crucial to London's role as a developed and cosmopolitan world city. The prosperity of the city depended to a very large extent on the quality of its transport system, which maximised the city's economic efficiency and quality of life of its citizens. However, its age and under-investment had meant that the Underground no longer provided a modern and satisfying service fit for use in the 21ˢᵗ century.

By the 1990s, the situation was critical and there was widespread public consensus about the need to modernise and regenerate London's Underground. The system was decrepit, inadequate and under-managed, having been insufficiently funded for years. Much of the Underground had tunnels that were only 3.5 metres in diameter, with no second set of bypass tracks for emergency or repair work. Workers had to wait for trains to stop running altogether before they could start maintenance work, meaning most major works could only be done over weekends. The costs and difficulties of investing in such a complex and enormous system with existing infrastructure over 100 years old in places could not be overestimated. Under outright state ownership the 'Tube' had received stop-go government investment as political parties and priorities changed over time, with little commitment to the long-term needs of the system as a whole. As LUL was effectively a transport operator and not a construction and maintenance specialist, the tendency had been to concentrate funding on the former, resulting in assets becoming run down, poor service performance, high operating costs and a need for premature replacement. The make-do-and-mend culture accounted for ineffectual investment despite substantial sums being involved. The average core investment for the two decades up to 1997 was £395 million per year.[2] Between 1997 and the start of the PPP operations in 2003, the level of investment increased to around £530 million per year, but was still not enough in light of the city's rapid economic development and the fast growth in tourism. Overcrowding had become a serious problem, with the number of passengers hovering around the 1 billion per year mark, yet LUL was still only able to generate 71% of its revenue from fares, relying on Treasury support for the majority of the balance (Hutton, n.d.). Cancellation of trains had become a common occurrence, with one in twenty peak hour trains not running. Breakdowns were frequent, repairs to the system were very slow, stations were neglected and severely under-maintained and projects undertaken were often late and over budget. The last major government-funded project had been the 16 km Jubilee Line extension which opened in 1999 and which, under LUL's management, had taken nine years to build (1.5 years longer than planned), coming in £1.5 billion over budget and with a faulty signalling system that almost immediately necessitated repair work (Saunders, 2006).

9.2.2 The Hong Kong MTRC

MTRC operates the Hong Kong mass transit railway (MTR), an important element in Hong Kong's transportation network, offering world-class performance in safety, reliability, service quality and efficiency. Trains are punctual, frequent and clean, stations bright with advertising displays and busy with commercial outlets and there

---

[2] £1.00 = US$1.64 = HK$12.

was an efficient, state of the art Octopus smart card system to ease the payment of enviably low fares. Since the first line was opened in 1979, the MTRC had planned and completed five major railway line construction projects, each on time and within budget. The company's commitment to the renewal and upgrade of the infrastructure and systems had ensured continuous improvements in customer service and operational efficiency. The MTR's performance pledge promised passenger journeys on time and train service delivery of 99.5%, and every year, the company surpassed its performance targets. With a consistent operating margin exceeding 50%, thus allowing the recovery of capital investment, the MTR is one of the few profitable metro systems in the world. By 2007 the MTR was carrying some 2.5 million passengers daily (in excess of 870 million per annum), and was one of the most intensively utilised, and efficient, mass transit systems in the world. The total route length ran over 91 km, including 121 km of tunnel (as some route sections had several tunnels). Out of the total 53 stations, 35 were under ground, at depths measuring between 12 to 37 metres below street level.

The MTRC, a public company listed on the Hong Kong Stock Exchange, although still 76% owned by the Hong Kong Government, had operated under a highly successful rail plus property development model for many years. Using the income from the commercial and residential property development rights at, and on top, of its stations to subsidise construction of rail lines, the MTRC had managed to build a top-tier rail network, earn consistent profits, list its shares and gain international recognition for its performance excellence. With a turnover of HK$9.541 billion[3] and operating profits of HK$5.201 billion (54.5%) in 2006, the MTRC had a strong financial position (MTR Corporation, 2007).[4]

The MTRC had been commercially quite active internationally, including providing design, construction and operation services to urban railway companies in major Chinese cities, and consulting services in China, Japan, other Asian countries, Australia, the USA and several European countries, through its consulting arm, MTR Consultancy Services (MTRCS). Recent consulting services it had provided in Europe and the USA included railway engineering design and peer technical reviews, operating systems advice, assistance in preparation of bids, assistance in due diligence exercises, numerous technical consultancies, and many more. However, despite extensive network involvements in China, and the extensive technical consultancy activities of MTRCS, MTRC had not yet established a bridgehead into actual urban railway operations in Europe. It did, however, have some experience of United Kingdom circumstances, having completed operations and maintenance consulting to the pre-PPP London Underground (LU), and systems review consultancy to Tube Lines, one of the private sector infrastructure companies (Infracos) that eventually secured a PPP to upgrade, maintain and operate the Jubilee, Northern and Piccadilly lines of the LU.

---

[3] US$1 = HK$7.8 on 31 December 2006.
[4] Operating profits stated were from the railways and related businesses, before depreciation, and did not include profits on property development.

However, neither the MTRC nor its consultancy arm had yet managed to secure the opportunity to design, construct and operate fully a railway line or system outside China, despite having attempted a number of times, and despite investigating a number of LUL station-related real estate development projects.

9.3   Public private partnership

9.3.1   PPPs in the UK

Private sector participation in the delivery of public infrastructure and services in the UK essentially followed two models: PPPs and private finance initiatives (PFIs). Both PPPs and PFIs could be characterised as attempts to attract private sector participation in the financing and delivery of public services, under circumstances of public sector budgetary constraints. These arrangements took various forms on a continuum ranging from outright sale of publicly owned assets with a commercially inspired mandate, through to simple outsourcing of services. While there was no standard model for PPPs/PFIs, it could be argued that clarity and simplicity (or specificity in economic terms) of the service being privatised were key to negotiating and executing success-fully any PPP/PFI transaction. Similarly, it could be argued that a clearly defined, dedicated cash flow stream was essential to attract private sector finance, where this was sought, particularly where a derivative of the project finance model was preferred in the provision of infrastructure such as toll road or tunnel facilities.

The UK's public services had suffered from a sustained legacy of under-investment. Public sector net investment fell by an average of more than 15% each year between 1991–92 and 1996–97, resulting in a back-log of repairs and maintenance of physical assets predominantly in schools, National Health Service buildings and the transport sector infrastructure (HM Treasury, 2006). The move towards engaging private sector participation in the delivery of public services in the United Kingdom began in the early 1990s, with the government citing this as a means to secure extra value for money from investments, viewing it also as an essential method by which to renew infrastructure with a higher level of investment on a longer term basis and at more predictable levels.

Private consortia, usually involving large construction firms, were contracted to design, build, and in some cases manage new projects. Contracts typically lasted for 30 years, during which time the associated building was leased by a public authority. By 2000, PPPs had delivered efficiency savings of 17% on average, compared with the traditional public sector procurement option (using private companies for specific major projects). By 2000, some 35 hospitals had been built or commissioned using PPP projects covering 373 schools were underway, and approval had been given to projects covering a further 281 schools. A government Ten Year Transport Plan prom-ised to deliver £180 billion of investment, with over £50 billion coming from the private sector. PPPs were boosting investment in National Air Traffic Services, the

Channel Tunnel Rail Link and modern tram schemes up and down the country. The partnerships between the public and private sectors had enabled programmes and projects to go ahead which, the government asserted, would not otherwise have been funded (Department for Transport, 2001).

9.3.2 Political sentiment for and against the LUL PPP

There were different views about the best way to fund the investment required on the Underground and the two main political parties, Conservative and Labour, proposed widely different strategies. In the 1997 election manifesto, the Labour party outlined its strategy in the following terms:

> *'The Conservative plan for the wholesale privatisation of the London Underground is not the answer. It would be a poor deal for the taxpayer and passenger alike. Yet again, public assets would be sold off at an under-valued rate. Much needed investment would be delayed. The core public responsibilities of the Underground would be threatened. Labour plans a new public/private partnership to improve the Underground, safeguard its commitment to the public interest and guarantee value for money to taxpayers and passengers.'*

In the latter half of 1997, as a result of the new Labour government's general commitment to PPP, and believing that the private sector would manage LUL's assets better than LUL itself, the government started to look seriously at a range of possible PPP structures for the Underground. Over the course of the next five years they worked out a structure which would divide the London Underground into four parts for the following 30 years, ultimately splitting the Underground's infrastructure from its operation (see Box 9.1 in which LUL attempts to explain the PPP to customers and gain support for it). Taking office in 2000, the first Mayor of London, Ken Livingstone, however was vehemently opposed to the PPP plan in the form proposed by central government. While he had always advocated improving public transport in London, he was opposed to the fragmentation of the system and believed a unified management structure had to be maintained. He brought in Bob Kiley as Commissioner of Transport for London (TfL), a public body with members appointed by the Mayor, to repeat his success with the New York City subway in raising money through bond issues backed by revenues from fares. Private contractors would still do the work, but without 30 year contracts. It was impossible for the Mayor to block the PPP process as the Underground was by that time still under full ownership of the government; nevertheless he, together with TfL, mounted a legal challenge against the power of LUL to proceed with the PPP. Central government (and especially finance minister Gordon Brown), however, was determined to see the process through, and, with one of the two actions brought by TfL against the PPP collapsing in July 2001 and the other one withdrawn in July 2002, the PPP was given the green light to proceed without being hampered by further legal proceedings.

Box 9.1 LUL PPP

- The public sector will continue to own and operate the Tube. The Tube map, travelcard, and integrated services will all remain. Private sector companies will be contracted to carry out maintenance and upgrades to track, signals, stations and other infrastructure.
- Services will be faster and more frequent – for example, a 20% increase in capacity on the Victoria Line.
- There will be fewer breakdowns and delays and more reliable services. The PPP will update technology – around half of all cancellations today are caused by equipment failures. The public sector will co-ordinate works – so you would not see, for example, the Northern and Victoria lines closed for upgrades at the same time.
- Every train on the Underground will be replaced or refurbished over the course of the whole PPP.
- Every London Underground-owned station will be refurbished or modernised in the next seven and a half years. In the first five years of the contract, 52 out of 255 stations will be fully modernised. Congestion problems will be tackled at overcrowded stations like Leicester Square and Brixton.
- Accessibility for all – including step-free access – will be provided at a number of major stations, as part of a long-term programme.
- Safety and security will be improved. London Underground will remain one of the safest mass transit systems in the world. The independent Health and Safety Executive will continue to ensure Underground safety. Many more stations will have security cameras. Every carriage will have CCTV for passenger safety.
- Fares will continue to be set by the elected Greater London Authority under the Mayor – the PPP does not require fares to rise faster than inflation.
- Other PPPs on the Tube are already delivering new trains on the Northern Line, and funding new ticketing, power and communications systems across the network.
- All these improvements are guaranteed under the PPP, locked in by contracts.

Source: Department for Transport (2001).

After much debate, TfL and the government finally reached agreement and LUL transferred from the control of the government to TfL on 15 July 2003 and, for the first time, came under the authority of the Mayor of London and the Greater London Authority. Under the terms of the agreement, TfL's new LUL management, led by Managing Director Tim O'Toole, would manage the Underground in the best interests of Londoners and take charge of negotiating the PPP contracts and monitoring their implementation to ensure that the improvements to be achieved by the infrastructure companies were delivered in a timely and cost-efficient manner. It was anticipated that the PPP would drive investment of £13 billion over 15 years, with £8.7 billion spent on enhancements and £4.3 billion on maintenance, a much higher level of sustained investment than before.

9.3.3 Structure of the LUL PPP project

The LUL PPP was intended to provide the Underground with a structure that would raise investment over time, promote value for money (VfM) and foster efficiency while retaining the social accountability that came with public ownership (see Figure 9.1 for the structure of the LUL PPP).

9.3.4 A public sector operating company

In March 1998, the central government announced that the best policy objective would be achieved by firstly retaining LUL as a public-facing operating company responsible for safely running the trains and stations, determining service patterns and setting fares, thus remaining in the public sector and retaining its public service focus.

9.3.5 Three private sector infrastructure companies

The second part of the policy was to set up three private sector infrastructure companies (Infracos), taking responsibility for, but not ownership of, the track, signal, stations, and rolling stock of a certain part of the network. In return, LUL would pay an annual infrastructure service charge (ISC). Some of the division of responsibilities depended upon the nature of the underground lines. The sub-surface lines in cut-and-cover tunnels were under Infraco SSL, while those lines running in deep tunnels were split between Infraco BCV and Infraco JNP. Infraco BCV was responsible for the Bakerloo, Central, Victoria and Waterloo & City lines. Infraco JNP was responsible for the Jubilee, Northern and Piccadilly lines. Infraco SSL was responsible for the Circle, District, Metropolitan, East London and Hammersmith & City lines (see Figure 9.2 to appreciate the complexity of this operational environment).

9.3.6 Union of expertise, division of duties

While LUL remained responsible for train operations, staffing, customer services at stations, fare collection and safety, the infrastructure companies took control of LUL's assets and were responsible for maintaining, renewing and upgrading London Underground's infrastructure under long-term contracts without actually owning the infrastructure itself. Six thousand LUL staff would be transferred to the private Infracos with certain undertakings and job protection assurances, while the remaining 8000 Underground staff working as train and station crews would continue to be employed by the public sector. Investors were invited to bid for the companies and their associated rights. The private sector was responsible for providing the underlying finance for

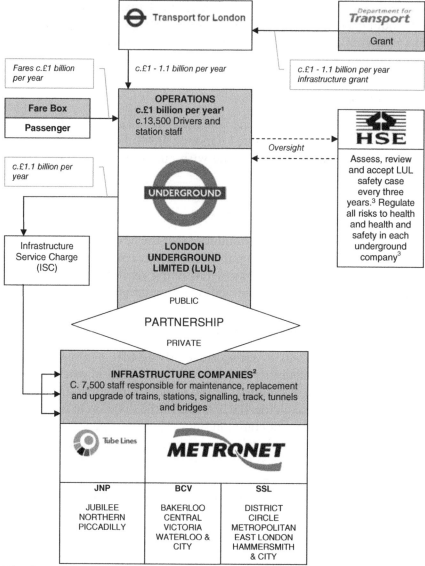

NOTES:

1   All monetary amounts are 2004 annual figures.

2   A partnership director, nominated by LUL, sits on all three Infraco boards.

3   Each Infraco is also required, under the PPP agreement, to satisfy safety requirements.

Source: National Audit Office (17 June 2004).

Figure 9.1   Structure of the LUL PPP.

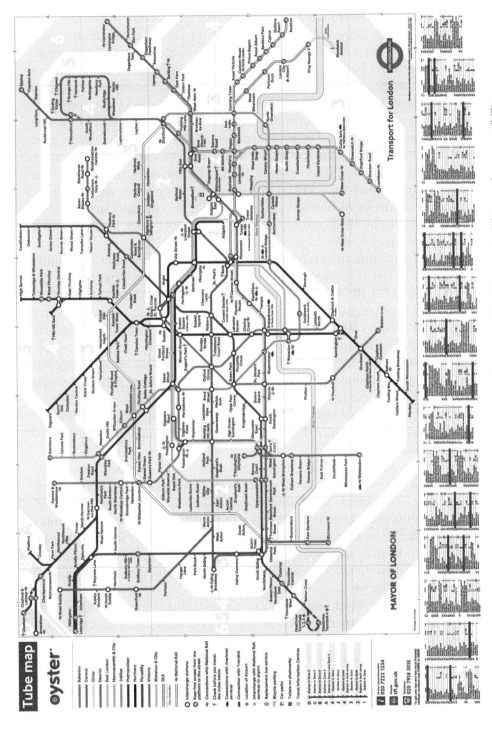

Figure 9.2 Map of the London Underground, showing overall context of the lines under Metronet's responsibility.
Source: Transport for London (2007).

the investments they would require to maintain and upgrade the facilities and services on the relevant tube lines, covering track, trains, tunnels, signals and stations.

The relationship between London Underground and the three infrastructure companies was based on an output specification of the Underground's infrastructure needs, defining exactly what had to be done by the three infrastructure companies in terms of agreed performance measures. The three infrastructure companies were to be paid for by delivering defined outputs which were linked to the key performance measures. The output specification considered what maintenance or other work was required and laid down which investment projects would be carried out. With a 30 year agreement timeframe, the intention was to focus on managing the assets on an efficient, whole-life basis. London Underground, being a public body, used social cost benefit analysis as a framework for decision making while the private sector, in contrast, used financial returns as its framework. Therefore it was deemed important to calibrate the payments to the three infrastructure companies in such a way that the two approaches would yield similar conclusions.

By this policy, the government intended to combine the strengths of the public sector's management of the train network and the private sector's expertise in the maintenance and renewal of the infrastructure, giving London Underground its first committed long-term investment programme in decades. The public's hope for better London Underground services was ostensibly being safeguarded through agreements which only paid the PPP companies for the results they achieved, with good performance rewarded and poor performance penalised. The intention was to encourage the private sector companies to reduce their risks and seek to avoid the penalties through better planning and the introduction of more reliable systems and newer technology. Undoubtedly, the introduction of significant private sector management and finance into the running and refurbishment of the world's oldest underground marked a milestone in the development of PPPs.

To assuage public concerns that separating the infrastructure from operations could jeopardise safety, the PPPs set down in detail safety responsibilities across the network for the infrastructure companies, promising that the London Underground would remain one of the safest mass transit systems in the world. To achieve maximum economic efficiency from the available funding, particular attention was paid to the way the assets were to be managed. Special attention was to be paid to performance and cost issues over the whole asset life-cycle of design, construction, maintenance, refurbishment and replacement.

In part, the practical realities of long-term asset management explained the split between operations and infrastructure. Whilst the private sector was bringing in a wide range of disciplines, skills and expertise driven by commercial incentives, innovation, management expertise and more, the public sector could focus on activities fundamental to the core of the role of the government, such as setting, monitoring and enforcing safety, quality and performance standards for the transport service, safeguarding the interests of the wider public and defining the level and quality of services to be provided.

9.3.7 Financial incentives

The public sector was limited in its ability to create powerful and consistent incentives to management and to take advantage of business opportunities. This was largely due to a multiplicity of policy objectives and the difficulty in defining clear measures of performance. Public opinion was another reason explaining why the public sector had a tendency to be risk averse. On the contrary, with commercial incentives, the private sector was more likely to accept challenges and to create innovative approaches in renewing and managing LUL assets. Incentives were created through contractual arrangements in which the private sector bore the financial risk involved in delivering specified services, where standards were enforced through regulation or payment by results. Putting their own capital at risk would lead the Infracos to make their own judgements about how best to deliver their contribution, balanced against heavy financial penalties if operating performance declined during the contract term. The intention was to provide a strong incentive for the Infracos to deliver the capital projects on time and within budget and to ensure that underlying assets were properly managed over their life-cycle.

9.3.8 Transfer of risk

Another contributing factor to the commercial incentives offered to the Infracos was the risk transfer involved. It was expected that by entering into contracts with the private Infracos, LUL could better manage its risks than it could alone. In practice LUL would specify the outputs that it required and leave the responsibility for many of the associated risks in delivering those outputs to the Infracos. This transfer would create incentives for the Infracos to increase their efficiency as they now bore the costs associated with those risks, such as non-delivery, cost over-runs and technical or asset management failures.

The fact that the private sector put its own capital at risk and was only paid based upon performance had been judged by the government to be an effective and efficient motivator on PPP projects. When this was coupled with the competitive pressures that came from bidding and innovation, PPPs had been able to outperform the public sector based upon recognised and established comparisons such as those described as public sector comparators (PSCs).

9.3.9 Management expertise

The government wanted to bring in the management expertise of the private sector since the mainstream business of many private firms involved complex investment projects. The private sector firms were seen to be more experienced and more likely to attract and retain staff equipped with the necessary skills to perform effectively, thus resulting in higher operational efficiencies. Nevertheless, routine and effective communications between the infrastructure companies and London Underground would be critical in relation to a wide range of technical, commercial and operational issues. To facilitate this, the PPP contract set out a detailed plan to manage the interactions between the public and private sectors.

9.3.10   Long term planning

Past annual rounds of funding debates had created continual uncertainty about any long term programme of works that the London Underground might have wished to undertake. In practice the rolling programme of works had frequently changed based upon things such as differing social benefit criteria, changes in the availability of funding, reassessments of engineering and customer requirements over both the short and long term, forecast of future trends and conflicting political objectives. These factors all led to uncertainty from a delivery point of view as the priorities changed.

The intention was for the LUL PPPs to provide a more stable funding environment for work to be planned and carried out, thereby achieving efficiencies through more predictable planning and delivery of the maintenance, operation and new investment programmes. The PPPs would bring supply-chain dynamics to the Underground.

The 30 year contract structure was expected to give the private sector time to make investment decisions and to procure and manage assets on an efficient whole life basis, unhindered by changes in scope or priorities. It also gave the private sector a better chance to recover its investments over the lifetime of the project.

9.3.11   The bidding process

The bidding process for the Infraco business was seen to be very important to the success of the PPPs. The city of London needed to have sufficient confidence in the bidding process that those bidders appointed to take over the three infrastructure companies would have the required technical and financial capability to do so. Thus the bidding process was long and involved with extensive evaluation of the bidder's submissions and their financial models. The bidding process started in late 1998 with a market consultation exercise, and then in June 1999 LU invited bidders to pre-qualify to bid for the PPP contracts, with the pre-qualified bidders announced in the following year and the invitations to tender issued, and best and final offers submitted in November 2000. In 2001 the preferred bidders for all three Infracos were announced. This was seen by many to have been done unnecessarily early in the negotiation process, while there were still a significant number of issues to be worked out, including the conditionality in the bids, outstanding technical issues and the impact of changes in scope. Deloitte & Touche documented their concern that negotiations undertaken without the benefit of competitive tension may well lead to material erosion of VfM (Deloitte & Touche, 2001). In reply to similar concerns voiced by TfL, LUL said that they considered that *'given the history of negotiations on the remaining open issues, they felt that nothing would be gained by keeping open the competition; that a preferred bidder would provide momentum to closure'*. The final appointment of those same preferred bidders was announced in the following year, and their operations started in 2003, two years later than originally planned, with the minutiae of the agreements being worked out into the most complex commercial contract Britain had witnessed (Hutton, 2002).

9.3.12   Tube Lines

Tube Lines, the asset management company which acquired Infraco JNP, was a collaboration between engineering groups Amey of the UK and Bechtel of the US. Logistics firm Jarvis was originally a member of Tube Lines but, facing bankruptcy, its 33% stake was bought out by Amey in 2005, giving Amey a 66% share of the company. Tube Lines was independent of equipment suppliers and would procure equipment globally, having ruled out the chance of contracts being awarded to member companies in less than competitive circumstances. Tube Lines was responsible for over 320 km of track, 100 stations, 251 trains, 2395 bridges and structures, 71 lifts and 227 escalators, spread across the three lines under its responsibility. These lines, carrying more than 1.5 million passengers a day, were the busiest and deepest on the network (Tube Lines, 2004/5). Initial debt financing comprised £1.795 billion in senior debt, parts of which were provided with financial guarantee 'wraps' by AAA-rated monoline insurer Ambac, and comprising £300 million in a European Investment Bank loan, a £1.23 billion term commercial syndicated loan, and smaller standby and liquidity facilities. Tube Lines successfully returned to the debt markets in August 2004, issuing £1.5 billion in long-term asset-backed securities to lengthen its overall liability profile and partly refinance the 2002 loans at more favourable interest costs (International Financing Review and Euroweek, passim).

9.3.13   Metronet

Metronet, a consortium of five equal partners, W S Atkins, Balfour Beatty, Bombadier Transportation, Seeboard and Thames Water, acquired the other two infrastructure companies, Infraco BCV and Infraco SSL. Unlike Tube Lines, Metronet's shareholders constituted the entire primary contractor supply chain, with Bombadier supplying trains and signalling, Balfour Beatty supplying track work and facility upgrades, and Balfour Beatty, Atkins, Seeboard and Thames Water forming a joint venture called Trans4m to handle infrastructure maintenance (Briginshaw, 2003). Metronet, with six lines under its responsibility, was contracted to modernise and refurbish the 150 stations by 2012, and to invest £17 billion over the course of the 30 year contract. Within its maintenance and capital project management remit it had over 750 km of track, 155 stations, 347 trains, 120 km of deep underground tunnels, over 2000 points, crossings, and bridges, 187 escalators, 2 travelators and 36 lifts (Metronet Rail, 2007). An initial £2.6 billion debt financing was arranged for Metronet's two corporate vehicles, Metronet Rail BCV Holdings Ltd and Metronet Rail SSL Holdings Ltd. This was a successful £1.3 billion issue together wrapped by two AAA-rated monoline insurers.

The total also included a matching £1.3 billion in bank loans, partly funded by the European Investment Bank. Unlike the Tube Lines 2002 fundraisings, Metronet's scheme included at inception a substantial bond issue: Tube Line's first bond was issued two years later when its project had become well understood by investors, allowing the borrower's interest costs to be reduced (International Financing Review and Euroweek, passim).

9.3.14 Contract commitment and dispute resolution

The size and complexity of the London Underground network and the amount of work required to modernise the system were undoubtedly going to have a bearing on the length and depth of negotiations, and the final contracts that emerged after five years of negotiations were indeed so complex that £455 million was paid in lawyers' and consultants' expenses just to have them drawn up. These 30 year, £16 billion contracts to modernise the tracks, stations and tunnels involved a unique and highly complex system of performance measurement entailing hundreds of mathematical formulae never previously tested in the world of transport (Vulliamy and Clark, 2005).

The contracts were based on four basic performance metrics: lost customer hours (availability and reliability of services); journey time (theoretical time taken to get from A to B); ambience (cleanliness); and service points (performance of services such as lighting and heating which did not affect the running of the trains) (Saunders, 2006). The private sector was to pay 25% towards the work, government grants 60% and fares 15%. From the private investors' point of view, the deals were almost risk-free and guaranteed them 30 years' work with periodic reviews every 7.5 years. The deals were expected to generate a good return for the Infracos in the last 12 years of the 30 year contracts, and offered guaranteed 18% returns on equity for 30 years, higher than in many other PFI contracts (Vulliamy and Clark, 2005). With benchmark performance targets set at 5% below existing levels, there was opportunity for substantial profits (Hutton, n.d.).

Charged with delivering an outstanding Underground for London, Tube Lines committed over £4.4 billion and Metronet around £7 billion to improve their respective lines over the first 7.5 years of the 30 year PPP agreements. The PPPs allowed for periodic reviews of the terms of the PPP service contracts every 7.5 years because it was judged to be unrealistic to expect the infrastructure companies to submit fixed prices for the whole contractual period, technological advances may affect project plans and also because it was not possible for LUL to predict its future service requirements. The periodic reviews would look at, for example, the output requirements, any changes to risk profile, any needs for new finance, and the level of infrastructure service charge to be made to the infrastructure companies. Given the size of the undertaking, the periodic reviews would avoid some of the drawbacks of typical PFI deals where schemes would be very rigidly devised within fixed and constrained budgets, yet the flexibility would also work in favour of the Infracos, theoretically allowing them to withhold investment if it was in their interest to do so. The ultimate price to be paid for the private finance investment was therefore not known, when the prices were only firm for the first 7.5 years of the 30 year contract.

According to an independent review of the LUL PPP commissioned by the Department for Transport in February 2002, Ernst & Young found that *'The contract structure does not allow the presentation of an indisputable quantified assessment of value for money'*, referring to the room for substantial adjustments permitted under the periodic reviews. It found that the risk transfer was complex and subject to various sharing and limitation procedures and that the levels of performance expected by the private and the public sector were subjective. Furthermore, it noted that the commercial leverage LU might apply during the periodic reviews would be potentially hampered by the costs of unwinding the long term relationships as set out in the contracts.

An independent Arbiter was appointed by the Secretary of State for Transport, responsible for ensuring that any financial disputes between the parties could be resolved quickly and cleanly. It was anticipated that there would be frequent referral to the Arbiter for interim guidance as well as those matters which might come up in the periodic reviews. The Arbiter was expected to act in a manner best calculated to promote an efficient and economic maintenance and upgrading of the London Underground, and was subject to judicial review by the courts should he, when discharging his duties, ignore matters brought to his attention by the various parties.

9.3.15   *PPP/PFI performance developments*

By 2007, government support for the PPP/PFI process remained very strong, using PPPs to deliver some 10% of its annual spending in public services, and fully expecting the process to pay for an increasing number of public sector projects in the future (HM Treasury, 2006). While admitting certain weaknesses in the process, the Treasury asserted that PFI would continue to be used where it was expected to deliver VfM and encouraged project teams to improve the process for both public and private sectors. PFI had grown in line with government investment in public services, and over 700 PFI transactions across all sectors had reached financial close by March 2006, with a total capital value of £46 billion. A successful city export industry had emerged off the back of PPPs and PFIs, as the private sector became more experienced in the requirements of the process and offered consultancy services overseas. Public sector skills however were still being honed, and the National Audit Office (NAO) report of March 2007 on *Improving the PFI Tendering Process* laid out very clearly the weaknesses attributed to the tendering stage, the area subject to harsh criticism in the case of LUL PPP. The report maintained that the public sector's skills in negotiating the deals remained below par, generally resulting in high costs and risking poor VfM and called for structured training and recycling of existing skills in complex procurement. A new procurement procedure known as competitive dialogue had been introduced to ensure the commercial basis of a PFI deal as well as key aspects of the detailed design being agreed with all bidders before a preferred bidder was selected, to maintain competitive tension for longer at the negotiating stage. The lengthy tendering process was thought to be a factor contributing to the poor response from the private sector to new PFI projects, another concern highlighted in the report which called for review on the scope of the project in question and the suitability of the bids on the table to ensure bidder interest did not weaken and the competition remained viable.

The LUL PPPs were already one-tenth of the way through their contracts, and while it was undeniable that more money was being spent on the Underground in a more considered and strategic way than had been the case for decades, the fundamental issue as to whether the LUL should have been funded by the private sector rather than the public purse remained controversial. Critics of the LUL PPP said a simpler public sector solution, unburdened by the need to make profits for shareholders, would have been more efficient but the PPPs had won through because of the Labour government's refusal to back down on its decision made years earlier, rather than accepting that while PFI might be the right approach in some projects, it might not be in others. Suspicions were rife about behind-the-scenes politicking pertaining to the transactions, and calls for the PPPs to be scrapped continued. The Mayor's London Plan expected the population to grow by 810 000 by 2016, of which over 500 000 would be of working age. Against such projections, the urgency of the requirement for a robust transport system (Figure 9.3) was more evident than ever (Greater London Authority, 2004).

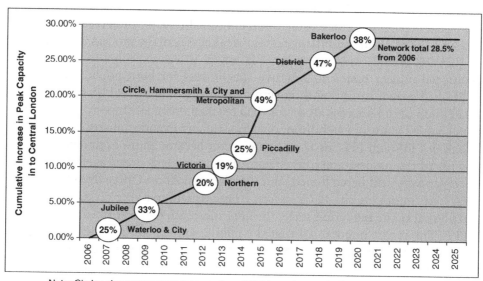

Note: Circles show percentage increase from 2006 in peak capacity for each individual line

Source: Transport for London (November 2006).

Figure 9.3   PPP line peak capacity increases into Central London.

9.3.16   *LUL performance under PPP*

Thrust under the public spotlight since starting operations, the private sector management expertise of the Infracos had been severely tested, not least their public relations skills to counter the barrage of criticism targeted at their performance. With the LUL PPP service charge representing 40% of all Underground expenditure, public expectation had been high, and while improvements had been seen, they were not to the levels expected. Three years on and the travelling public continued to suffer through massive overcrowding at peak times, made worse in summer by the record temperatures. While both organisations had received praise for their swift actions following the London terror attacks in July 2005, the ongoing service disruptions, delays, repeated track, signal and rolling stock breakdowns all remained commonplace and certain high profile failures had only served to emphasise the level of turnaround required.

> *'Availability is the most important factor for Tube travellers. All the Infracos needed to do to meet their availability benchmarks was to perform only a little worse than in the past. On most lines they did not even manage that.'*
>
> (House of Commons Transport Committee, 2005)

The wrath of the public and the press was mostly directed against the larger Infraco, Metronet.

LUL had repeatedly voiced its displeasure with the Infraco, citing lack of confidence in the management. It had issued Corrective Action Notices to both Metronet BCV and Metronet SSL for their repeated failures in delivering the contracted improvements, and in May 2005 the chief executive of Metronet had been sacked after it was revealed that the company had made £50 million profit despite being behind on all its major works (see Table 9.1 for Metronet's financial information). This payment for poor service outraged the public and the blame was laid squarely on the doorstep of the government for appearing to bend over backwards to get the deals done. Professor Stephen Glaister of Imperial College, London, who also sat on the TfL board, said that the infrastructure firms;

> *'were able to do what they did because the government was so committed to the PPP, and therefore over a barrel. You have got to be prepared to walk away from a negotiation to make it work – but they were clear to the Infracos that they were not going to walk away from it.'*
>
> (Vulliamy and Clark, 2005).

It remained highly unlikely that there would be any government turnaround on the policy, not least because of the costs and difficulty for the public sector to get out of the deals, but it was a possibility that could not be disregarded.

In its third annual review of the PPP in 2006, LUL reported that Metronet appeared to be making progress on line upgrades, but were far behind schedule on their station programme. Tube Lines had come off rather better in the report, having succeeded in

Table 9.1   Metronet Infracos financial information.

Statutory accounts year ended 31 March 2005 £million	Metronet BCV	Metronet SSL
Reported profits		
Turnover	324	351
EBITDA	54	60
EBITDA/turnover	17%	17%
Profit after tax – year	14	20
Profit after tax – since contract start	30	39
Net debt		
Net debt B/F	(97)	(35)
Change in year	(140)	(160)
Net debt C/F	(237)	(195)

Total bonus/abatement (year to 31 March) £million	Metronet BCV	Metronet SSL
2003	n/a	n/a
2004	(7.4)	1.8
2005	(3.7)	5.8
2006	1.8	(1.8)
Total	**(9.3)**	**5.8**
Infraco expectation at transfer	17.3	26.3
Above/(below) Infraco expectation	**(26.6)**	**(20.5)**

Cumulative to 31 March 2006 £million	Metronet BCV	Metronet SSL
Baseline ISC	998.8	1,042.6
Capability	3.6	2.9
Availability	(8.2)	16.8
Ambience	1.9	(0.6)
Service points	(3.6)	(7.6)
Specific project adjustments	(3.1)	(5.7)
Total bonus/(abatement)	**(9.3)**	**5.8**
Usage	(2.6)	(0.0)
Performance adjusted ISC	**986.8**	**1,048.4**

Cumulative capital investment to 31 March 2006 £million	Metronet BCV	Metronet SSL
Track and infrastructure	222.3	164.4
Trains, depots and signalling	313.5	318.0
Stations, lifts and escalators	231.7	209.3
Total cumulative investment	**767.5**	**691.7**
Infraco expectation at transfer	796.1	712.9
Above/(below) Infraco expectation	**(28.6)**	**(21.2)**

Table 9.1   *Continued*

Cumulative maintenance expenditure to 31 March 2006 £million	Metronet BCV	Metronet SSL
Track and infrastructure	89.8	208.0
Trains, depots and signalling	238.4	223.4
Stations, lifts and escalators	130.9	149.6
Total asset maintenance	**459.1**	**581.0**
Administration and overheads	256.3	243.3
Total	**715.4**	**824.3**

Sources of funding cumulative to 31 March 2005 £million	Metronet BCV		Metronet SSL	
ISC	626	60%	710	59%
Debt draw down (net of investment of surplus bond proceeds)	369	35%	443	37%
Equity paid and loans	56	5%	51	4%
Total	**1,051**		**1,024**	

Source: Transport for London.

turning around performance on the Piccadilly Line. However, performance on the notoriously unreliable Northern Line (dubbed the 'misery line' by commuters) remained significantly worse than benchmark (see Figure 9.4 for a comparative performance of the lines). Tube Lines CEO Terry Morgan had admitted that the task had been '*much, much more difficult than we ever appreciated*' with costs higher than expected and value for money hard to achieve (Railway Gazette International, 2006). Both Infracos defended their poor performance, talking of the years of underinvestment, and certainly their limited prior knowledge of the existence or condition of the less accessible infrastructure had played a part in the slow progress of the improvements (Saunders, 2006).

In November 2006, the Arbiter's independent report on Metronet, the largest of the two Infracos, was published, wherein he concluded that '*there had been poor delivery of maintenance and renewals*'. Metronet, from 2003 to 2006, had not carried out its activities in an overall efficient and economic manner and in accordance with good industry practice, further detailing assessments of a long list of criteria against agreed benchmarks, and citing examples of gross underperformance. The company's bid document had estimated that station modernisations on the sub-surface lines would cost £2 m each, whereas in fact they had cost an average of £7.5 million. Track renewals on those lines cost double the bid estimate. Deep underground tube reconditioning, estimated to cost £3 million per km, actually cost £5.7 million. In signing the contracts, the Infracos were supposed to be working under incentives to perform, facing penalties for poor performance. The Arbiter's report showed that Metronet had

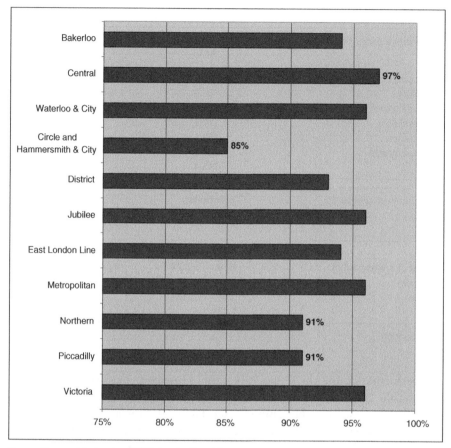

Source: Transport for London (2006), *London Travel Report 2006*.

Figure 9.4   London Underground service reliability, comparative performance by line (2005/6).

received £3 billion in service charges between 2003 and 2006, while the penalties it had paid for poor performance were minimal in comparison (Travers, 2006).

The Arbiter's report on Metronet was for guidance only and did not lead to any financial penalty for the company. The ball was in the court of the consortium's shareholder companies to take the necessary action and deliver on their promises. In response, the firm defended its record, saying the report found it generally performed *'at or better than benchmark'* and promised to ensure that appropriate improvements were made in good time to take effect before the end of the first contract period in 2010 (BBC, 2006).

LUL lamented the work entailed in monitoring the performance of its Infracos. The contracts stipulated a level of performance monitoring that required hundreds of LUL staff to evaluate more than 2000 aspects of the Underground's operation every month, in order to be able to pay to the Infracos the performance-related bonuses and penalties as set out under the terms of the contract. The bureaucracy involved was taking its toll on LUL staff and management, and diverting manpower from the operations. *'With the PPP an awful lot of time and energy is spent just keeping score,'* Tim O'Toole, MD London Underground (Saunders, 2006).

9.3.17 Weighing up the risks

The Metronet Board, being familiar with the MTRC's record as consultant to the Tube Lines Infraco and aware of its wide international urban railway consulting experience, as well as being mindful of the MTRC's superb record and performance on the systems and lines it operates in Hong Kong and China, considered offering MTRC a 20% stake in the PPP and had put out initial overtures to assess the MTRC's interest in such an investment. It seemed clear that the world-famous rail operation efficiency record of MTRC might be extremely useful in improving its overall performance on its LUL PPP obligations. Metronet's Board was convinced that there would be no obstacle to such a transaction from regulatory authorities, given that it would not be relinquishing control of Metronet and its commitments through a secondary transaction, but in effect that it was attempting to secure additional expertise for Metronet in fulfilling its PPP obligations to upgrade and operate its two Infracos under the LUL PPP. This was considered essential, given the importance to London of the approaching 2012 Olympic Games and the critically important role the LU would play in this event, and the public pressure Metronet had been under for its apparently under-performing Infraco BCV and Infraco SSL. Further, Metronet was cautiously non-specific about how it saw the MTRC participating in the PPP if it acquired a stake, but was also aware that without offering the MTRC a reasonably meaningful opportunity to participate in operational activities, it was unlikely that the MTRC would be interested.

Having gone through partial privatisation itself in recent years, the MTRC had managed to blend the best of public and private sector skills, avoiding the trap of chronic government under-funding that had blighted London's Underground. The company's experience was largely gained through greenfield projects and existing system overhauls, certainly not on the scale required by the LUL PPP. Yet the company's success in building, running and maintaining the Hong Kong MTR, reputedly the world's most intensively used urban underground rail system, at a good profit would be an invaluable resource to Metronet in its remit under the PPP. But under the terms and conditions of one of the most complex contracts imaginable, and against the backdrop of widespread public, professional and industrial dissatisfaction, political controversy, media scrutiny, and a litany of independent and painstaking examinations of the LUL PPP and the Infracos, it was understandable that the MTRC was concerned about its decision whether or not it should enter negotiations at all with Metronet shareholders. It was one thing to complete discrete, relatively well-defined consulting assignments internationally, but quite another to operate a railway that it had not designed or constructed, with rolling stock it was unfamiliar with, in a state of repair that it was uncertain about, and in a political environment it was not immersed in. It became increasingly clear to the MTRC Board that it would have to outline clearly what the critical issues were that could influence most a potential investment in the Metronet LUL PPP. What were these critical issues, and given that it had to decide very soon if it would enter into preliminary discussions with Metronet, how long would it take to obtain sufficient insight into the issues to make any decisions? How would it evaluate its own proposal? Would there be any benefit to appointing an independent consultant to identify and analyse the issues? Which issues would require priority analysis? How would the MTRC brief such a consultant?

9.4   Endnote

After a lengthy period of financial uncertainty and disputes with TfL, Metronet BCV and Metronet SSL entered PPP administration on 18 July 2007. The long-term result of this development is unknown as this book goes to press but it seems likely that the administrators, whose appointment was sanctioned by the Mayor of London under the PPP agreements and the Greater London Authority Act 1999, will seek a short-term accommodation with Metronet's financiers, prior to preparing the company and its subsidiaries for sale. Tube Lines is unaffected by this development. Furthermore, Tube Lines has apparently not experienced Metronet's operating and cash flow problems, in spite of being responsible for rail lines that are no less complex to maintain or in need of restorative investment.

Metronet's descent into quasi-receivership followed immediately upon a decision of the statutory PPP arbitrator to award Metronet BCV only 21.9% of its claim for a £551 million increase after 2007 in the Infrastructure Service Charge to cover additional or remedial work required by TfL. The award was insufficient to ensure Metronet adequate liquidity to continue functioning. For some months, its operating performance and acrimony with TfL had induced breaches of covenant and technical defaults on most of its financing agreements, denying the company access to its established bank facilities. With lenders refusing additional funding, neither of the two Metronet vehicles was able to continue operations and the group thus sought protection from its creditors.

While the arbitration result was ostensibly the cause of Metronet's fall, and coincided with shareholder Bombardier writing off its entire equity investment in the venture, the causes of Metronet's problems had been subject to widespread informal discussion in financial and specialist media circles for some time, not least due to the declining value of outstanding unwrapped Metronet debt. There has been speculation that Metronet's long-term performance was hindered by its favouring shareholders as principal operating contractors, which may be an agency problem to be addressed in future PPP agreements.

At the time of publication, the view among financiers was that Metronet's interest in the project would be sold (International Financing Review, 2007). This was no surprise to the financial markets, which had concerns as to Metronet's early operating performance in mid-2005 (International Financing Review, 2005), nor to parliamentary scrutineers (House of Commons Select Committee of Public Accounts, 2006). Most observers have believed that the complex London Underground PPP documentation was nevertheless sufficiently robust to provide comfort to financial investors, that is, it approached a high degree of contractual completeness in providing for uncertain events. However, not all aspects of the financing have been tested in this respect. For example, doubts exist at the time of writing that 'Spens costs' are fully covered by monoline insurance wraps, and may cause losses for investors and a blow to confidence in the system's transactional integrity. The UK Treasury uses Spens costs as a term for break or refinancing costs associated with the early termination of fixed rate financing transactions.

This chapter reflects events as they had unfolded to early 2007. It is intended to illustrate the risks of applying a PPP model to cases as demanding as the London Underground. In particular, the upkeep and renovation of London's rail systems have an inherent complexity requiring the most intricate institutional arrangements, both to initiate the PPP schemes and govern their operation. This is seen notably in arrays of performance covenants in both the PPP and external financing agreements. Our study shows not only the need for contractual completeness among all parties and stakeholders but highlights differences that arose from inception in the managerial approach and performance of the two principal PPP operators. Only much later will it become clear how complete are the Underground PPP contracts in specifying performance and required outcomes.

References

4Ps (2000) *Output Specifications for PFI Projects – 4Ps Guide.* Public Private Partnerships Programme, London.

4Ps (2004) *A Map of the PFI Process.* Public Private Partnerships Programme, London.

Akintoye, A., Taylor, C. and Fitzgerald (1998) Risk Analysis and Management of Private Finance Initiative Projects. *Engineering, Construction & Architectural Management,* 5 (1):9–21.

Akintoye, A., Beck, M. and Hardcastle, C. (eds.) (2003) *Public-Private Partnerships: Managing Risks and Opportunities.* Blackwell, Oxford.

Allison, T. (2001) Enron's Eight-Year Power Struggle in India. *Asia Times Online,* 18 January, retrieved December 29, 2006, from http://www.atimes.com/reports/CA13Ai01.html

Anderson, F. (2001) Free and Clear of Enron's Woes. *Business Week,* 26 November, retrieved January 26, 2007, from http://www.businessweek.com/magazine/content/01_48/c3759014.htm

Aoki, M., Gustavson, B. and Williamson, O.E. (eds) (1990) *The Firm as a Nexus of Treaties.* Sage, London.

Arner D. (2002) Development of the American Law of Corporations to 1832, *SMU Law Review,* 55:23–57.

Arner D. (2007) *Financial Stability, Economic Growth and the Role of Law.* Cambridge University Press, New York.

BBC (2002) *Indian Receivers Take Control of Dabhol,* 2 April, http://news.bbc.co.uk/1/hi/business/1907254.stm

BBC (2006) *Tube firm 'has performed poorly',* 16 November, retrieved March, 2007, from http://news.bbc.co.uk/2/hi/uk_news/england/london/6153414.stm

Benninga, S. (2000) *Financial Modelling,* 2nd edn. MIT Press, Cambridge, MA.

Benninga, S. (2006) *Principles of Finance with Excel.* OUP, New York.

Berle A.A. and Means, G.C. (1991) *The Modern Corporation and Private Property.* Transaction Publishers, US [1932].

Blockley, D. and Godfrey, P. (2000) *Doing it Differently: Systems for Rethinking Construction.* Thomas Telford, London.

Bodie, Z. and Merton, R.C. (1998) *Finance.* Prentice Hall, US.

Brauchli, M.W. (1995) Enron Project is Scrapped by India State. *The Wall Street Journal,* 4 August.

Brealey, R.A., Myers, S.C. and Allen, F. (2006) *Principles of Corporate Finance,* 8th edn. McGraw-Hill Irwin.

Briginshaw, D. (2003) Will PPPs work for London? *Railway Age,* 204 (5):64.

Chapman, C.B. (1997) Project Risk Analysis and Management – PRAM the Generic Process. *International Journal of Project Management,* 15 (5):273–281.

Checkland, P. (1999) *System Thinking, Systems Practice.* Wiley, England.

Cheung Kong Infrastructure Holdings Limited (2000) *Annual Report,* retrieved October 2, 2006, from http://www.cki.com.hk/english/PDF_file/annualReport/2000/2000_AR_projectProfile.pdf

CIPFA (2001) *Risk Management in the Public Services.* Chartered Institute of Public Finance and Accountancy, London.

Coase, R.H. (2000) The Acquisition of Fisher Body by General Motors. *Journal of Law and Economics,* 43 (1):15–31.

Commission of the European Communities (2004) *Green Paper on Public-Private Partnerships and Community Law on Public Contracts and Concessions.* Brussels.

Copeland, T. and Antikarov, V. (2001) *Real Options; A Practitioner's Guide.* Texere, New York.

Copeland, T., Koller, T. and Murrin, J. (2000) *Valuation: Measuring and Managing the Value of Companies*, 3rd edn. John Wiley & Sons, New York.

Crane, D.B., Froot, K.A., Mason, S.P., Perold, A.F., Merton, R.C., Bodie, Z., Sirri, E.R. and Tufano, P. (1995) *The Global Financial System: A Functional Perspective.* Harvard Business School Press, Cambridge, MA.

Cross City Motorway Pty Ltd, *Construction.* retrieved August 30, 2006, from http://www.crosscity.com.au/DynamicPages.asp?cid=7&navid=7

Cross City Tunnel Parliamentary Notice, retrieved October 7, 2006 from http://www.lee.greens.org.au/campaigns/crosscity.htm

Deloitte & Touche Corporate Finance for Transport for London (2001) *London Underground public private partnership: emerging findings.*

Denton Wilde Sapte International Projects Group (2000) *A Guide to Project Finance.*

Department for Transport (2001) *London Underground Public Private Partnership: the offer to Londoners*, retrieved March 15, 2007, from http://www.dft.gov.uk/transportforyou/london/londonunderground/londonundergroundpublicpriva6176

Diamond, D. (1984) Financial Intermediation and Delegated Monitoring. *Review of Economic Studies*, **51** (3):393.

Dixit, A. and Pindyck, R. (1994) *Investment Under Uncertainty.* Princeton University Press, New Jersey.

Douma, S. and Schreuder, H. (2002) *Economic Approaches to Organisations*, 3rd edn. Prentice Hall, New Jersey.

Dutta, S. (2002) *The Enron Saga.* ICFAI Centre for Management Research, Hydarabad.

Economist Intelligence Unit (1995) India Business Intelligence: Dabhol is Heading for a Happy Conclusion, 29 November.

Edwards, B. and Shukla, M. (1995) The Mugging of Enron, *Euromoney*, October.

Euroweek (1995) Financing in place for $920m Indian Project, 3 March.

Euroweek (2003) Australasia, Syndicated Loans, 28 January.

Fernandes, N. (1995) State Halts India's Largest Foreign Investment, Enron's $2.8 Billion, *Associated Press*, 3 August.

Fidler, S. (2002) Enron chief scorned asset division, *Financial Times*, 11 February, retrieved January 28, 2007, from http://specials.ft.com/enron/FT3AB0FQKXC.html

Fislage, B. and Heymann, E. (2003) Road Operation Projects: Lucrative for Institutional Investors, *Deutsche Bank Research*, 10 June.

Fox, J. and Tott, N. (1999) *The PFI Handbook.* Jordans, Bristol.

Freud, D. (2006) *Freud in the City.* Bene Factum Publishing, London.

G.E. to Help Restart Power Plant in India (2005) *New York Times*, 30 May.

Goode, R. (2004) *Commercial Law*, 3rd edn. Penguin, London.

Greater London Authority (2001) *The Mayor's Transport Strategy*, July, retrieved March 13, 2007, from http://www.london.gov.uk/mayor/strategies/transport/pdf/final_execsumm.pdf

Greater London Authority (2004) *The Mayor's London Plan*, February.

Greenbaum, S.I. and Thakor, A.V. (1995) *Contemporary Financial Intermediation.* The Dryden Press, Texas.

Haley, G. (1999) PFI Contractors: Some Emerging Principles. *Construction Law Journal*, **15** (3):220–230.

Handy, C. (1999) *Understanding Organisations*, 4th edn. Penguin Books, London.

Harris, S. (2004) *Public Private Partnerships: Delivering Better Infrastructure Services.* Inter-American Development Bank, Working Paper, Washington.

Hassett, S. (2006) NSW Government Baulks at Tunnel Bid. *The Weekend Australian*, 16 November retrieved January 6, 2007, from http://www.theaustralian.news.com.au/printpage/0,5942,20769578,00.html

Hill, C.W.L. (2005) Enron International in India. Additional Case at McGraw Hill Online Learning Centre with PowerWeb, *International Business: Competition in the Global Marketplace*, 5th edn., retrieved January 28, 2007, from http://highered.mcgraw-hill.com/sites/0072873957/student_view0/additional_cases.html

HM Treasury (1997) *'The Green Book' Economic Appraisal in Central Government: A Technical Guide for Government Departments.* HMSO, London.

HM Treasury (2003) *PFI: Meeting the Investment Challenge.* HMSO, London.

HM Treasury Task Force (1998a) *A Step-by-Step Guide to the PFI Procurement Process.* Office of Government Commerce, London.

HM Treasury Task Force (1998b) *How to Construct a Public Sector Comparator, Technical Note 5.* Office of Government Commerce, London.

HM Treasury Task Force (2003), *Standardisation of PFI Contract Terms.* HMSO, London.

HM Treasury (2006) *PFI: strengthening long-term partnerships.* HMSO, London.

Hong Kong Census and Statistics Department (2006a) *Gross Domestic Product*, retrieved December 24, 2006, from http://www.censtatd.gov.hk/hong_kong_statistics/key_economic_and_social_indicators/index.jsp#prices

Hong Kong Census and Statistics Department (2006b) *Consumer Price Index*, retrieved December 24, 2006, from http://www.censtatd.gov.hk/hong_kong_statistics/key_economic_and_social_indicators/index.jsp#prices

Hong Kong SAR Government (2006) *Economic Situation in the Third Quarter of 2006 and Latest GDP and Price Forecasts for 2006*, retrieved December 24, 2006, from http://www.info.gov.hk/gia/general/200611/21/P200611210182.htm

Hong Kong SAR Government Press Release (2006) *Proponents respond to West Kowloon development parameters*, 26 January, retrieved January 20, 2007, from http://www.info.gov.hk/gia/general/200601/27/P200601270256.htm

House of Commons Transport Select Committee (2005) *The Performance of the London Underground*, 18 March, The Stationery Office, London.

House of Commons Select Committee of Public Accounts (2006), Update on PFI debt refinancing and the PFI equity market, minutes of evidence, 11 December.

Hutton, W., *The London Underground Public Private Partnership: An Independent Review.* The Industrial Society, London.

Hutton, W. (2002) A vote for capital punishment, *The Observer*, 10 February.

Inkpen, A. (2002) *Enron and the Dabhol Power Company.* Thunderbird School of Global Management Arizona.

In *re Denton's Estate*, Licenses Insurance Corporation And Guarantee Fund, Limited v Denton, Court of Appeal [1904] 2 Ch 178.

Institution of Civil Engineers, Faculty and Institute of Actuaries (1998) *RAMP: Risk Analysis and Management of Projects.* Thomas Telford, London.

International Financing Review (1995) Power project propelled – turn up and go, 18 March.

International Financing Review (2003), Australia: Finally launched, 11 January.

International Financing Review (2005), Divisions stall Metronet, 16 July.

International Financing Review (2007), Mind the gap: administrators confident of Metronet sale, 21 July.

Jensen, M.C. (2000) *A Theory of the Firm: Governance, Residual Claims, and Organisational Forms.* Harvard University Press, Massachusetts.

Jensen, M.C. and Meckling, W.H. (1976) Theory of the Firm: Managerial Behaviour, Agency Costs and Ownership Structure. *Journal of Financial Economics*, 3:305–360.

Joshi, P. (2002) Dabhol: A Case Study of Restructuring Infrastructure Projects. *Journal of Structural and Project Finance*, Spring, 27–34.

Kramer, N. and DeSmit, J. (1977) *Systems Thinking – concepts and notions.* M Nijhoff Social Sciences Division, Leiden.

Laffont, J.J. and Martimort, D. (2002) *The Theory of Incentives: The Principal-Agent Model.* Princeton University Press, New Jersey.

Lamb, D. and Merna, T. (2004) *A Guide to the Procurement of Privately Financed Projects: An Indicative Assessment of the Procurement Process,* Thomas Telford, London.

Lang, L. (1998) *Project Finance in Asia,* Elsevier, Amsterdam, 257–271.

Lau, C.K. (2001) Plight at end of the tunnel. *South China Morning Post,* 19 December.

Lee, R. *Cross City Tunnel Parliamentary Notice,* retrieved October 7, 2006, from http://www. lee.greens.org.au/campaigns/crosscity.htm

Lindrup, G. and Godfrey, E. (1998) *Butterworths PFI Manual.* Butterworths Tolley, London.

McLean, B. (2001) Is Enron overpriced? *Fortune* **143**: 5 5 March.

McLean, B., et al (2001) Why Enron went bust, *Fortune* **144**(13): 24 December.

Merna, T. and Njiru, C. (2002) *Financing Infrastructure Projects,* Thomas Telford, London.

Merna, T. and Smith, N. (1999) Privately financed infrastructure in the 21ˢᵗ century. *Proceedings of the Institution of Civil Engineering,* **132** (Nov):166–173.

Metronetrail 2007, retrieved March 22, 2007, from http://www.Metronetrail.com

Milgrom, P. and Roberts J. (1992) *Economics, Organisation and Management.* Prentice Hall International, Englewood Cliffs.

Modigliani, F. and Miller, M.H. (1958) The Cost of Capital, Corporate Finance and the Theory of Investment. *American Economic Review,* **48:**261–297.

Modigliani, F. and Miller, M.H. (1963) Corporate Income Taxes and the Cost of Capital: A Correction. *American Economic Review,* **53:**433–443.

Morgan Stanley (2004) *Cheung Kong Infrastructure,* 28 April, Hong Kong.

MTR Corporation Limited (2007) *2006 annual results,* retrieved March 26, 2007, from http://www. mtr.com.hk/eng/homepage/e_corp_index.html

Naik, N. (2003) *Dabhol Power Company,* case study, London Business School, retrieved December 28, 2006, from http://www.london.edu/assets/documents/PDF/2.3.4.1.6_dabhol.pdf

Nainan, M. (1995) Enron Submits Renegotiation Proposals for Power Project. *Agence France Presse,* 19 September.

National Audit Office (1999) *Examining the Value for Money of Deals under the Private Finance Initiative.* HMSO, London.

National Audit Office (2004) *Managing Risks to Improve Public Services.* HMSO, London.

Nevitt, P. and Fabozzi, F. (2000) *Project Financing.* Euromoney Books, London.

New South Wales Government (2001) *Working with Government Guidelines for Privately Financed Projects,* November, New South Wales Government, Sydney.

Nicholson, M. (1995) Lights Go Out for India's Foreign Investors. *Financial Times,* 4 August.

Noel, M. and Wladyslaw, J.B. (2004) *Mobilizing Private Finance for Local Infrastructure in Europe and Central Asia: An Alternative Public Private Partnership Framework.* World Bank Publications, Washington.

North, D.C. (1990) *Institutions, Institutional Change and Economic Performance.* Cambridge University Press, Cambridge.

NSW Audit Office (2006) *Performance Reports 2006: The Cross City Tunnel Project – Executive Summary*, retrieved January 6, 2007, from http://www.audit.nsw.gov.au/publications/reports/performance/2006/cross_city_tunnel/execsum.htm

Owen, G. and Merna, A. (1997) The Private Finance Initiative. *Engineering, Construction and Architectural Management*, **4** (3):163–177.

Parliament of New South Wales (2006) *Joint Select Committee on the Cross City Tunnel, First Report*, February, New South Wales Government, Sydney.

Paterson, C. (2006) Investor-to-State Dispute Settlement in Infrastructure Projects. *OECD Working Papers on International Investment*, March 2006/2, retrieved January 1, 2007, from http://www.sourceoecd.org/10.1787/416335763425

Pesta, J. (2000) Indian State Looks Again at Enron Project – Calling Power Princes High, Official Wants New Talks. *The Wall Street Journal*, 7 December.

Porter, M. (1980) *Competitive Strategy: Techniques for analysing industries and competitors*. The Free Press, New York.

Porter, M. (1985) *Competitive Advantage: Creating and sustaining competitive advantage*. The Free Press, New York.

Posner, R.A. (1992) *Economic Analysis of Law*, 4th edn. Little, Brown and Company, Boston.

Pretorius, F., Chau, K.W. and Walker A. (2003) Exploitation, Expropriation and Capital Assets: the Economics of Commercial Real Estate Leases. *Journal of Real Estate Literature*, **11** (1):3–34.

Price, A.D.F. and Shawa, H.H. (1997) Risk and Risk Management in Project Related Finance. *Journal of Construction Procurement*, **3** (3):27–46.

Price Waterhouse Coopers (2005) *PPP Global Trends*. London, unpublished.

Private Finance Panel (1993) *The PFI: Breaking New Ground*. HMSO, London.

Private Finance Panel (1996) *Risk and Reward in PFI Contracts: Practical Guidance*. HMSO, London.

Project Finance (2000) *Dabhol – The Big One*, January.

Public Private Partnerships: UK Expertise for International Markets (2003) *International Financial Services, London*.

Railway Gazette International (2006) Three years on, and the PPP presents a mixed picture, **162** (10): 669.

Rao, K. (2001) Foreign Creditors Feel the Pinch. *Euromoney*, October.

Reiss, G. (1996) *Project Management Demystified*, 2nd edn. E&F N Spon, London.

Report of the Energy Review Committee (2001) *Report of the Energy Review Committee*. Government of Maharashtra, Mumbai, India, 10 April, retrieved January 4, 2007, from http://www.prayaspune.org/energy/erc_godbole_report_part-I.htm

Report of the Maharashtra State Cabinet Sub-committee to Review the Dabhol Project (1995).

Rescher, N. (1998) *Complexity: A Philosophical Overview*. Transaction Publishers, New Jersey.

Robinson, P. (2001) *PFI and the Public Finances*. Institute for Public Policy Research, London.

RTA (2003) *Cross City Tunnel Summary of Contracts*. June.

RTA (2002) *The Cross City Tunnel Supplementary Environmental Impact Statement*, August: 46.

Sant, G., Dixit, S. and Wagle, S. (1995) The Enron Controversy: Techno-Economic Analysis and Policy Implications, September, retrieved December 28, 2006, from http://www.prayaspune.org/energy/04_Enron_Controversy%20.pdf

Sarkar, S.K. and Sharma, V. (2006) *Encouraging Investment in Infrastructure Services: Political and Regulatory Risks*. The Energy and Resource Institute, retrieved December 18, 2006, from http://www.teriin.org/pub/papers/ft10.pdf

Saunders, A. (1997) *Financial Institutions Management: A Modern Perspective*, 2nd edn. McGraw-Hill, New York.

Saunders, A. (2006) Fixing the Tube. *Management Today*, October at http://managementtoday. co.uk/search/article/598618/fixing_tube.

Scott, S. and Allen, L. (2005) City Tunnel Contract in Doubt. *Australasian Business Intelligence*, **23**, Sydney.

Senge, P. (1990) *The Fifth Discipline*. Doubleday, New York.

Senn, M.A. (1990) *Commercial Real Estate Leases: Preparation and Negotiation*, 2nd edn. Wiley & Sons, New York.

Simon, P., Hillson, D. and Newland, K. (eds.) (1997) *Project Risk Analysis and Management Guide*. The Association for Project Management, Oxford.

Sridhar, V. (2005) Reviving Dabhol. *Frontline*, 30 July–12 August.

The Economist (1994) Infrastructure in Asia: the trillion-dollar dream, 26 February.

The Economist (2001a) Generation gaps, 31 January.

The Economist (2001b) Enron, and on, and on, 21 April.

The Economist (2004) Can Dabhol be Fired-Up Again? Energy in India, 1 May.

Tirole, J. (2006) *The Theory of Corporate Finance*. Princeton University Press, New Jersey.

Transport for London (2007) *London Underground and the PPP – the third year 2005/06*, retrieved March 20, 2007 from http://www.tfl.gov.uk/pppreport

Travers, T. (2006) Money down the Tube? *Public Finance*, 24–30 November: 4.

Trigeorgis, L. (2000) *Real Options*. MIT Press, Massachusetts.

Tube Lines (2004/5) *Corporate Review*, retrieved March 22, 2007 from http://www.tubelines.com/ aboutus/investorrel/Corporate_review_2004_5.pdf

Tunnels and Tunnelling International (2003) Cross City Contract Signed, **35** (1):15.

Vulliamy, E. and Clark, A. (2005) Down the tube: how PPP deal is costing London. *The Guardian*, 21 February.

Waldorp, M.M. (1992) *Complexity: The Emerging Science at the Edge of Order and Chaos*. Touch-stone, New York.

Weil, J. (2000) Energy traders cite gains, but some math is missing, *Wall Street Journal*, 20 September.

Western Harbour Crossing Ordinance, Cap. 436, ss. 44–45, retrieved from http://www.hklii.org/ hk/legis/en/ord/436/.

Western Harbour Tunnel Company Limited (2006) *Annual Report*.

Williamson, O.E. (1985) *The Institutions of Capitalism*. The Free Press, New York.

Wilson, R. (2004) The high toll of price distortion, *South China Morning Post*, 27 July.

Woodward, D.G. (1997) Risk Analysis and Allocation in Project Financing. *Accounting and Business Review*, **4** (1):117–141.

World Bank (1997) *Dealing with Public Risk in Private Infrastructure*. World Bank, Washington.

World Bank (2000) *Private Infrastructure: A Review of Projects with Private Participation 1990–2000*. World Bank, Washington.

Index

Printed and bound in the UK by
CPI Antony Rowe, Eastbourne

Printed and bound by CPI Group (UK) Ltd, Croydon, CR0 4YY

16/04/2025

14658830-0003